Peptide Synthesis Protocols

Methods in Molecular Biology

John M. Walker, SERIES EDITOR

35. **Peptide Synthesis Protocols,** edited by ***Michael W. Pennington and Ben M. Dunn,*** 1994
34. **Immunocytochemical Methods and Protocols,** edited by ***Lorette C. Javois,*** 1994
33. ***In Situ*** **Hybridization Protocols,** edited by ***K. H. Andy Choo, 1994***
32. **Basic Protein and Peptide Protocols,** edited by ***John M. Walker, 1994***
31. **Protocols for Gene Analysis,** edited by ***Adrian J. Harwood, 1994***
30. **DNA–Protein Interactions,** edited by ***G. Geoff Kneale, 1994***
29. **Chromosome Analysis Protocols,** edited by ***John R. Gosden, 1994***
28. **Protocols for Nucleic Acid Analysis by Nonradioactive Probes,** edited by ***Peter G. Isaac, 1994***
27. **Biomembrane Protocols:** *II. Architecture and Function,* edited by ***John M. Graham and Joan A. Higgins, 1994***
26. **Protocols for Oligonucleotide Conjugates,** edited by ***Sudhir Agrawal, 1994***
25. **Computer Analysis of Sequence Data:** *Part II,* edited by ***Annette M. Griffin and Hugh G. Griffin, 1994***
24. **Computer Analysis of Sequence Data:** *Part I,* edited by ***Annette M. Griffin and Hugh G. Griffin, 1994***
23. **DNA Sequencing Protocols,** edited by ***Hugh G. Griffin and Annette M. Griffin, 1993***
22. **Optical Spectroscopy, Microscopy, and Macroscopic Techniques,** edited by ***Christopher Jones, Barbara Mulloy, and Adrian H. Thomas, 1994***
21. **Protocols in Molecular Parasitology,** edited by ***John E. Hyde, 1993***
20. **Protocols for Oligonucleotides and Analogs,** edited by ***Sudhir Agrawal, 1993***
19. **Biomembrane Protocols:** *I. Isolation and Analysis,* edited by ***John M. Graham and Joan A. Higgins, 1993***
18. **Transgenesis Techniques,** edited by ***David Murphy and David A. Carter, 1993***
17. **Spectroscopic Methods and Analyses,** edited by ***Christopher Jones, Barbara Mulloy, and Adrian H. Thomas, 1993***
16. **Enzymes of Molecular Biology,** edited by ***Michael M. Burrell, 1993***
15. **PCR Protocols,** edited by ***Bruce A. White, 1993***
14. **Glycoprotein Analysis in Biomedicine,** edited by ***Elizabeth F. Hounsell, 1993***
13. **Protocols in Molecular Neurobiology,** edited by ***Alan Longstaff and Patricia Revest, 1992***
12. **Pulsed-Field Gel Electrophoresis,** edited by ***Margit Burmeister and Levy Ulanovsky, 1992***
11. **Practical Protein Chromatography,** edited by ***Andrew Kenney and Susan Fowell, 1992***
10. **Immunochemical Protocols,** edited by ***Margaret M. Manson, 1992***
9. **Protocols in Human Molecular Genetics,** edited by ***Christopher G. Mathew, 1991***
8. **Practical Molecular Virology,** edited by ***Mary K. L. Collins, 1991***

Earlier volumes are still available. Contact Humana for details.

Methods in Molecular Biology • 35

Peptide Synthesis Protocols

Edited by

Michael W. Pennington

Bachem Bioscience, King of Prussia, PA

and

Ben M. Dunn

University of Florida, Gainesville, FL

Humana Press **Totowa, New Jersey**

999 Riverview Drive, Suite 208
Totowa, New Jersey 07512

This publication is printed on acid-free paper. ∞
ANSI Z39.48-1984 (American National Standards Institute)
Permanence of Paper for Printed Library Materials

Printed in the United States of America. 10 9 8 7 6 5 4 3 2 1

Library of Congress Cataloging in Publication Data

Peptide synthesis protocols / edited by Michael W. Pennington and Ben M. Dunn
p. cm.—(Methods in molecular biology; 35)
Companion v. to : Peptide analysis protocols.
Includes index.
ISBN 0-89603-273-6
1. Peptides—Synthesis—Laboratory manuals. I. Pennington, Michael W. II. Dunn, Ben M. III. Peptide analysis protocols IV. Series: Methods in molecular biology (Totowa, NJ); 35
[DNLM: 1. Peptides—chemical synthesis. W1 ME9616J v.35 1994 / QU 68 P424 1994]
QP552.P4P458 1994
574.19'296—dc20
DNLM/DLC 94-26326
for Library of Congress CIP

Preface

Peptide synthesis has emerged as one of the most powerful tools in biochemical, pharmacological, immunological, and biophysical laboratories. Recent improvements include general solid-phase methodology, new protecting groups, and automated equipment. These advances have allowed the facile synthesis of increasingly more complex peptides. Many of these new and improved methods for the synthesis of peptides and peptide-related substances have been reported in various publications, but never compiled in a convenient handbook. Like other volumes in this series, *Peptide Synthesis Protocols* concentrates on the practical aspects of these procedures, providing the researcher with detailed descriptions and helpful tips about potential problems. This volume is not intended to serve as a basic guide to standard Merrifield-type solid-phase strategy, but rather to provide the researcher with some of the most recent applications in the field of peptide science. A companion volume, *Peptide Analysis Protocols*, will detail methodology for the charaterization of new synthetic peptides.

Development of new methods and applications has continued actively even as this volume was in preparation. Owing to the number of contributors to this volume, it was necessary to establish a cutoff for publication purposes. We feel that all of the protocols presented are timely and up-to-date. Several promising new strategies, such as allyloxycarbonyl-based syntheses, were being developed at the time this volume was in the editing stages and will be included in future editions.

The editors would like to thank all of the contributing authors and their colleagues for their many helpful suggestions regarding new applications and methods. Special appreciation must be given to R. B. Merrifield, whose original conceptualization and implementation of the solid-phase approach to peptide synthesis has now been extended to the preparation of peptide libraries, oligonucleotides, and organic molecule libraries. With these new applications, we are certain that the next decade will be a truly exciting time for the peptide chemist.

Michael W. Pennington
Ben M. Dunn

Contents

Contents of the companion volume:

Peptide Analysis Protocols

(Note: Cross references to these chapters will include the notation PAP)

Contributors

FERNANDO ALBERICIO • *Department of Organic Chemistry, University of Barcelona, Barcelona, Spain; Millipore Corporation, Bedford, MA; Department of Chemistry, University of Minnesota, Minneapolis, MN*

DAVID ANDREU • *Department of Organic Chemistry, University of Barcelona, Barcelona, Spain*

ANATOL ARENDT • *Department of Ophthalmology, University of Florida School of Medicine, Gainesville, FL*

GEORGE BARANY • *Department of Chemistry, University of Minnesota, Minneapolis, MN*

G. GREGG BONNER • *Department of Chemistry, University of Arizona, Tucson, AZ*

MICHAEL E. BYRNES • *Bachem Bioscience, King of Prussia, PA*

FRITZ DICK • *Bachem Feinchemikalien AG, Bubendorf, Switzerland*

MARC FERRER • *Department of Chemistry, University of Minnesota, Minneapolis, MN*

CYNTHIA G. FIELDS • *Department of Laboratory Medicine and Pathology, The Biomedical Engineering Center, University of Minnesota, Minneapolis, MN*

GREGG B. FIELDS • *Department of Laboratory Medicine and Pathology, The Biomedical Engineering Center, University of Minnesota, Minneapolis, MN*

PAUL A. HARGRAVE • *Department of Biochemistry and Molecular Biology, University of Florida School of Medicine, Gainesville, FL*

VICTOR J. HRUBY • *Department of Chemistry, University of Arizona, Tucson, AZ*

MONIKA MERGLER • *Bachem Feinchemikalien AG, Bubendorf, Switzerland*

MARK C. MUNSON • *Department of Chemistry, University of Minnesota, Minneapolis, MN. Present Address: Amgen Inc., Boulder, CO*

ROLF NYFELER • *Bachem Feinchemikalien AG, Bubendorf, Switzerland*

MICHAEL W. PENNINGTON • *Bachem Bioscience, King of Prussia, PA*

XINHUA QIAN • *Department of Chemistry, University of Arizona, Tucson, AZ*

FRANK A. ROBEY • *Peptide and Immunohistochemistry Unit, Laboratory of Cellular Development and Oncology, National Institute of Dental Research, National Institutes of Health, Bethesda, MD*

NÚRIA A. SOLÉ • *Department of Chemistry, University of Minnesota, Minneapolis, MN. Present Address: Millipore Corporation, Bedford, MA*

CHAPTER 1

Procedures to Improve Difficult Couplings

Michael W. Pennington and Michael E. Byrnes

1. Introduction

The successful coupling of amino acid derivatives during the synthesis of a peptide by either solution or solid-phase procedures depends on both the reactivity of the carboxyl group of the N-protected amino acid and the steric accessibility of the reactive nucleophile (either a primary or secondary amine). Activation of the carboxyl group is a requisite for the synthesis of an amide bond. Many activation procedures have been developed to accomplish this, and ultimately, the reactivity of the activated species is crucial in determining the coupling yield.

Improvements in solid-phase assembly techniques now permit the routine synthesis of long (>40 residues) complex peptides. However, as the ability to assemble these longer molecules on a solid-phase matrix improved, new problems were encountered. Successful synthesis was hampered by steric factors of the bulky protected derivatives *(1)*, intermolecular aggregation of the protected peptide chain *(2,3)*, formation of hydrogen bonding structures, such as β-sheet *(4–7)*, premature termination, or cyclization on the resin *(8–10)*.

Our laboratory routinely synthesizes large quantities of many peptides. We employ a semiautomated procedure where each individual coupling is monitored for completeness prior to the next deblocking/elongation step. As a result of this type of strategy, we encounter many couplings

From: *Methods in Molecular Biology, Vol. 35: Peptide Synthesis Protocols*
Edited by: M. W. Pennington and B. M. Dunn Copyright ©1994 Humana Press Inc., Totowa, NJ

that do not proceed to completeness using either a single carbodiimide/HOBT coupling *(11)* or double coupling employing the same carbodiimide/HOBT strategy. During the past several years, we have evaluated many of the methods described in the literature to improve the coupling yield. It is important to point out that every peptide presents its own unique set of complications. Thus, it is impossible to give a universal procedure that will work for every peptide. It is the purpose of this chapter to present several of these protocols, which we have found to be very useful.

2. Materials

1. All materials and reagents are purchased from commercial sources and used as such.
2. Synthesis solvents, such as 1-methyl-2-pyrrolidinone (NMP), *N,N*-dimethylformamide (DMF), and dichloromethane (DCM), may be obtained from commercial sources, such as Burdick Jackson (Baxter, McGaw Park, IL), Baker (Phillipsburg, NJ), or Fisher (Fair Lawn, NJ).
3. Coupling agents, such as dicyclohexylcarbodiimide (DCC), diisopropylcarbodiimide (DIC), 1-hydroxybenzotriazole (HOBT), and *N,N*-diisopropylethylamine (DIEA), may be obtained from Chem Impex International (Wood Dale, IL), Aldrich (Milwaukee, WI), or other commercial sources.
4. The following reagents are available from Aldrich, unless otherwise noted: 2,2,2-trifluoroethanol (TFE) 99+% toxic, 1,4-dioxane (anhydrous, 99%), and 4-dimethylaminopyridine (DMAP). Benzotriazol-1-yl-oxy-tris (dimethylamino) phosphonium hexafluorophosphate (BOP reagent), and 2-(1H-benzotriazol-1-yl)-1,1,3,3-tetramethyluronium hexafluorophosphate (HBTU), as well as the related compound 2-(1H-benzotriazol-1-yl)-1,1,3,3-tetramethyluronium tetrafluoroborate (TBTU) may be obtained from Richelieu Biotechnologies (QC. Canada).
5. Chaotropic salts, such as potassium thiocyanate and sodium perchlorate (anhydrous 99%, oxidizer, hygroscopic, irritant, Aldrich), are also commercially available.

3. Methods

The general strategy of this section is to detail several techniques that promote accessibility of the reactive amino group, increase reactivity of the activated carboxyl group, or both. The following techniques have been reported in the literature and successfully employed in our laboratory where a problematic residue or sequence has been encountered.

3.1. Difficult Couplings

Ideally, the coupling reaction of a deprotected amino group and an activated carboxyl group proceeds to near 100% completion. However, because of the factors mentioned earlier, this is sometimes rather difficult to accomplish. Incomplete couplings quickly destroy the fidelity of the synthesis causing an increase in deletion sequences. Capping protocols *(12)* help to eliminate these deletion sequences and are essential in longer syntheses. During a long synthesis, each incomplete coupling is magnified sufficiently so as to reduce the yield of the desired product and increase the levels of deletion sequences and capped, truncated peptidyl-sequences.

As a general rule, difficult couplings are usually sequence-dependent and not residue-specific. It has been observed that many difficulties arise in the synthesis as peptides are elongated through residues 12–20 of their sequences *(2)*. This phenomenon has been attributed to the propensity to form β-structure aggregates on the resin *(3–7)*. Examples of this are peptides with known β-structure (M. W. P., personal communication), as well as peptides rich in hydrogen bonding residues, such as Asn and Gln, which in Boc synthesis are generally incorporated with unprotected side chains *(13)*. It is possible to incorporate both Asn and Gln with protected side chains in a Boc strategy using one of the TFA-stable substituted mono- or bisbenzylamides *(14)*. However, these derivatives are not routinely commercially available. When a fluorenyl methoxycarbonyl (Fmoc) strategy is employed, Asn and Gln side chain protection is possible with trityl *(15)* and methyltrityl *(16)* groups. These protecting groups help prevent the aggregation phenomenon *(16)*.

An incomplete coupling may be identified by the reaction of a portion of the peptidyl resin with ninhydrin as described by Kaiser et al. *(17)* and elsewhere in this volume (*see* Chapter 8). This is a colorimetric reaction that yields a purple, blue, or blue-green color following incubation at an elevated temperature with ninhydrin if any primary amines are present. Secondary amines, such as Pro and *N*-methyl amino acids, usually are less reactive with ninhydrin and result in a reddish-brown color as a positive reaction. Such a positive result indicates an incomplete coupling reaction. When a manual strategy is employed, a recoupling should be performed.

In automated synthesizers using a Boc strategy, a recoupling protocol may be programmed prior to synthesis, but this may not be practical. In most cases, a failed synthesis during a Boc scheme will be identified after the peptide has been completed by analysis of resin samples taken by the instrument, such as the ABI 430, during synthesis *(18)*. Many technicians opt to employ a double-coupling scheme routinely throughout a specific region (residues 8–18, for example) or an entire synthesis, even when this is not necessary, so as to avoid having to resynthesize the molecule if it fails during a single coupling strategy.

On-line acylation and Fmoc removal monitoring by UV spectroscopy have significantly increased the appeal of Fmoc synthesis *(19)*. This feature has been exploited mostly by continuous flow synthesizers, which employ a microprocessor that controls the acylation and deblocking steps by directly interpreting the data. This interpretation allows immediate recoupling during the synthesis much like that during a manual synthesis.

3.2. Resin Substitution

Use of low-substitution resins (0.1–0.4 mmol/g) may increase α-amine accessibility by decreasing steric interactions as well as interchain aggregation. Many commercial resins are supplied with substitutions of 1 mmol/g or greater. For small peptides of 8–20 residues, this may be acceptable. However, for longer peptides, this high degree of substitution can present difficulties later in the synthesis *(20)*. We routinely lower the substitution in these cases during the first cycle of synthesis. This is easily accomplished by performing the first coupling with a limiting amount of protected amino acid. Following this coupling, the remaining free amino groups are capped, thus eliminating any further reactivity at these sites.

3.2.1. Example Method: Reduction of Substitution of mBHA Resin

1. Place 10 g of mBHA resin (substitution value 1.1 mmol/g) in 125-mL flask. Swell the resin with 100 mL of DCM. Filter the solvent away over a scintered glass funnel. Repeat this procedure twice.
2. In a separate flask, preactivate 5 mmol of Boc-amino acid with 10 mmol of DCC and 15 mmol of HOBT in 100 mL of NMP for 30 min.
3. Filter the activated amino acid solution over a separate scintered glass funnel to remove the DCU that has formed during the activation.
4. Add this filtered solution to the swollen mBHA resin, and gently mix for 2 h at room temperature.
5. Terminate the reaction by filtering the activated amino acid solution away from the resin.

6. Wash the resin beads repetitively with 2 × 100 mL DMF, followed by 2 × 100 mL DCM, followed by 2 × 100 mL MeOH, and lastly 2 × 100 mL DCM again.
7. Monitor a sample of the resin by Kaiser analysis (*see* Chapter 8) for positive amino groups. The beads should still turn very dark blue.
8. Initiate a capping procedure by reacting the unreacted primary amino groups with 100 mL of a 20% solution of acetic anhydride in DMF with 2 Eq of DIEA for 1 h.
9. Repeat steps 6 and 7. The Kaiser test should now give a clear yellow (negative test) solution, indicating all unreacted amino groups have been capped.
10. Following a standard TFA deblocking step and subsequent solvent and base washes, a Kaiser test of the resin beads should show a positive result, either blue or reddish brown color (only for Pro). Accurate determination of the actual substitution can be determined by amino acid analysis. A rough approximation can be determined by performing a quantitative ninhydrin test as described by Sarin et al. *(21)*.

3.3. Elevated Temperature

Coupling efficiencies may be increased in a temperature-dependent manner because of thermal disruption of interchain aggregates, although extensive studies on racemization and other peptide modifications must be performed in order to quantify its benefits fully *(22,23)*. **Note: Coupling reactions maintained above the recommended temperature may result in significant amounts of dehydrated material when performed on peptides containing Asparagine and Glutamine** *(23)*.

1. Elevated temperature coupling reactions should be maintained at 35–50°C.
2. Temperature elevation is accomplished by wrapping the reaction vessel in Thermolyne heating tape (Fisher) and regulated with a reostat.
3. The reaction temperature must be checked manually with a thermometer to ensure against variations in temperature.
4. This procedure should be tested experimentally on a small scale until the optimized conditions are found.
5. Alternatively, this procedure may be performed in 5-min intervals every 15 min during a 2-h coupling reaction in order to minimize the deleterious effects of heating.

3.4. Carboxyl Activation Procedures

Peptide bond formation is facilitated by activation of the carboxyl group by addition of a condensing agent to a mixture of the amine component of the existing peptide chain and the carboxyl component of the amino acid being introduced to the synthesis. The earliest procedures, and still today among the most common, incorporated the use of dicyclohexylcarbodiimide (DCC) *(24)*. Also, diisopropylcarbodiimide (DIC) may be substituted in order to allow the formation of diisopropylurea, which is more readily soluble than the dicyclohexylurea formed with DCC use.

The activation procedure may take place *in situ*. However, reaction of the activating reagent with the amino as well as the carboxyl component is possible. External activation permits activation in a nonpolar medium, as well as avoiding contact of the amino group with the reactive carbodiimide or the coproduct urea. This procedure, however, requires the fresh preparation of solutions before each use.

In situ activation is also possible with the phosphonium (BOP and PyBOP) and the uronium (TBTU and HBTU) type activators. These have the unique advantage of generating the activated species without generating the insoluble urea byproducts (*see* Section 3.4.3.).

3.4.1. HOBT Active Esters

Although addition of HOBT to DCC-mediated couplings has been reported to improve coupling reactions, the preformed HOBT ester is widely held to be extremely effective *(11)*, and is especially useful for Asn, Gln, Arg, and His derivatives.

1. For a 1.0-mmol synthesis (1.0 mmol of theoretical amino groups), 5 mmol of amino acid, 0.77 g (5 mmol) of HOBT (153 g/mol), and 1.03 g (5 mmol) of DCC (206 g/mol) are dissolved in 25–30 mL of cold DMF.
2. The prepared solution is allowed to warm up to room temperature and stand at room temperature for approx 30 min before adding to the washed peptide resin. We routinely protect this solution from moisture by keeping the solution under an N_2 atmosphere.
3. Add this solution to the deblocked peptidyl resin.
4. After approx 30 min of coupling, an additional 20 mL of DMF may be added to the resin in order to facilitate wetting and mixing of the resin.
5. Active esters may racemize slowly in DMF. Therefore, it is advised to recouple after an initial positive ninhydrin test, rather than extend the reaction time *(11)*.
6. NMP or other appropriate solvents may also be used during the coupling reaction. Additionally, DIC (126 g/mol; 0.806 g/mL) may be substituted for DCC. Many automated synthesizers successfully use this type of chemistry for activation and do not use cold DMF.

or for peptides rich in β-structural elements to a substitution value of 0.25–0.4 mmol/g of resin.

References

1. Kent, S. B. H. (1988) Chemical synthesis of peptides and proteins. *Ann. Rev. Biochem.* **57,** 957–989.
2. Meister, S. M. and Kent, S. B. H. (1984) Sequence-dependent coupling problems in stepwise solid-phase peptide synthesis: occurrence, mechanism, and correction, in *Peptides: Structure and Function, Proceedings of the 8th American Peptide Symposium* (Hruby, V. J. and Rich, D. H., eds.), Pierce Chem. Co., Rockford, IL, pp. 103–106.
3. Kent, S. B. H. (1985) Difficult sequences in stepwise peptide synthesis: common molecular origins in solution and solid phase, in *Peptides: Structure and Function, Proceedings of the 9th American Peptide Symposium* (Deber, C. M., Hruby, V. J., and Kopple, K. D., eds.), Pierce Chem. Co., Rockford, IL, pp. 407–414.
4. Mutter, M., Altmann, K.-H., Bellot, D., Florsheimer, A., Herbert, J., Huber, M., Klein, B., Strauch, L., and Vorherr, T. (1985) The impact of secondary structure formation in peptide synthesis, in *Peptides: Structure and Function, Proceedings of the 9th American Peptide Symposium* (Deber, C. M., Hruby, V. J., and Kopple, K. D., eds.), Pierce Chem. Co., Rockford, IL, pp. 397–405.
5. Baron, M. H., Deloze, C., Toniolo, C., and Fasman, G. D. (1978) Structure in solution of protected homo-oligopeptides of L-Valine, L-Isoleucine and L-Phenylalanine: an infrared adsorption study. *Biopolymers* **17,** 2225–2239.
6. Pillai, V. and Mutter, M. (1981) Conformational studies of poly(oxyethylene)-bound peptides and protein sequences. *Acc. Chem. Res.* **14,** 122–130.
7. Narita, M., Chen, J. Y., Sato, H., and Lim, Y. (1985) Critical peptide size for insolubility caused by β-sheet aggregation and solubility improvement by replacement of alanine residues with α-aminoisobutyric acid residues. *Bull. Chem. Soc. Jpn.* **58,** 2494–2501.
8. Gisin, B. F. and Merrifield, R. B. (1972) Carboxyl-catalyzed intramolecular aminolysis: a side reaction in solid-phase peptide synthesis. *J. Am. Chem. Soc.* **94,** 3102–3106.
9. Barany, G., Kneib-Cordonier, N., and Mullen, D. G. (1987) Solid-phase peptide synthesis: a silver anniversary report. *Int. J. Peptide Protein Res.* **30,** 705–739.
10. Fields, G. B. and Noble, R. L. (1990) Solid-phase peptide synthesis utilizing fluorenylmethoxycarbonyl amino acids. *Int. J. Peptide Protein Res.* **35,** 161–214.
11. Konig, W. and Geiger, R. (1970) Eine neue zur synthese von peptiden: aktivierung der carboxylgruppe mit dicyclohexyl-carbodiimid unter zusatz von 1-hydroxybenzotriazolen. *Chem. Ber.* **103,** 788–798.
12. Barany, G. and Merrifield, R. B. (1979) Solid-phase peptide synthesis, in *The Peptides: Analysis, Synthesis and Biology*, vol. 2 (Gross, E. and Meinenhofer, J., eds.), Academic, New York, pp. 1–284.
13. Marglin, A. and Merrifield, R. B. (1966) Synthesis of bovine insulin by the solid-phase method. *J. Am. Chem. Soc.* **88,** 5051,5052.

14. Pietta, P. G., Biondi, P. A., and Brenna, O. (1976) Comparative acidic cleavage of methoxybenzyl protected amino acids. *J. Org. Chem.* **41,** 703,704.
15. Sieber, P. and Riniker, B. (1991) Protection of carboxamide functions by the trityl residue: application to peptide synthesis. *Tet. Lett.* **32,** 739–742.
16. Sax, B., Dick, F., Tanner, R., and Gosteli, J. (1992) 4-Methyltrityl (Mtt): a new protecting group for the side chains of Asn and Gln in solid-phase peptide synthesis. *Peptide Res.* **5,** 245,246.
17. Kaiser, E., Colescott, R. C., Bossinger, C. D., and Cook, P. I. (1970) Color test for the detection of free terminal amino groups in the solid-phase synthesis of peptides. *Anal. Biochem.* **34,** 595–598.
18. Kent, S. B. H., Hood, L. E., Beilar, H., Meister, S., and Geiser, T. (1984) High yield chemical synthesis of biologically active peptides on an automated peptide synthesizer of novel design, in *Peptides 1984: Proceedings of the 18th European Peptide Symposium* (Ragnarsson, E., ed.), Almqvist and Wiksell, Stockholm, Sweden, pp. 185–188.
19. Atherton, E. and Sheppard, R. C. (1989) Analytical and monitoring techniques in solid-phase peptide synthesis, in *Solid-Phase Peptide Synthesis: A Practical Approach,* IRL, New York, pp. 112–130.
20. Kent, S. B. H. and Merrifield, R. B. (1981) The role of crosslinked resin support in enhancing the solvation and reactivity of self-aggregating peptides: solid-phase synthesis of acyl carrier protein (65-74), in *Peptides 1980: Proceedings of the 16th European Peptide Symposium* (Brunfeldt, K., ed.), Scriptor, Copenhagen, pp. 328–333.
21. Sarin, V. K., Kent, S. B. H., Tam, J. P., and Merrifield, R. B. (1981) Quantitative monitoring of solid-phase peptide synthesis by the ninhydrin reaction. *Anal. Biochem.* **117,** 147–157.
22. Tam, J. P. (1985) Enhancement of coupling efficiency in solid-phase peptide synthesis by elevated temperature, in *Peptides: Structure and Function, Proceedings of the 9th American Peptide Symposium* (Deber, C. M., Hruby, V. J., and Kopple, K. D., eds.), Pierce Chem. Co., Rockford, IL, pp. 423–425.
23. Lloyd, D. H., Petrie, G. M., Noble, R. L., and Tam, J. P. (1990) Increased coupling efficiency in solid phase peptide synthesis using elevated temperature, in *Peptides: Chemistry, Structure and Biology, Proceedings of the 11th American Peptide Symposium* (Rivier, J. E. and Marshall, G. R., eds), Escom, Leiden, Netherlands, pp. 909,910
24. Sheehan, J. C. and Hess, G. P. (1955) A new method of forming peptide bonds. *J. Am. Chem. Soc.* **77,** 1067.
25. Stewart, J. M. and Young, J. D. (1984) *Solid Phase Peptide Synthesis,* Pierce Chem. Co., Rockford, IL, pp. 81–83.
26. Knorr, R., Trezciak, A., Bannwarth, W., and Gillessen, D. (1989) New coupling reagents in peptide chemistry. *Tet. Lett.* **30,** 1927–1930.
27. Dourtoglou, V., Ziegler, J. C., and Gross, B. (1978) L'Hexafluoro-phosphate de *O*-benzotriazolyl-*N-N-N'N'*-tetramethyluronium: un reactif de couplage petidique nouveau et efficace. *Tet. Lett.* **15,** 1269–1272.

28. Fields, C. G., Lloyd, D. H., Macdonald, R. L., Otteson, K. M., and Noble, R. L. (1991) HBTU activation for automated solid-phase peptide synthesis. *Peptide Res.* **4,** 95–101.
29. Castro, B., Dormoy, J. R., Evin, G., and Selvy, C. (1975) Peptide coupling reactions with benzotriazol-1-yl-tris (dimethylamino) phosphonium hexafluorophosphate (BOP). *Tet. Lett.* **14,** 1219–1222.
30. Fournier, A., Wang, C. T., and Felix, A. M. (1988) Applications of BOP reagent in solid phase peptide synthesis. *Int. J. Peptide Protein Res.* **31,** 86–97.
31. Forest, M. and Fournier, A. (1990) BOP reagent for the coupling of pGlu and Boc-His(Tos) in solid phase peptide synthesis. *Int. J. Peptide Protein Res.* **35,** 89–94.
32. Coste, J., Le Nguyen, D., and Castro, B. (1990) PyBOP: a new peptide coupling reagent devoid of toxic by-product. *Tet. Lett.* **31,** 205–208.
33. Yamashiro, D., Blake, J., and Li, C. H. (1976) The use of trifluoroethanol for improved coupling in solid-phase peptide synthesis. *Tet. Lett.* **18,** 1469–1472.
34. Milton, S. C. F. and De L. Milton, R. C. (1990) An improved solid-phase synthesis of a difficult sequence peptide using hexafluoro-2-propanol. *Int. J. Peptide Protein Res.* **36,** 193–196.
35. Ogunjobi, O. and Ramage, R. (1990) Ubiquitin: preparative chemical synthesis, purification and characterization. *Biochem. Soc. Trans.* **18,** 1322–1333.
36. Nozaki, S. (1990) Solid phase synthesis of steroidogenesis-activator polypeptide under continuous flow conditions. *Bull. Chem. Soc. Jpn.* **63,** 842–846.
37. Steinauer, R., Chen, F. M. F., and Benoiton, N. L. (1989) Studies on racemization associated with the use of benzotriazol-1-yl-tris (dimethylamino)phosphonium hexafluorophosphate (BOP). *Int. J. Peptide Protein Res.* **34,** 295–298.
38. Klis, W. A. and Stewart, J. M. (1990) Chaotropic salts improve solid-phase peptide synthesis coupling reactions, in *Peptides: Chemistry, Structure and Biology, Proceedings of the 11th American Peptide Symposium* (Rivier, J. E. and Marshall, G. R., eds.), Escom, Leiden, Netherlands, pp. 904–906.
39. Thaler, A., Seebach, D., and Cardinaux, F. (1991) Lithium salt effects in peptide synthesis, part II. Improvement of degree of resin swelling and efficiency in solid-phase peptide synthesis. *Helv. Chim. Acta* **74,** 628–643.
40. Steinauer, R., Chen, F. M. F., and Benoiton, N. L. (1990) Studies on racemization associated with the coupling of activated hydroxyamino acids, in *Peptides: Chemistry, Structure and Biology, Proceedings of the 11th American Peptide Symposium* (Rivier, J. E. and Marshall, G. R., eds.), Escom, Leiden, Netherlands, pp. 967,968.
41. Atherton, E., Hardy, P. M., Harris, D. E., and Matthews, B. H. (1991) Racemisation of C-terminal cysteine during peptide assembly, in *Peptides 1990: Proceedings of the 21st European Peptide Symposium* (Giralt, E. and Andreu, D., eds.), Escom, Leiden, Netherlands, pp. 243, 244.
42. Wang, S. S., Tam, J. P., Wang, B. S. H., and Merrifield, R. B. (1981) Enhancement of peptide coupling reactions by 4-Dimethylaminopyridine. *Int. J. Peptide Protein Res.* **18,** 459–467.
43. Crest, M., Jacquet, G., Gola, M., Zerrouk, H., Benslimane, A., Rochat, H., Mansuelle, P., and Martin-Eauclaire, M.-F. (1992) Kaliotoxin, a novel peptidyl

inhibitor of neuronal BK-type Ca^{+2}-activated K^+ channels characterized from *Androctonus mauretanicus mauretanicus* venom. *J. Biol. Chem.* **267,** 1640–1647.

44. Fuller, W. D., Cohen, M. P., Shabankareh, M., and Blair, R. K. (1990) Urethane-protected amino acid *N*-carboxyanhydrides and their use in peptide synthesis. *J. Am. Chem. Soc.* **112,** 7414–7416.
45. Carpino, L. A., Cohen, B. J., Stephens, K. E., Sadat-Aalee, D., Tien, J. H., and Landridge, D. C. (1986) ((9-Fluorenylmethyl)-oxy)carbonyl (Fmoc) acid chlorides. Synthesis, characterization and application to the rapid synthesis of short peptide segments. *J. Org. Chem.* **51,** 3732–3734.
46. Bentho, J. N., Loffet, A., Pinel, C., Reuther, F., and Sennyey, G. (1991) Amino acid fluorides: their preparation and use in peptide synthesis. *Tet. Lett.* **32,** 1303–1306.
47. Spencer, J. R., Antonenko, V. V., Delaet, N. G. J., and Goodman, M. (1992) Comparative study of methods to couple hindered peptides. *Int. J. Peptide Protein Res.* **40,** 282–293.

CHAPTER 2

Methods for Removing the Fmoc Group

Gregg B. Fields

1. Introduction

The electron withdrawing fluorene ring system of the 9-fluorenylmethyloxycarbonyl (Fmoc) group renders the lone hydrogen on the β-carbon very acidic and, therefore, susceptible to removal by weak bases *(1,2)*. Following the abstraction of this acidic proton at the 9-position of the fluorene ring system, β-elimination proceeds to give a highly reactive dibenzofulvene intermediate *(1–5)*. Dibenzofulvene can be trapped by excess amine cleavage agents to form stable adducts *(1,2)*. The stability of the Fmoc group to a variety of bases *(6–10)* is reported in Table 1. The Fmoc group is, in general, rapidly removed by primary (i.e., cyclohexylamine, ethanolamine) and some secondary (i.e., piperidine, piperazine) amines, and slowly removed by tertiary (i.e., triethylamine [Et_3N], *N,N*-diisopropylethylamine [DIEA]) amines. Removal also occurs more rapidly in a relatively polar medium (*N,N*-dimethylformamide [DMF] or *N*-methylpyrrolidone [NMP]) compared to a relatively nonpolar one (dichloromethane [DCM]). During solid-phase peptide synthesis (SPPS), the Fmoc group is removed typically with piperidine, which in turn scavenges the liberated dibenzofulvene to form a fulvene-piperidine adduct. Standard conditions for removal include 30% piperidine-DMF for 10 min *(11)*, 20% piperidine-DMF for 10 min *(12,13)*, 55% piperidine-DMF for 20 min *(14)*, 30% piperidine in toluene-DMF (1:1) for 11 min *(11,15–17)*, 23% piperidine-NMP for 10 min *(9)*, and 20% piperidine-NMP for 18 min *(18)*. Piperidine-DCM should not be utilized, since an amine salt precipitates after relatively brief stand-

From: *Methods in Molecular Biology, Vol. 35: Peptide Synthesis Protocols*
Edited by: M. W. Pennington and B. M. Dunn Copyright ©1994 Humana Press Inc., Totowa, NJ

Table 1
Removal of the Fmoc Group

Compound	Base	Solvent	Time, min	Deprotection, %	Reference
Fmoc-Gly-PS	10% Morpholine	DCM	240	18[a]	*6*
Fmoc-Gly-PS	10% Morpholine	DMF	240	75[a]	*6*
Fmoc-Gly-PS	50% Morpholine	DCM	240	100[a]	*6*
Fmoc-Val	50% Morpholine	DMF	1	50[b]	*7*
Fmoc-Ala-O*t*Bu	50% Morpholine	DCM	120	100[c]	*8*
Fmoc-Gly-PS	10% Piperidine	DCM	240	100[a]	*6*
Fmoc-Val	20% Piperidine	DMF	0.1	50[b]	*7*
Fmoc-Gly-HMP-PS	23% Piperidine	NMP	0.25	50[d]	*9*
Fmoc-Ala-O*t*Bu	50% Piperidine	DCM	<5	100[c]	*8*
Fmoc-Val	5% Piperazine	DMF	0.33	50[b]	*7*
Fmoc-Ala-O*t*Bu	50% Piperazine	DCM	60	100[c]	*8*
Fmoc-PCA	59% 1,4-bis-(3-aminopropyl)piperazine	$CDCl_3$	2	100[e]	*10*
Fmoc-Val	50% Dicyclohexylamine	DMF	35	50[b]	*7*
Fmoc-Ala-O*t*Bu	50% Dicyclohexylamine	DCM	>1080	100[c]	*8*
Fmoc-Val	50% DIEA	DMF	606	50[b]	*7*
Fmoc-Ala-O*t*Bu	50% DIEA	DCM	>1080	100[c]	*8*
Fmoc-Val	10% 4-Dimethylaminopyridine	DMF	85	50[b]	*7*

Fmoc-Ala-O*t*Bu	50% DBU	DCM	<5	100[c]	*8*
Fmoc-Ala-O*t*Bu	50% Pyrrolidine	DCM	<5	100[c]	*8*
Fmoc-Ala-O*t*Bu	50% Cyclohexylamine	DCM	<5	100[c]	*8*
Fmoc-Ala-O*t*Bu	50% Ethanolamine	DCM	<5	100[c]	*8*
Fmoc-Ala-O*t*Bu	50% Diethylamine	DCM	180	100[c]	*8*
Fmoc-Ala-O*t*Bu	50% Triethylamine	DCM	>1080	100[c]	*8*
Fmoc-Ala-O*t*Bu	50% Ammonia	DCM	>1080	100[c]	*8*
Fmoc-Ala-O*t*Bu	50% Tributylamine	DCM	>1080	100[c]	*8*
Fmoc-Ala-O*t*Bu	1.0 m*M* triethylenediamine	DCM	>1080	100[c]	*8*
Fmoc-Ala-O*t*Bu	1.0 m*M* Hydroxylamine HCl	DCM	>1080	100[c]	*8*
Fmoc-Ala-O*t*Bu	0.5 mmol Proton sponge	DCM	>1080	100[c]	*8*
Fmoc-Ala-O*t*Bu	2.0 mmol NaOH	30% CH_3OH-*p*-dioxane	<5	100[c]	*8*
Fmoc-PCA	50% Tris(2-aminoethyl)amine	$CDCl_3$	2	100[e]	*10*
Fmoc-PCA	59% 1,3-Cyclohexanebis-(methylamine)	$CDCl_3$	2	100[e]	*10*

[a]Deprotection of Fmoc-Gly-PS was quantitated spectrophotometrically at 273 nm *(6)*.

[b]Deprotection of Fmoc-Val was quantitated by amino acid analysis *(7)*.

[c]Deprotection of Fmoc-Ala-O-*t*Bu was quantitated by thin-layer chromatography *(8)*.

[d]Deprotection of Fmoc-Gly-HMP-PS was quantitated by ninhydrin analysis *(9)*.

[e]Deprotection of 9-fluorenylmethyl *N*-*p*-chlorophenyl carbamate (Fmoc-PCA) was quantitated by ^{1}H-NMR *(10)*. Dibenzofulvene was scavenged in 2 min by tris(2-aminoethyl)amine, 15 min by 1,3-cyclohexanebis-(methylamine), and 50 min by 1,4-bis-(3-aminopropyl)piperazine.

ing *(11)*. An inexpensive alternative to piperidine for Fmoc removal is diethylamine, with standard conditions being 60% diethylamine-DMF for 180 min *(19,20)* or 10% diethylamine-*N,N*-dimethylacetamide (DMA) for 120 min *(21,22)*.

2. Monitoring

Fmoc removal can be monitored spectrophotometrically because of the formation of dibenzofulvene or fulvene-piperidine adducts. Monitoring is especially valuable in "difficult" sequences, where Fmoc removal may be slow or incomplete *(17,23,24)*. Slow deprotection has been correlated to a broad fulvene-piperidine peak detected at 312 nm *(24–26)*. Monitoring of a broad fulvene-piperidine peak at 365 nm has been used to demonstrate slow deprotection from Fmoc-$(Ala)_5$-Val-4-hydroxymethylphenoxy (HMP)-copoly(styrene-1%-divinylbenzene)-resin (PS); in turn, detection of a narrow fulvene-piperidine peak demonstrated efficient deprotection of the same sequence on a different solid support (HMP-polyethylene glycol-PS) *(27)*. Monitoring of fulvene-piperidine at 313 nm was utilized during the successful synthesis of the entire 76-residue sequence of ubiquitin *(28)*. Dibenzofulvene formation has been monitored at 270 or 304 nm *(29)*.

3. Side Reactions

Repetitive piperidine treatments can result in a number of deleterious side reactions, such as diketopiperazine and aspartimide formation and racemization of esterified Cys derivatives. Base-catalyzed cyclization of resin-bound dipeptides to diketopiperazines is especially prominent in sequences containing Pro, Gly, D-amino acids, or *N*-methyl amino acids. For continuous-flow Fmoc SPPS, diketopiperazine formation is suppressed by deprotecting for 1.5 min with 20% piperidine-DMF at an increased flow rate (15 mL/min), washing for 3 min with DMF at the same flow rate, and coupling the third Fmoc-amino acid *in situ* with benzotriazolyl *N*-oxytrisdimethylaminophosphonium hexafluorophosphate (BOP), 4-methylmorpholine, and 1-hydroxybenzotriazole (HOBt) in DMF *(30)*. For batch-wise SPPS, rapid (a maximum of 5 min) treatments by 50% piperidine-DMF should be used, followed by DMF washes and then *in situ* acylations mediated by BOP or 2-(1H-benzotriazole-1-yl)-1,1,3,3-tetramethyluronium hexafluorophosphate (HBTU) *(31)*. Piperidine catalysis of aspartimide formation from side-chain-protected Asp residues can be rapid, and is dependent on the side-chain-protecting group. Treatment of Asp(OBzl)-Gly, Asp(OcHex)-Gly, and Asp(O*t*Bu)-Gly with 20% piperidine-DMF for 4 h resulted in 100, 67.5, and 11% aspartimide formation, respectively *(32)*, whereas treatment of Asp(OBzl)-Phe with 55% piperidine-DMF for 1 h resulted in 16% aspartimide formation *(33)*. The racemization of C-terminal-esterified Cys derivatives by 20% piperidine-DMF is also problematic, with D-Cys formed to the extent of 11.8% from Cys(Trt), 9.4% from Cys(Acm), 5.9% from Cys(*t*Bu), and 36.0% from Cys(S*t*Bu) after 4 h of treatment *(34)*.

Some piperidine-catalyzed side-reactions may be minimized by using other bases to remove the Fmoc group. Two percent 1,8-diazabicyclo[5.4.0]undec-7-ene (DBU)-DMF, at a flow rate of 3 mL/min for 10 min, is used to minimize monodealkylation of either Tyr(PO_3Me_2) or Tyr(PO_3Bzl_2) *(29)*. For example, 50% monodealkylation of Tyr(PO_3Me_2) occurred in 7 min with 20% piperidine-DMF, but required 5 h with 1*M* DBU in DMF, whereas 50% monodealkylation of Tyr(PO_3Bzl_2) occurred in 12 min with 20% piperidine-DMF and 14 h with 1*M* DBU in DMF *(29)*. Racemization of esterified Cys(Trt) was reduced from 11.8% with 20% piperidine-DMF to only 2.6% with 1% DBU-DMF after 4 h of treatment *(29,34)*. Unfortunately, aspartimide formation of Asp(O*t*Bu)-Asn is worse with DBU compared to piperidine *(35)*. This reagent is recommended for continuous-flow syntheses only, since the dibenzofulvene intermediate does not form an adduct with DBU and thus must be washed rapidly from the peptide resin to avoid reattachment of dibenzofulvene *(29)*. However, a solution of DBU-piperidine-DMF (1:1:48) is effective for batch syntheses, since the piperidine component scavenges the dibenzofulvene.

4. Glycopeptide Synthesis

The mild conditions of Fmoc chemistry are, in general, more suited for glycopeptide syntheses than Boc chemistry, because repetitive acid treatments can be detrimental to sugar linkages *(36)*. However, some researchers prefer morpholine to piperidine as an Fmoc removal agent during glycopeptide SPPS, because the pK_a of morpholine (8.3) is lower than that of piperidine (11.1), and is thus less detrimental to side-chain glycosyls *(36,37)*. Side-chain Ser and Thr glycosyls are stable to base deprotection by neat morpholine *(38,39)* for 30 min *(40)* and 50%

morpholine-DMF for 20–30 min *(41–43)*. A 4-h treatment of Cys(Trt) with 50% morpholine-DMF resulted in 3.8% D-Cys, which is considerably less racemization than that seen with piperidine *(34)*.

5. Solution Syntheses

For rapid solution-phase synthesis, it is desirable to use an Fmoc removal agent that forms a dibenzofulvene adduct that can be extracted in phosphate buffer (pH 5.5). Such an adduct is obtained when either 4-(aminomethyl)piperidine *(44)* or tris(2-aminoethyl)amine is used for Fmoc removal *(10)*. Precipitates or emulsions can form during 4-(aminomethyl)piperidine-fulvene adduct extraction from a DCM layer, so tris(2-aminoethyl)amine is preferred *(10)*. Complete deprotection and scavenging of 9-fluorenylmethyl *N*-*p*-chlorophenyl carbamate (Fmoc-PCA) (0.14 mmol) was achieved in 2 min with 2 mL of tris(2-aminoethyl)amine (100 Eq) in 2 mL $CDCl_3$ *(10)*. Polymeric-bound amines, such as piperazine-PS (2.4 mEq/g) *(45)* and a copolymer of styrene, 2,4,5-trichlorophenyl acrylate, and *N,N'*-dimethyl-*N,N'*-bisacryloylhexamethylene diamine, with subsequent replacement of activated ester groups by 1-(2-aminoethyl)piperazine (3.3 mEq/g) *(46)*, also efficiently remove the Fmoc group in solution-phase syntheses. The use of polymeric-bound amines allows for the isolation of the free amino component by simple filtration of the resin, since the polymer traps the dibenzofulvene *(45,46)*.

6. Notes

1. Amine impurities that could possibly remove the Fmoc group include dimethylamine found in DMF *(47)* and methylamine found in NMP *(48)*. Fmoc-Gly was found to be deprotected after 7 d in DMA, DMF, and NMP to the extent of 1, 5, and 14%, respectively *(49)*. Although these rates of decomposition are considered extremely low, it is recommended that these solvents be freshly purified before use *(26,47)*. The presence of HOBt (0.001–0.1*M*) greatly reduces the detrimental effect of methylamine *(48,50)* whereby Fmoc-Gly-HMP-PS was <1% deprotected after 20 h in NMP *(48)*.
2. The primary and secondary amine lability of the Fmoc group also prompted an investigation of Fmoc removal by esterified or resin-bound amino acids. Fmoc-Ala and Fmoc-Gly (in DMF) were labile to Pro-O*t*Bu, where $t_{1/2} \sim 9$ and 7 h, respectively *(51)*. Fmoc liberation was less rapid by Pro-Lys(4-NO_2-Z)-Gly-OET ($t_{1/2} \sim 40$ and 35 h for Fmoc-Ala and Fmoc-Gly, respectively, in the presence of 1 Eq DIEA), and greatly reduced by the presence of HOBt (1 Eq) and 2,4-dinitrophenol (2 Eq) *(51)*. The Fmoc group was

less labile to primary amino acid esters, even in the presence of DIEA *(51)*. Fmoc-Leu (in DCM) was deprotected very slowly by Gly-PS, with $t_{1/2}$ = 300 and 1500 h in the presence of 1.8 and 1.2 Eq of DIEA, respectively *(8)*. These rates of Fmoc removal by Gly-PS are insignificant in SPPS.

3. There are several alternatives to base removal of the Fmoc group, such as fluoride ion or hydrogenation. Fmoc-Phe was rapidly deprotected (~ 2 min) by 0.05–0.1*M* tetrabutylammonium fluoride trihydrate (TBAF) in DMF *(52)*. Continuous-flow Fmoc SPPS of Leu-Ala-Gly-Val, carried out with 20-min deprotections of 0.02*M* TBAF in DMF, resulted in a highly homogeneous crude product *(52)*. Adding 100 Eq of MeOH to TBAF-DMF solutions could inhibit readdition of dibenzofulvene to the peptide resin and diketopiperazine formation *(52)*. Succinimide formation from Asn, glutarimide formation from Gln, and the instability of benzyl ester groups are potential problems of TBAF deprotection *(53,54)*. Complete deprotection of Fmoc-Ala (in CH_3OH), Fmoc-Gly (in 95% ethanol), and Fmoc-Leu (in 75% aqueous ethanol) by hydrogenation with 10% Pd-on-charcoal catalyst in the presence of acetic acid (two drops) occurred in 4, 22, and 4 h, respectively *(55)*. Deprotection was solvent-dependent, with generation of Gly from Fmoc-Gly occurring with $t_{1/2}$ ~ 30 h in 20% acetic acid–CH_3OH, $t_{1/2}$ ~ 17 h in DMF, and $t_{1/2}$ ~ 7 h in DMF containing 2 Eq of DIEA by hydrogenation with 10% Pd-on-charcoal catalyst *(49)*. Fairly rapid Fmoc-Gly deprotection in DMF ($t_{1/2}$ ~ 2.5 h) was found when $Pd(OAc)_2$ was used as the catalyst instead of Pd-on-charcoal *(49)*. Studies with Fmoc-Gly-OBzl showed selective removal of the benzyl ester in the presence of the Fmoc group by hydrogenation in CH_3OH with 10% Pd-$BaSO_4$ catalyst for ~ 1 h *(56)*.

References

1. Carpino, L. A. and Han, G. Y. (1972) The 9-fluorenylmethoxycarbonyl amino-protecting group. *J. Org. Chem.* **37,** 3404–3409.
2. Carpino, L. A. (1987) The 9-fluorenylmethyloxycarbonyl family of base-sensitive amino-protecting groups. *Acc. Chem. Res.* **20,** 401–407.
3. O'Ferrall, R. A. M. and Slae, S. (1970) β-elimination of 9-fluorenylmethanol in aqueous solution: an *E1cB* mechanism. *J. Chem. Soc.* **(B),** 260–268.
4. O'Ferrall, R. A. M. (1970) β-elimination of 9-fluorenylmethanol in solutions of methanol and *t*-butyl alcohol. *J. Chem. Soc.* **(B),** 268–274.
5. O'Ferrall, R. A. M. (1970) Relationships between *E2* and *E1cB* mechanisms of β-elimination. *J. Chem. Soc.* **(B),** 274–277.
6. Merrifield, R. B. and Bach, A. E. (1978) 9-(2-Sulfo)fluorenylmethyloxycarbonyl chloride, a new reagent for the purification of synthetic peptides. *J. Org. Chem.* **43,** 4808–4816.
7. Atherton, E., Logan, C. J., and Sheppard, R. C. (1981) Peptide synthesis, part 2: procedures for solid-phase synthesis using N^{α}-fluorenylmethoxycarbonylamino-

acids on polyamide supports: synthesis of substance P and of acyl carrier protein 65-74 decapeptide. *J. Chem. Soc. Perkin Trans.* **I,** 538–546.

8. Chang, C.-D., Waki, M., Ahmad, M., Meienhofer, J., Lundell, E. O., and Haug, J. D. (1980) Preparation and properties of N^{α}-9-fluorenylmethyloxycarbonyl-amino acids bearing *tert.*-butyl side chain protection. *Int. J. Peptide Protein Res.* **15,** 59–66.
9. Harrison, J. L., Petrie, G. M., Noble, R. L., Beilan, H. S., McCurdy, S. N., and Culwell, A. R. (1989) Fmoc chemistry: synthesis, kinetics, cleavage, and deprotection of arginine-containing peptides, in *Techniques in Protein Chemistry* (Hugli, T. E., ed.), Academic, San Diego, pp. 506–516.
10. Carpino, L. A., Sadat-Aalaee, D., and Beyermann, M. (1990) Tris(2-aminoethyl)amine as a substitute for 4-(aminomethyl)piperidine in the FMOC/polyamine approach to rapid peptide synthesis. *J. Org. Chem.* **55,** 1673–1675.
11. Albericio, F., Kneib-Cordonier, N., Biancalana, S., Gera, L., Masada, R. I., Hudson, D., and Barany, G. (1990) Preparation and application of the 5-(4-(9-fluorenylmethyloxycarbonyl)aminomethyl-3,5-dimethoxyphenoxy)valeric acid (PAL) handle for the solid-phase synthesis of C-terminal peptide amides under mild conditions. *J. Org. Chem.* **55,** 3730–3743.
12. Atherton, E., Fox, H., Harkiss, D., Logan, C. J., Sheppard, R. C., and Williams, B. J. (1978) A mild procedure for solid phase peptide synthesis: use of fluorenylmethoxycarbonylamino-acids. *J. Chem. Soc., Chem. Commun.* 537–539.
13. Atherton, E., Fox, H., Harkiss, D., and Sheppard, R. C. (1978) Application of polyamide resins to polypeptide synthesis: an improved synthesis of β-endorphin using fluorenylmethoxycarbonylamino-acids. *J. Chem. Soc., Chem. Commun.* 539,540.
14. Chang, C.-D., Felix, A. M., Jimenez, M. H., and Meienhofer, J. (1980) Solid-phase peptide synthesis of somatostatin using mild base cleavage of N^{α}-fluorenylmethyloxycarbonylamino acids. *Int. J. Peptide Protein Res.* **15,** 485–494.
15. Hudson, D. (1988) Methodological implications of simultaneous solid-phase peptide synthesis 1: comparison of different coupling procedures. *J. Org. Chem.* **53,** 617–624.
16. Otvös, L., Jr., Urge, L., Hollosi, M., Wroblewski, K., Graczyk, G., Fasman, G. D., and Thurin, J. (1990) Automated solid-phase synthesis of glycopeptides: incorporation of unprotected mono- and disaccharide units of *N*-glycoprotein antennae into T cell epitopic peptides. *Tetrahedron Lett.* **31,** 5889–5892.
17. Fontenot, J. D., Ball, J. M., Miller, M. A., David, C. M., and Montelaro, R. C. (1991) A survey of potential problems and quality control in peptide synthesis by the fluorenylmethoxycarbonyl procedure. *Peptide Res.* **4,** 19–25.
18. Fields, G. B. and Fields, C. G. (1991) Solvation effects in solid-phase peptide synthesis. *J. Am. Chem. Soc.* **113,** 4202–4207.
19. Sivanandaiah, K. M., Gurusiddappa, S., and Babu, V. V. S. (1988) Peptides related to leucine-/methionine-enkephalinamides: synthesis and biological activities. *Indian J. Chem.* **27B,** 645–648.
20. Sivanandaiah, K. M., Gurusiddappa, S., Channe Gowda, D., and Suresh Babu, V. V. (1989) Improved solid phase synthesis of luteinizing hormone releasing hormone analogues using 9-fluorenylmethyloxycarbonyl amino acid active esters and

catalytic transfer hydrogenation with minimal side-chain protection and their biological activities. *J. Biosci.* **14,** 311–317.
21. Butwell, F. G. W., Haws, E. J., and Epton, R. (1988) Advances in ultra-high load polymer supported peptide synthesis with phenolic supports 1: a selectively-labile C-terminal spacer group for use with a base-mediated N-terminal deprotection strategy and Fmoc amino acids. *Makromol. Chem., Macromol. Symp.* **19,** 69–77.
22. Butwell, F. G. W., Epton, R., McLaren, J. V., Small, P. W., and Wellings, D. A. (1985) Gel phase ^{13}C n.m.r. spectroscopy as a method of analytical control in ultra-high load solid (gel) phase peptide synthesis with special reference to LH-RH, in *Peptides: Structure and Function* (Deber, C. M., Hruby, V. J., and Kopple, K. D., eds.), Pierce Chemical Co., Rockford, IL, pp. 273–276.
23. Larsen, B. D., Larsen, C., and Holm, A. (1991) Incomplete Fmoc-deprotection in solid phase synthesis, in *Peptides 1990* (Giralt, E. and Andreu, D., eds.), Escom, Leiden, The Netherlands, pp. 183–185.
24. Atherton, E. and Sheppard, R. C. (1985) Detection of problem sequences in solid phase synthesis, in *Peptides: Structure and Function* (Deber, C. M., Hruby, V. J., and Kopple, K. D., eds.), Pierce Chemical Co., Rockford, IL, pp. 415–418.
25. Sheppard, R. C. (1988) True automation of peptide synthesis. *Chem. Britain* **24,** 557–562.
26. Atherton, E. and Sheppard, R. C. (1989) *Solid Phase Peptide Synthesis: A Practical Approach.* IRL, Oxford, UK.
27. Barany, G., Solé, N. A., Van Abel, R. J., Albericio, F., and Selsted, M. E. (1992) Recent advances in solid-phase peptide synthesis, in *Innovation and Perspectives in Solid Phase Synthesis—1992* (Epton R., ed.), Intercept, Andover, UK, pp. 29–38.
28. Ogunjobi, O. and Ramage, R. (1990) Ubiquitin: Preparative chemical synthesis, purification and characterization. *Biochem. Soc. Trans.* **18,** 1322–1333.
29. Wade, J. D., Bedford, J., Sheppard, R. C., and Tregear, G. W. (1991) DBU as an N^{α}-deprotecting reagent for the fluorenylmethoxycarbonyl group in continuous flow solid-phase peptide synthesis. *Peptide Res.* **4,** 194–199.
30. MilliGen/Biosearch (1990) Flow rate programming in continuous flow peptide synthesizer solves problematic synthesis. *MilliGen/Biosearch Report* **7,** 4–5,11. MilliGen/Biosearch Division of Millipore, Bedford, MA.
31. Pedroso, E., Grandas, A., de las Heras, X., Eritja, R., and Giralt, E. (1986) Diketopiperazine formation in solid phase peptide synthesis using *p*-alkoxybenzyl ester resins and Fmoc-amino acids. *Tetrahedron Lett.* **27,** 743–746.
32. Nicolás, E., Pedroso, E., and Giralt, E. (1989) Formation of aspartimide peptides in Asp-Gly sequences. *Tetrahedron Lett.* **30,** 497–500.
33. Schön, I., Colombo, R., and Csehi, A. (1983) Effect of piperidine on benzylaspartyl peptides in solution and in the solid phase. *J. Chem. Soc., Chem. Commun.* 505–507.
34. Atherton, E., Hardy, P. M., Harris, D. E., and Matthews, B. H. (1991) Racemization of C-terminal cysteine during peptide assembly, in *Peptides 1990* (Giralt, E. and Andreu, D., eds.), Escom, Leiden, The Netherlands, pp. 243,244.
35. Kitas, E. A., Knorr, R., Trzeciak, A., and Bannwarth, W. (1991) Alternative strategies for the Fmoc solid-phase synthesis of O^4-phospho-L-tyrosine-containing peptides. *Helv. Chim. Acta* **74,** 1314–1328.

36. Kunz, H. (1987) Synthesis of glycopeptides: partial structures of biological recognition components. *Angew. Chem. Int. Ed. Engl.* **26,** 294–308.
37. Paulsen, H., Adermann, K., Merz, G., Schultz, M., and Weichert, U. (1988) Synthese von *O*-glycopeptiden. *Starch/Stärke* **40,** 465–472.
38. Hoogerhout, P., Guis, C. P., Erkelens, C., Bloemhoff, W., Kerling, K. E. T., and van Boom, J. H. (1985) Synthesis of the glycotripeptide alanyl-(*O*-β-D-galactopyranosyl-seryl)-alanine using (9-fluorenylmethoxy)carbonyl and esters of 2-(hydroxymethyl)-9,10-anthraquinone as temporary protecting groups. *Recl. Trav. Chim. Pays-Bas* **104,** 54–59.
39. Paulsen, H., Merz, G., and Weichert, U. (1988) Solid-phase synthesis of *O*-glycopeptide sequences. *Angew. Chem. Int. Ed. Engl.* **27,** 1365–1367.
40. Paulsen, H., Merz, G., Peters, S., and Weichert, U. (1990) Festphasensynthese von *O*-glycopeptiden. *Liebigs Ann. Chem.*, 1165–1173.
41. Jansson, A. M., Meldal, M., and Block, K. (1990) The active ester *N*-Fmoc-3-*O*-[Ac_4-α-D-Man*p*-(1→2)-Ac_3-α-D-Man*p*-1-]-threonine-*O*-Pfp as a building block in solid-phase synthesis of an *O*-linked dimannosyl glycopeptide. *Tetrahedron Lett.* **31,** 6991–6994.
42. Peters, S., Bielfeldt, T., Meldal, M., Block, K., and Paulsen, H. (1991) Multiple column solid phase glycopeptide synthesis. *Tetrahedron Lett.* **32,** 5067–5070.
43. Meldal, M. and Jensen, K. J. (1991) Pentafluorophenyl esters for the temporary protection of the α-carboxy group in solid phase glycopeptide synthesis. *J. Chem. Soc., Chem. Commun.* 483–485.
44. Beyermann, M., Bienert, M., Niedrich, H., Carpino, L. A., and Sadat-Aalaee, D. (1990) Rapid continuous peptide synthesis via FMOC amino acid chloride coupling and 4-(aminomethyl)piperidine deblocking. *J. Org. Chem.* **55,** 721–728.
45. Carpino, L. A. and Williams, J. R. (1978) Polymeric de-blocking agents for the fluoren-9-ylmethoxycarbonyl (FMOC) amino-protecting group. *J. Chem. Soc., Chem. Commun.* 450,451.
46. Arshady, R., Atherton, E., and Sheppard, R. C. (1979) Basic polymers for the cleavage of fluorenylmethoxycarbonyl amino-protecting groups in peptide synthesis. *Tetrahedron Lett.*, 1521–1524.
47. Stewart, J. M. and Young, J. D. (1984) *Solid Phase Peptide Synthesis*, 2nd ed., Pierce Chemical Co., Rockford, IL.
48. Otteson, K. M., Harrison, J. L., Ligutom, A., and Ashcroft, P. (1989) Solid phase peptide synthesis with *N*-methylpyrrolidone as the solvent for both Fmoc and Boc synthesis, in *Poster Presentations at the Eleventh American Peptide Symposium,* Applied Biosystems, Inc., Foster City, CA, pp. 34–38.
49. Atherton, E., Bury, C., Sheppard, R. C., and Williams, B. J. (1979) Stability of fluorenylmethoxycarbonylamino groups in peptide synthesis: cleavage by hydrogenolysis and by dipolar aprotic solvents. *Tetrahedron Lett.*, 3041,3042.
50. Albercio, F. and Barany, G. (1987) Mild, orthogonal solid-phase peptide synthesis: use of N^{α}-dithiasuccinoyl (Dts) amino acids and *N*-(iso-propyldithio) carbonylproline, together with *p*-alkoxybenzyl ester anchoring linkages. *Int. J. Peptide Protein Res.* **30,** 177–205.

51. Bodanszky, M., Deshmane, S. S., and Martinez, J. (1979) Side reactions in peptide synthesis. 11. Possible removal of the 9-fluorenylmethyloxycarbonyl group by the amino components during coupling. *J. Org. Chem.* **44,** 1622–1625.
52. Ueki, M. and Amemiya, M. (1987) Removal of 9-fluorenylmethyloxycarbonyl (Fmoc) group with tetrabutylammonium fluoride. *Tetrahedron Lett.* **28,** 6617–6620.
53. Sieber, P. (1977) Der 2-trimethylsilyläthyl-rest als selektiv abspaltbare carboxy-schutzgruppe. *Helv. Chim. Acta* **60,** 2711–2716.
54. Mullen, D. G. and Barany, G. (1988) A new fluoridolyzable anchoring linkage for orthogonal solid-phase peptide synthesis: design, preparation, and application of the *N*-(3 or 4)-[[4-(hydroxymethyl)phenoxy]-*tert*-butylphenylsilyl]phenyl pentanedioic acid monoamide (Pbs) handle. *J. Org. Chem.* **53,** 5240–5248.
55. Martinez, J., Tolle, J. C., and Bodanszky, M. (1979) Side reactions in peptide synthesis. 12. Hydrogenolysis of the 9-fluorenylmethyloxycarbonyl group. *J. Org. Chem.* **44,** 3596–3598.
56. Rabanal, F., Haro, I., Reig, F., and Garcia-Anton, J. M. (1990) Estudio de la estabilidad del 9-fluorenilmetoxicarbonil frente a la hidrogenacion catalitica. *Anal. Quim.* **86,** 84–88.

CHAPTER 3

Solvents for Solid-Phase Peptide Synthesis

Cynthia G. Fields and Gregg B. Fields

1. Introduction

Effective solvation of the peptide resin is perhaps the most crucial condition for efficient chain assembly during solid-phase peptide synthesis (SPPS) *(1)*. ^{1}H-, ^{2}H-, ^{13}C-, and ^{19}F-nuclear magnetic resonance (NMR) experiments have shown that, under proper solvation conditions, the linear polystyrene chains of copoly(styrene-1%-divinylbenzene)-resin (PS) are nearly as accessible to reagents as if free in solution *(2–6)*. When PS is well solvated, diffusion of reagents is not rate-limiting *(7–9)*. PS swelling tests are thus recommended strongly prior to synthesis *(1)*. The swelling capability of peptidyl-PS increases with increasing peptide length owing to a net decrease in free energy from solvation of the linear peptide chains *(10)*. Under proper solvent conditions, there was no decrease in synthetic efficiency of the model peptide (Leu-Ala-Gly-Val)$_n$ up to a length of 60 amino acids *(11)*.

^{13}C- and ^{19}F-NMR studies of Pepsyn (copolymerized dimethylacrylamide, *N,N'*-bisacryloylethylenediamine, and acryloylsarcosine methyl ester *[12]*) have shown similar mobilities at resin reactive sites as PS *(6)*. Additional supports created by grafting polyethylene glycol (polyoxyethylene) onto PS, either by controlled anionic polymerization of ethylene oxide on tetraethylene glycol-PS (POE-PS) *(13,14)* or coupling N^{α}-Boc or Fmoc-polyethylene glycol acid or polyethylene glycol diacid to amino-functionalized PS (PEG-PS) *(15–17)*, have the potential to combine the advantages of liquid-phase synthesis (homogeneous reaction

From: *Methods in Molecular Biology, Vol. 35: Peptide Synthesis Protocols*
Edited by: M. W. Pennington and B. M. Dunn

environment) and solid-phase synthesis (insoluble support) *(13–18)*. ^{13}C-NMR measurements of POE-PS showed the polyoxyethylene chains to be more mobile than the PS matrix *(13,19)*, with the highest T_1 spin-lattice relaxation times seen for POE of mol wt 2000–3000 *(14)*. Each of these solid supports (PS, Pepsyn, POE-PS, and PEG-PS) has its own distinct solvation behavior *(20,21)*.

In practice, obtaining proper solvation conditions of resins and peptide resins is not always straightforward. Peptide resins are expected to have physicochemical properties that differ considerably from the initial resin because of the addition of the polar peptide backbone. Difficult couplings have often been attributed to poor solvation of the growing chain by dichloromethane (DCM) *(22–24)*. The peptide backbone would require a polar solvent to ensure optimum solvation and, hence, accessibility *(22–26)*. Electron microscopy *(10)* and ESR spectroscopy *(27)* have shown increased solvent polarity (*N,N*-dimethylformamide [DMF] vs DCM) to provide increased peptide-resin solvation when no side-chain-protecting groups were present. Subsequently, it was demonstrated that the increased solvent polarity of *N*-methylpyrrolidone (NMP), *N,N*-dimethylacetamide (DMA), DMF, or *N,N'*-dimethyl-*N,N'*-ethyleneurea (DMEU) vs DCM was extremely beneficial for synthetic efficiency and peptide-resin solvation when benzyl (Bzl)- or *tertiary*-butyl (*t*Bu)-based side-chain protection was present *(24,28–36)*.

2. Problems in Peptide Synthesis

Specific sequences that contain difficult couplings and/or deprotections during SPPS have been characterized for 20 yr, the best known example being the acyl carrier protein (ACP) 65–74 sequence *(22)*. Several studies have correlated difficult couplings to the identity of the activated and N-terminal amino acids *(8,28,37–39)* or to the average coil parameter (P_c) for a segment of amino acids *(29,40–42)*. Difficult sequences have been shown by infrared and NMR spectroscopies to have a common molecular origin as interchain β-sheets *(29,41,43–50)*, which are responsible for lowering coupling *(5,30)* and Fmoc deprotection *(51)* efficiencies. The formation of interchain β-sheets is believed to have the same effect as increased resin crosslinking *(30,52)*. β-sheet formation occurs primarily at chain lengths of <20 residues *(5,28,30,31,44,46)*. Solvents of increasing polarity (DMF or NMP vs DCM; DMSO vs DMF) are favored because of inhibition of β-sheet formation *(29,30,43–48,50,53)* as well as the previously discussed enhancement of solvation of the polar peptide backbone. A scale of β-sheet structure-stabilizing potential has been developed for Boc-amino acid derivatives *(32)*.

3. Mixed-Solvent Systems

Mixed-solvent systems may optimize peptide-resin solvation by combining relatively polar and nonpolar solvents *(23,26,29,33–35,41,42,47, 48)*. Several mixed-solvent systems used successfully in SPPS include 2,2,2-trifluoroethanol (TFE)-DCM *(26,33,35,54)*, dimethyl sulfoxide (DMSO)-NMP *(33,35,55)*, DMSO-DMF *(56–58)*, *p*-dioxane-DMF *(59,60)*, 1,1,1,3,3,3-hexafluoro-2-propanol (HFIP)-DCM *(42,48)*, and urea-DMF *(23)* (Table 1). Recently, theory incorporating solvent electron donor (DN) and acceptor (AN) numbers *(61)* has been used to create mixed-solvent systems that minimize intermolecular β-sheet formation *(29,47,62)*. As solvent AN and DN values become larger, the β-sheet disrupting potential becomes greater *(62)*. Mixed-solvent systems should be homogeneous in terms of electron potential, i.e., only acceptor or donor solvents should be used together *(47)*. Strong electron acceptor solvents, such as HFIP or TFE, are mixed with DCM, whereas electron donor solvents, such as hexamethylphosphoric triamide (HMPA) or DMSO, are mixed with DMA, DMF, or NMP *(61)*.

The solvation of PS and peptidyl-PS can be correlated well to solvent Hildebrand solubility (δ) and hydrogen-bonding solubility (δ_h) parameters *(33)*; δ is either experimentally measured or calculated using *(63)*:

$$\delta = [(-\Sigma z \Delta U / \Sigma z V)]^{1/2} \tag{1}$$

where z is the number of each group type (e.g.,—OH,—CH_2—), ΔU is the molar vaporization energy, and V is the molar volume; δ_h is either experimentally measured or calculated using *(64)*:

$$\delta_h = [(5000N / \Sigma z V)]^{1/2} \tag{2}$$

where N is the number of hydroxyl groups. By constructing contour plots that compare solvent δ and δ_h values to PS and peptidyl-PS solvation, optimized mixed-solvent systems can be designed *(34)*. To correlate the solvation effects of mixed-solvent systems to solvent properties, solubility parameters are estimated using *(63)*:

$$\delta_{1+2} = \phi_1 \delta_1 + \phi_2 \delta_2 \tag{3}$$

Table 1
Mixed Solvent Systems in Solid-Phase Peptide Synthesis

Mixed solvent	Properties	Reference
30% DMF-DCM	Improved coupling yields during Boc SPPS of SRFGSWGAEGQSTFGK	*23*
50% DMF-*p*-dioxane	Improved couplings during Fmoc SPPS of ubiquitin-HMP resin[a]	*59*
50% DMSO-DMF	Improved couplings of Boc-2-Aib HOBt ester[b]	*56*
	Improved couplings during Boc SPPS of α helical peptides	*57*
20% DMSO-NMP	Enhanced solvation of ACP 65-74-PAM resin[c]	*35*
	Disaggregation of ACP 65-74-PAM resin[c]	*58*
	Enhanced solvation of Ala[15]-GRF 11-29- and ACP 65-74-PAM resin[c]	*33*
	Allowed for single coupling SPPS of insulin-like growth factor I	*55*
10% HFIP-DCM	Improved coupling yields for Boc SPPS of D-Ala[17]-phGnRH (14–36)[d]	*42*
2% HFIP-DCM	Improved coupling yields during Boc SPPS of resin-bound proinsulin protected peptide fragments–PAM resin[c]	*48*
20% HFIP-DCM	Disaggregation of resin-bound protected peptides	*41*
6% HFIP-THF	Disaggregation of resin-bound protected peptides	*47*
17% HMPA-DCM	Disaggregation of resin-bound proinsulin protected peptide fragments–PAM resin[c]	*29*
20% TFE-DCM	Enhanced solvation of AAH toxin II 44-52-Nbb-resin[e]	*53*
	Improved coupling of Boc-Lys(2-BrZ) during SPPS of KKKKKEEELLWF[f]	*26*
	Enhanced solvation of [*t*Bu]-Lys9-conotoxin–Rink resin[g]	*34*
	Enhanced solvation of ALA[15]-GRF 11-29- and ACP 65-74-PAM resins[c,h]	*33*
2% TFE-DCM	Disaggregation of resin-bound proinsulin protected peptide fragments–PAM resin[c]	*29*
35% THF-NMP	Enhanced solvation of [*t*Bu] and [Bsl]-Lys[9]-conotoxin–Rink resin[g]	*34*
7% Urea-DMF	Improved coupling yield of Boc-Gln in SRFGSWGAEGQSPFGK	*23*
0.4*M* KSCN in DCM-DMF (1:1)	Improved coupling yields during Boc SPPS of RNase 1-13-MBHA resin[i]	*71*
2.0*M* LiBr in THF	Disaggregation of resin-bound, protected amyloid-β protein fragments	*73*
0.4*M* LiCl in NMP	Improved coupling yield of Fmoc-Ala[6] to (Ala)[5]-Phe-HMP resin and enhanced solvation[a]	*71*
0.4*M* $NaClO_4$ in DCM-DMF (1:1)	Improved coupling yields during Boc SPPS of RNase 1-13-MBHA resin[i]	*72*

[a]HMP = 4-hydroxymethylphenoxy; [b]2-Aib = 2-aminoisobutyric acid; [c]PAM = phenylacetamidomethyl; [d]phGnRH = human gonadotropin-releasing hormone precursor; [e]AAH = *Androctonus australis Hector*, Nbb = nitrobenzamidobenzyl; [f]2-BrZ = 2-bromobenzyloxycarbonyl; [g]Rink = 4-(2',4'-dimethoxyphenylaminomethyl)phenoxy; [h]GRF = growth hormone releasing factor; [i]MBHA = 4-methylbenzhydrylamine.

where ϕ is the volume fraction of the solvent. From comparison of peptidyl-PS solvation to solvent δ and δ_h values, mixed-solvent systems, such as 20% TFE-DCM and 35% THF-NMP, were proposed *(34)*. These mixed-solvent systems provided excellent peptide-resin solvation regardless of side-chain protection *(34)*. It should be noted that solvation is **not** the same as β-sheet disrupting potential.

4. Overcoming Hydrophobic Collapse

Interchain aggregation may occur in regions of apolar side-chain protecting groups resulting in a collapsed gel structure *(52,65)*. Transfer-free energies of amino acid side-chain groups from cyclohexane to water or 1-octanol show a Bzl side chain to be considerably more polar than branched alkyl (e.g., *t*Bu) side chains *(66,67)*, whereas enhanced peptide-resin solvation by polar solvents is favored with Bzl-based side-chain protection rather than with *t*Bu-based side-chain protection *(34)*. During the Fmoc SPPS of the 66-104 fragment of cytochrome C, the accumulation of nonpolar, *t*Bu-based side chains resulted in a collapsed gel that was not solvated by polar solvents, such as DMA *(65)*. Replacement of apolar side chains [Lys(Boc), Met] by more polar side chains [Lys(Tfa), Met(O)] was required for efficient chain assembly *(52,65)*. It has been noted that among Fmoc-Cys-protecting groups, Acm is considerably more polar than *t*Bu, S*t*Bu, or Trt *(68)*. Cys(Acm) may therefore be used to reduce the overall hydrophobicity of a growing peptide chain in a similar fashion to Lys(Tfa) and Met(O) for the cytochrome C synthesis mentioned above. The use of solvent mixtures containing both a polar and nonpolar component, such as 35% THF-NMP or 20% TFE-DCM, is recommended to alleviate the problem of side-chain-induced resin collapse *(34)*. The partial substitution or complete replacement of *t*Bu-based side-chain-protecting groups for carboxyl, hydroxyl, and amino side chains by more polar groups would also aid peptide-resin solvation *(34,65,69)*. Unfortunately, changes in side-chain-protecting group structure are not correlated easily to β-sheet disrupting potential *(70)*.

5. Acylation Conditions

The use of mixed solvents requires careful consideration of acylation conditions. Dioxane-DMF is compatible with carbodiimide activation, and thus has been used with preformed species or *in situ (59,60)*. Twenty percent TFE-DCM and 10% HFIP-DCM are used by preforming an activated species in DCM with a carbodiimide, and then adding TFE or HFIP after filtering the urea *(26,42)*. For Fmoc amino acids that are not entirely soluble in DCM, a small amount of DMF is used for solubilization, followed by the addition of DCM and the carbodiimide. For 20% DMSO-NMP, HOBt esters have been preformed in NMP with carbodiimide, with subsequent addition of DMSO *(55)*.

Chaotropic salts have been shown to inhibit interchain β-sheet aggregates and, hence, improve peptide-resin solvation and coupling efficiencies *(71–73)* (Table 1). The efficiency of acylation reactions in organic solvents containing chaotropic salts is highly dependent on the nature of the salt and/or solvent *(53,71–73)*; 0.4*M* $NaClO_4$, KSCN, or LiBr was helpful for several couplings in DCM-DMF (1:1) during the Boc SPPS of RNase 1-13-MBHA-PS *(72)*; 2*M* LiBr in THF was shown to be excellent at inhibiting interchain aggregation and enhancing peptide-resin solvation, but coupling yields were considerably reduced in this solvent system compared to DCM *(73)*; 0.4*M* LiCl in NMP, but not in DMF, was advantageous for coupling Fmoc-Ala-OPfp to $(Ala)_5$-Phe-HMP-PS *(71)*. The same coupling was only slightly improved by 0.4*M* LiCl in NMP when POE-PS or encapsulated Pepsyn resins were used *(71)*. Problems of complexes between salts and amino acid side-chain functionalities, such as Thr(Bzl) and His(Bom), must be considered, and preactivation in the presence of the salt avoided *(72)*.

An interesting alternative approach to the study of synthetic efficiency as a function of solvent composition was an evaluation of synthetic efficiency as a function of peptide-resin charge state *(74–76)*. The use of 50% DCM-DMF or DMF for couplings of difficult sequences was effective for trifluoroacetic acid-treated peptide resins, but not for neutralized peptide resins *(75,76)*. Couplings are by "*in situ* neutralization" methods *(74–78)*.

References

1. Pugh, K., York, E. J., and Stewart, J. M. (1992) Effects of resin swelling and substitution on solid phase synthesis. *Int. J. Peptide Protein Res.* **40,** 208–213.
2. Manatt, S. L., Horowitz, D., Horowitz, R., and Pinnell, R. P. (1980) Solvent swelling for enhancement of carbon-13 nuclear magnetic resonance spectral information from insoluble polymers: chloromethylation levels in crosslinked polystyrenes. *Anal. Chem.* **52,** 1529–1532.
3. Ford, W. T. and Balakrishnan, T. (1981) Carbon-13 nuclear magnetic resonance relaxation in cross-linked polystyrene gels. *Macromolecules* **14,** 284–288.

4. Live, D. and Kent, S. B. H. (1982) Fundamental aspects of the chemical applications of cross-linked polymers, in *Elastomers and Rubber Elasticity* (Mark, J. E., ed.), American Chemical Society, Washington, DC, pp. 501–515.
5. Ludwick, A. G., Jelinski, L. W., Live, D., Kintanar, A., and Dumais, J. J. (1986) Association of peptide chains during Merrifield solid-phase peptide synthesis: a deuterium NMR study. *J. Am. Chem. Soc.* **108,** 6493–6496.
6. Albericio, F., Pons, M., Pedroso, E., and Giralt, E. (1989) Comparative study of supports for solid-phase coupling of protected-peptide segments. *J. Org. Chem.* **54,** 360–366.
7. Rudinger, J. and Buetzer, P. (1975) Some rate measurements in solid phase synthesis, in *Peptides 1974* (Wolman, Y., ed.), Halsted, New York, pp. 211–219.
8. Hetnarski, B. and Merrifield, R. B. (1988) Kinetics of coupling reactions in solid phase peptide synthesis, in *Peptides: Chemistry and Biology* (Marshall, G. R., ed.), Escom, Leiden, The Netherlands, pp. 220–222.
9. Pickup, S., Blum, F. D., and Ford, W. T. (1990) Self-diffusion coefficients of Boc-amino acid anhydrides under conditions of solid phase peptide synthesis. *J. Polym. Sci. A: Polym. Chem.* **28,** 931–934.
10. Sarin, V. K., Kent, S. B. H., and Merrifield, R. B. (1980) Properties of swollen polymer networks: solvation and swelling of peptide-containing resins in solid-phase peptide synthesis. *J. Am. Chem. Soc.* **102,** 5463–5470.
11. Sarin, V. K., Kent, S. B. H., Mitchell, A. R., and Merrifield, R. B. (1984) A general approach to the quantitation of synthetic efficiency in solid-phase peptide synthesis as a function of chain length. *J. Am. Chem. Soc.* **106,** 7845–7850.
12. Arshady, R., Atherton, E., Clive, D. L. J., and Sheppard, R. C. (1981) Peptide synthesis, part 1: preparation and use of polar supports based on poly(dimethylacrylamide). *J. Chem. Soc. Perkin Trans.* **I,** 529–537.
13. Bayer, E., Hemmasi, B., Albert, K., Rapp, W., and Dengler, M. (1983) Immobilized polyoxyethylene, a new support for peptide synthesis, in *Peptides: Structure and Function* (Hruby, V. J. and Rich, D. H., eds.), Pierce Chemical Co., Rockford, IL, pp. 87–90.
14. Bayer, E. and Rapp, W. (1986) New polymer supports for solid-liquid-phase peptide synthesis, in *Chemistry of Peptides and Proteins*, vol. 3 (Voelter, W., Bayer, E., Ovchinnikov, Y. A., and Ivanov, V. T., eds.), Walter de Gruyter & Co., Berlin, pp. 3–8.
15. Zalipsky, S., Albericio, F., and Barany, G. (1985) Preparation and use of an aminoethyl polyethylene glycol-crosslinked polystyrene graft resin support for solid-phase peptide synthesis, in *Peptides: Structure and Function* (Deber, C. M., Hruby, V. J., and Kopple, K. D., eds.), Pierce Chemical Co., Rockford, IL, pp. 257–260.
16. Zalipsky, S. and Barany, G. (1986) Preparation of polyethylene glycol derivatives with two different functional groups at the termini. *Polymer Preprints* **27,** 1–2.
17. Barany, G., Albericio, F., Biancalana, S., Bontems, S. L., Chang, J. L., Eritja, R., Ferrer, M., Fields, C. G., Fields, G. B., Lyttle, M. H., Solé, N. A., Tian, Z., Van Abel, R. J., Wright, P. B., Zalipsky, S., and Hudson, D. (1992) Biopolymer synthesis on novel polyethylene glycol-polystyrene (PEG-PS) graft supports, in *Peptides:*

Chemistry and Biology (Smith, J. A., and Rivier, J. E., eds.), Escom, Leiden, The Netherlands, pp. 603,604.

18. Hellermann, H., Lucas, H.-W., Maul, J., Pillai, V. N. R., and Mutter, M. (1983) Poly(ethylene glycol)s grafted onto crosslinked polystyrenes, 2: multidetachably anchored polymer systems for the synthesis of solubilized peptides. *Makromol. Chem.* **184,** 2603–2617.
19. Bayer, E., Albert, K., Willisch, H., Rapp, W., and Hemmasi, B. (1990) ^{13}C NMR relaxation times of a tripeptide methyl ester and its polymer-bound analogues. *Macromolecules* **23,** 1937–1940.
20. Bayer, E. (1991) Towards the chemical synthesis of proteins. *Angew. Chem. Int. Ed. Engl.* **30,** 113–129.
21. Fields, G. B. and Fields, C. G. (1992) Optimization strategies for Fmoc solid-phase peptide synthesis: synthesis of triple-helical collagen-model peptides, in *Innovation and Perspectives in Solid Phase Synthesis—Peptides, Polypeptides and Oligonucleotides—1992* (Epton, R., ed.), Intercept, Andover, UK, pp. 153–162.
22. Hancock, W. S., Prescott, D. J., Vagelos, P. R., and Marshall, G. R. (1973) Solvation of the polymer matrix: source of truncated and deletion sequences in solid phase synthesis. *J. Org. Chem.* **38,** 774–781.
23. Westall, F. C. and Robinson, A. B. (1970) Solvent modification in Merrifield solid-phase peptide synthesis. *J. Org. Chem.* **35,** 2842–2844.
24. Kent, S. B. H. and Merrifield, R. B. (1981) The role of crosslinked resin supports in enhancing the solvation and reactivity of self-aggregating peptides: solid phase syntheses of acyl carrier protein (65–74), in *Peptides 1980* (Brunfeldt, K., ed.), Pierce Chemical Co., Rockford, IL, pp. 328–333.
25. Sheppard, R. C. (1973) Solid phase peptide synthesis—an assessment of the present position, in *Peptides 1971* (Nesvadba, H., ed.), North-Holland Publishers, Amsterdam, pp. 111–125.
26. Yamashiro, D., Blake, J., and Li, C. H. (1976) The use of trifluoroethanol for improved coupling in solid-phase peptide synthesis. *Tetrahedron Lett.* 1469–1472.
27. Nakaie, C. R., Marchetto, R., Schreier, S., and Paiva, A. C. M. (1988) Synthetic and physicochemical studies of benzhydrylamine resins with different substitution levels: implications for solid phase peptide synthesis, in *Peptides: Chemistry and Biology* (Marshall, G. R., ed.), Escom, Leiden, The Netherlands, pp. 249–251.
28. Meister, S. M. and Kent, S. B. H. (1983) Sequence-dependent coupling problems in stepwise solid phase peptide synthesis: occurrence, mechanism, and correction, in *Peptides: Structure and Function* (Hruby, V. J. and Rich, D. H., eds.), Pierce Chemical Co., Rockford, IL, pp. 103–106.
29. Narita, M., Umeyama, H., and Yoshida, T. (1989) The easy disruption of the β-sheet structure of resin-bound human proinsulin C-peptide fragments by strong electron-donor solvents. *Bull. Chem. Soc. Jpn.* **62,** 3582–3586.
30. Live, D. H. and Kent, S. B. H. (1983) Correlation of coupling rates with physicochemical properties of resin-bound peptides in solid phase synthesis, in *Peptides: Structure and Function* (Hruby, V. J. and Rich, D. H., eds.), Pierce Chemical Co., Rockford, IL, pp. 65–68.

31. Kent, S. B. H. (1985) Difficult sequences in stepwise peptide synthesis: common molecular origins in solution and solid phase?, in *Peptides: Structure and Function* (Deber, C. M., Hruby, V. J., and Rich, D. H., eds.), Pierce Chemical Co., Rockford, IL, pp. 407–414.
32. Narita, M. and Kojima, Y. (1989) The β-sheet structure-stabilizing potential of twenty kinds of amino acid residues in protected peptides. *Bull. Chem. Soc. Jpn.* **62,** 3572–3576.
33. Fields, G. B., Otteson, K. M., Fields, C. G., and Noble, R. L. (1990) The versatility of solid phase peptide synthesis, in *Innovation and Perspectives in Solid Phase Synthesis* (Epton, R., ed.), Solid Phase Conference Coordination, Ltd., Birmingham, UK, pp. 241–260.
34. Fields, G. B. and Fields, C. G. (1991) Solvation effects in solid-phase peptide synthesis. *J. Am. Chem. Soc.* **113,** 4202–4207.
35. Geiser, T., Beilan, H., Bergot, B. J., and Otteson, K. M. (1988) Automation of solid-phase peptide synthesis, in *Macromolecular Sequencing and Synthesis: Selected Methods and Applications* (Schlesinger, D. H., ed.), Liss, New York, pp. 199–218.
36. Kitas, E. A., Knorr, R., Trzeciak, A., and Bannwarth, W. (1991) Alternative strategies for the Fmoc solid-phase synthesis of O^4-phospho-L-tyrosine-containing peptides. *Helv. Chim. Acta* **74,** 1314–1328.
37. Young, J. D., Huang, A. S., Ariel, N., Bruins, J. B., Ng, D., and Stevens, R. L. (1990) Coupling efficiencies of amino acids in solid phase synthesis of peptides. *Peptide Res.* **3,** 194–200.
38. van Woerkom, W. J. and van Nispen, J. W. (1991) Difficult couplings in stepwise solid phase peptide synthesis: predictable or just a guess? *Int. J. Peptide Protein Res.* **38,** 103–113.
39. Wang, S. and Foutch, G. L. (1989) Reaction rates for the production of selected hormones by solid-phase peptide synthesis. *Biotechnol. Prog.* **7,** 111–115.
40. Narita, M., Ishikawa, K., Chen, J. -Y., and Kim, Y. (1984) Prediction and improvement of protected peptide solubility in organic solvents. *Int. J. Peptide Protein Res.* **24,** 580–587.
41. Narita, M., Isokawa, S., Honda, S., Umeyama, H., Kakei, H., and Obana, S. (1989) Individuality of amino acid residues in protected peptides. *Bull. Chem. Soc. Jpn.* **62,** 773–779.
42. Milton, S. C. F. and Milton, R. C. de L. (1990) An improved solid-phase synthesis of a difficult-sequence peptide using hexafluoro-2-propanol. *Int. J. Peptide Protein Res.* **36,** 193–196.
43. Narita, M., Isokawa, S., Tomotake, Y., and Nagasawa, S. (1983) Synthesis and the solid-state conformations of cross-linked resin-bound oligo(leucine)s. *Polymer J.* **15,** 25–32.
44. Narita, M., Tomotake, Y., Isokawa, S., Matsuzawa, T., and Miyauchi, T. (1984) Syntheses and properties of resin-bound oligopeptides 2. *Macromolecules* **17,** 1903–1906.
45. Narita, M., Isokawa, S., Matsuzawa, T., and Miyauchi, T. (1985) Liquid-phase peptide synthesis by fragment condensation on soluble polymer support 7. *Macromolecules* **18,** 1363–1366.

46. Mutter, M., Altmann, K. H., Bellof, D., Flörsheimer, A., Herbert, J., Huber, M., Klein, B., Strauch, L., Vorherr, T., and Gremlich, H. U. (1985) The impact of secondary structure formation in peptide synthesis, in *Peptides: Structure and Function* (Deber, C. M., Hruby, V. J., and Kopple, K. D., eds.), Pierce Chemical Co., Rockford, IL, pp. 397–405.
47. Narita, M., Umeyama, H., Isokawa, S., Honda, S., Sasaki, C., and Kakei, H. (1989) The electron donor-acceptor interaction between mixed solvents and its influence on their β-sheet structure-disrupting potential. *Bull. Chem. Soc. Jpn.* **62,** 780–785.
48. Narita, M., Umeyama, H., and Yoshida, T. (1989) Peptide segment separation by tertiary peptide bonds. *Bull. Chem. Soc. Jpn.* **62,** 3577–3581.
49. Deber, C. M., Lutek, M. K., Heimer, E. P., and Felix, A. M. (1989) Conformational origin of a difficult coupling in a human growth hormone releasing factor analog. *Peptide Res.* **2,** 184–188.
50. Milton, R. C. de L., Milton, S. C. F., and Adams, P. A. (1990) Prediction of difficult sequences in solid-phase peptide synthesis. *J. Am. Chem. Soc.* **112,** 6039–6046.
51. Larsen, B. D., Larsen, C., and Holm, A. (1991) Incomplete Fmoc-deprotection in solid phase synthesis, in *Peptides 1990* (Giralt, E. and Andreu, D., eds.), Escom, Leiden, The Netherlands, pp. 183–185.
52. Atherton, E. and Sheppard, R. C. (1985) Detection of problem sequences in solid phase synthesis, in *Peptides: Structure and Function* (Deber, C. M., Hruby, V. J., and Kopple, K. D., eds.), Pierce Chemical Co., Rockford, IL, pp. 415–418.
53. Hyde, C., Johnson, T., and Sheppard, R. C. (1992) Internal aggregation during solid phase peptide synthesis: dimethyl sulfoxide as a powerful dissociating solvent. *J. Chem. Soc., Chem. Commun.*, 1573–1575.
54. Giralt, E., Eritja, R., Pedroso, E., Granier, C., and van Rietschoten, J. (1986) Convergent solid phase peptide synthesis III: synthesis of the 44-52 protected segment of the toxin II of *Androctonus australis Hector. Tetrahedron* **42,** 691–698.
55. Bagley, C. J., Otteson, K. M., May, B. L., McCurdy, S. N., Pierce, L., Ballard, F. J., and Wallace, J. C. (1990) Synthesis of insulin-like growth factor I using *N*-methyl pyrrolidinone as the coupling solvent and trifluromethane sulphonic acid cleavage from the resin. *Int. J. Peptide Protein Res.* **36,** 356–361.
56. Hoeprich, P. D., Jr. and Hugli, T. E. (1986) Helical conformation at the carboxy-terminal portion of human C3a is required for full activity. *Biochemistry* **25,** 1945–1950.
57. Ho, S. P. and DeGrado, W. F. (1987) Design of a 4-helix bundle protein: synthesis of peptides which self-associate into a helical protein. *J. Am. Chem. Soc.* **109,** 6751–6758.
58. Mapelli, C. and Swerdloff, M. D. (1991) Monitoring of conformational and reaction events in resin-bound peptides by ^{13}C NMR spectroscopy in various solvents, in *Peptides 1990* (Giralt, E. and Andreu, D., eds.), Escom, Leiden, The Netherlands, pp. 316–319.
59. Ogunjobi, O. and Ramage, R. (1990) Ubiquitin: preparative chemical synthesis, purification and characterization. *Biochem. Soc. Trans.* **18,** 1322–1333.

60. Nozaki, S. (1990) Solid-phase synthesis of steroidogenesis-activator polypeptide under continuous flow conditions. *Bull. Chem. Soc. Jpn.* **63,** 842–846.
61. Gutmann, V. (1978) *The Donor-Acceptor Approach to Molecular Interactions,* Plenum, New York.
62. Narita, M., Honda, S., and Obana, S. (1989) The β-sheet structure-disrupting potential of electron-donor and -acceptor solvents and role of mixed solvents in solvation of peptides. *Bull. Chem. Soc. Jpn.* **62,** 342–344.
63. Barton, A. F. M. (1983) *CRC Handbook of Solubility Parameters and Other Cohesion Parameters,* CRC, Boca Raton, FL.
64. Barton, A. F. M. (1975) Solubility parameters. *Chem. Rev.* **75,** 731–753.
65. Atherton E., Woolley, V., and Sheppard, R. C. (1980) Internal association in solid phase peptide synthesis: synthesis of cytochrome C residues 66-104 on polyamide supports. *J. Chem. Soc., Chem. Commun.* 970–971.
66. Guy, H. R. (1985) Amino acid side-chain partition energies and distribution of residues in soluble proteins. *Biophys. J.* **47,** 61–70.
67. Radzicka, A. and Wolfenden, R. (1988) Comparing the polarities of the amino acids: Side-chain distribution coefficients between the vapor phase, cyclohexane, 1-octanol, and neutral aqueous solution. *Biochemistry* **27,** 1664–1670.
68. Atherton, E., Pinori, M., and Sheppard, R. C. (1985) Peptide synthesis, part 6: Protection of the sulphydryl group of cysteine in solid-phase synthesis using N^{α}-fluorenylmethoxycarbonylamino acids: linear oxytocin derivatives. *J. Chem. Soc. Perkin Trans.* **I,** 2057–2064.
69. Bedford, J., Johnson, T., Jun, W., and Sheppard, R. C. (1992) Rate-slowing effects and side reactions in the solid phase peptide-resin matrix, in *Innovation and Perspectives in Solid Phase Synthesis—Peptides, Polypeptides and Oligonucleotides—1992* (Epton, R., ed.), Solid Phase Conference Coordination, Ltd., Birmingham, UK, pp. 213–219.
70. Bedford, J., Hyde, C., Johnson, T., Jun, W., Owen, D., Quibell, M., and Sheppard, R. C. (1992) Amino acid structure and "difficult sequences" in solid phase peptide synthesis. *Int. J. Peptide Protein Res.* **40,** 300–307.
71. Thaler, A., Seebach, D., and Cardinaux, F. (1991) Lithium-salt effects in peptide synthesis, part II: improvement of degree of resin swelling and of efficiency of coupling in solid-phase synthesis. *Helv. Chim. Acta* **74,** 628–643.
72. Stewart, J. M. and Klis, W. A. (1990) Polystyrene-based solid phase peptide synthesis: the state of the art, in *Innovation and Perspectives in Solid Phase Synthesis* (Epton, R., ed.), Solid Phase Conference Coordination, Ltd., Birmingham, UK, pp. 1–9.
73. Hendrix, J. C., Halverson, K. J., Jarrett, J. T., and Lansbury, P. T., Jr. (1990) A novel solvent system for solid-phase synthesis of protected peptides: the disaggregation of resin-bound antiparallel β-sheet. *J. Org. Chem.* **55,** 4517,4518.
74. Jezek, J. and Houghten, R. A. (1991) A comparative study of BOP as a coupling agent using simultaneous multiple peptide synthesis (SMPS), in *Peptides 1990* (Giralt, E. and Andreu, D., eds.), Escom, Leiden, The Netherlands, pp. 74,75.
75. Beyermann, M. and Bienert, M. (1992) Synthesis of difficult peptide sequences: a comparison of Fmoc- and Boc-technique. *Tetrahedron Lett.* **33,** 3745–3748.

76. Schnölzer, M., Alewood, P., Jones, A., Alewood, D., and Kent, S. B. H. (1992) *In situ* neutralization in Boc-chemistry solid phase peptide synthesis: rapid, high yield assembly of difficult sequences. *Int. J. Peptide Protein Res.* **40,** 180–193.
77. Suzuki, K., Nitta, K., and Endo, N. (1975) Suppression of diketopiperazine formation in solid phase peptide synthesis. *Chem. Pharm. Bull.* **23,** 222–224.
78. Gairi, M., Lloyd-Williams, P., Albericio, F., and Giralt, E. (1990) Use of BOP reagent for the suppression of diketopiperazine formation in Boc/Bzl solid-phase peptide synthesis. *Tetrahedron Lett.* **31,** 7363–7366.

CHAPTER 4

HF Cleavage and Deprotection Procedures for Peptides Synthesized Using a Boc/Bzl Strategy

Michael W. Pennington

1. Introduction

On completion of chemical synthesis of the peptide chain, the final step requires the removal from the solid-phase support and liberation of the protected side chains of the trifunctional amino acids *(1)*. Many different approaches to this problem have been established, but the procedure most widely used for all Boc/Bzl-based peptides has been treatment with liquid HF.

Anhydrous HF is a strongly protonating acid ($H_0 = -10.8$) *(2)* with the additional properties of being a nonoxidizing, excellent solvent for peptides *(3)*. In conventional strong-acid cleavage methods, one major drawback of using HF is the number of deleterious side reactions that are promoted by the S_N1 removal of protecting groups. The solvolysis reaction generates benzyl carbocations that are stabilized by HF. These carbocations are potent alkylating species of amino acids with nucleophilic side chains, such as cysteine, methionine, tryptophan, and tyrosine *(4–9)*. The addition of compounds, such as anisole, dimethyl sulfide, and *p*-cresol, to the reaction mixture helps to minimize these reactions by scavenging these carbocations *(10)*. Several of the most commonly used scavengers in Boc/Bzl-based strategies are shown in Table 1.

An additional consequence of the strong protonating ability of HF is the generation of an acylium ion from the dehydration of the side chains of aspartic acid and glutamic acid. The acylium ion then leads to irre-

From: *Methods in Molecular Biology, Vol. 35: Peptide Synthesis Protocols*
Edited by: M. W. Pennington and B. M. Dunn

Table 1
Commonly Used Scavengers in HF Deprotection Schemes

	Trp	Met	Tyr	Cys
Anisole	–	–	o	–
m- or *p*-Cresol	o	o	+	o
Dimethylsulfide	o	+	o	o
p-Thiocresol	+	+	+	+
1,2 Ethanedithiol	+	+	–	+
4-Methylindole	+	o	–	o

+ = good.
o = fair.
– = poor.

versible side products of anisylated glutamic acid *(11)* and aspartimide formation in the case of aspartic acid *(12,13)*.

More recently, an improvement over the conventional strong-acid S_N1 deprotection process has been developed by Tam and coworkers *(14)*. This is a two-part procedure that incorporates an S_N2 process as an initial step followed by an S_N1 step to remove the more resistant protecting groups. In the initial step, this procedure effectively lowers the acidity function of the HF while incorporating a nucleophile to promote an S_N2 mechanism. In an S_N2 process, reactions proceed with less carbocation character in the transition state and are aided by the participation of a nucleophile in the heterolysis of the C—O bond. By incorporating a weak base, such as a sulfide or mercaptan, to serve as a solvent for the HF, the acidity function of the HF is lowered to an H_0 between –4.6 and –5.2 *(14)*. This acid strength is strong enough to protonate most of the weak base protecting groups, such as benzyl carbamates, benzyl esters, benzyl ethers, and benzyl carbonates. Furthermore, this weak base owing to the lowered acidity of the HF remains partially unprotonated and, thus, can act as a nucleophile assisting in the removal of the protonated protecting groups *(14)*. The byproducts generated from the cleaved protecting groups instead of being stable carbocations are relatively inert sulfonium salts or sulfides.

An additional step is required when using this procedure to remove the more stable protecting groups, such as the methylbenzyl group for cysteine, the tosyl group for arginine, and the cyclohexyl group for aspartic and glutamic acid. This is accomplished by removal of the volatile weak base as well as the HF. The reaction can then be resumed as an S_N1 process using the high concentration of HF used in the conventional procedure *(14)*. The amount of benzyl carbocations will be greatly reduced and will minimize most of the potential problems described above. This two-step procedure has been named the "low-high HF deprotection procedure."

It is the purpose of this chapter to describe a variety of different cleavage techniques that we have used successfully to cleave and deprotect relatively simple as well as complex peptides. Descriptions of a number of precleavage solid-phase deprotection procedures will be given as well. Although alternative strategies exist, such as a number of different variations of the TFMSA-TFA procedure *(15–17)*, in this chapter we will focus on the HF-based procedure.

2. Materials

1. HF cylinders may be purchased from Matheson Gas Company (Newark, NJ). This material is highly corrosive and toxic. HF dissolves glass and must be handled in an HF-resistant apparatus in a scrubber equipped fume hood. It is mandatory that the user wear eye protection, rubber gloves, rubber apron, and face shield.
2. A suitable HF-resistant cleavage apparatus may be purchased from Immuno-dynamics, Inc. (La Jolla, CA), Multiple Peptide Systems (San Diego, CA), or Peninsula Laboratories, Inc. (Belmont, CA). These instruments are made from the HF-resistant polymer Kel-F.
3. It is recommended that all users keep the HF antidote calcium gluconate in the laboratory where HF is handled. This product is sold by Peptides International (Louisville, KY).
4. A wide variety of scavengers are routinely used in HF reactions. The following list may be obtained from Aldrich: anisole, *p*-cresol, *p*-thiocresol, dimethylsulfide, 1,2 ethanedithiol, *m*-cresol, ethylmethylsulfide, and thiophenol.
5. The solvents used for extraction may be obtained from Fisher Scientific. These solvents include diethyl ether, ethyl acetate, and glacial acetic acid.
6. The other reagents and solvents required for solid-phase deprotection reactions may be obtained from Fisher Scientific (Fair Lawn, NJ), Aldrich (Milwaukee, WI), J. T. Baker (Phillipsburg, NJ), or most other scientific supply sources. These include trifluoroacetic acid, dichloromethane, dimethyl formamide, piperidine, methanol, ethanol, ethanolamine, cyclohexylamine, and 2-mercaptoethanol. Piperidine is a DEA-controlled substance, and proper license will be required to purchase this reagent. If this presents

a problem, cyclohexylamine or ethanolamine may be substituted. Urea, acetic acid, and Tris may also be obtained from Fisher Scientific.

7. Fritted-glass funnels (medium and coarse porosity) and side-arm filtration flasks are necessary for pre- and post-HF work-up.

3. Methods

Successful cleavage and deprotection of the final peptide product represent one of the most crucial steps in the synthetic cycle. Poor cleavage procedures greatly increase the work-up time required to isolate the product because of the increased complexity of the crude product mixture. In general, several protecting groups may be removed directly from the resin-bound peptide prior to cleavage. Selection of the proper scavenger cocktail is also crucial to minimize deleterious side reactions during the HF cleavage. Finally, the time and temperature are two variables that can be easily modified if incomplete deprotection of the peptide appears to be a problem. This chapter will describe each of these topics, as well as present general protocols for standard and low-high HF cleavage procedures.

3.1. Precleaving Procedures

3.1.1. Removal of N-terminal Boc Group

Removal of the N-terminal Boc group eliminates the deleterious *t*-butyl carbocation from being generated in the HF reaction. This carbocation is a potent alkylator of amino acids with nucleophilic side chains.

1. Add the peptide resin to either a round-bottom or Erlenmeyer flask with a stir bar.
2. Add 50% (v/v) trifluoroacetic acid/dichloromethane to the flask (approx 20 mL/g of resin). For peptides containing reduced Met or unprotected Trp, dithiothreitol or methylindole may be added at a concentration of 1–5% wt/vol to the deblocking solution to prevent oxidation of these side chains.
3. Allow reaction to evolve CO_2 for several minutes before capping. Place the flask on a magnetic stir plate and allow to mix for 30 min. This reaction should be performed in a fume hood to prevent noxious trifluoroacetic acid fumes from entering the laboratory.
4. Filter the resin through a fritted-glass funnel to remove the trifluoroacetic acid, and wash the resin several times with dichloromethane. Some researchers also perform a base-wash following the deblocking to remove the TFA salt using 10% TEA in DCM.

5. Wash the resin with methanol several times followed by dichloromethane and then methanol last. Some laboratories prefer a final DCM wash to leave the resin "fluffy" as a means of improving the HF and anisole permeation of the resin.
6. Dry the resin overnight in a vacuum desiccator. Additionally, the resin may be dried on most automated synthesizers by passing N_2 through a vessel containing the peptide resin.

3.1.2. Deformylation of Trp(For)

Peptides containing formylated tryptophan can be conveniently deformylated prior to HF cleavage. This step is generally performed after removal of the final Boc group in order to eliminate exposure of the indole side chain to deleterious trifluoroacetic acid. Although deformylation of the peptide resin can be accomplished *in situ* during the HF reaction with the appropriate scavenger (1,2 EDT) *(18)*, many prefer to remove it prior to cleavage.

1. Suspend peptide resin in a solution of 20–25% piperidine (ethanolamine or cyclohexylamine) in dimethylformamide (20 mL/g of resin).
2. Stir or mix this suspension for 1 h at room temperature.
3. Filter on fritted-glass funnel, and wash resin with dimethyl formamide, followed by isopropanol, followed by dichloromethane.
4. Dry the resin overnight in a vacuum desiccator. When combining this procedure with other deblocking procedures, it is not necessary to dry the peptide resin prior to performing another step. It is only necessary to dry the resin prior to final HF cleavage.

3.1.3. Removal of Dnp from Histidine

The Dnp (dinitrophenyl)-protecting group is stable to HF as well as TFMSA cleavage procedures. This protecting group is most easily removed directly from the resin prior to cleavage. **Removal of the Dnp group should precede the deblocking of the N-terminus because of the reactivity that Dnp has for primary amines.**

1. Suspend resin in 10 mL of dimethylformamide/g of resin, and mix for several minutes. Filter off the DMF, and replace with fresh DMF.
2. Add 40 mmol of 2-mercaptoethanol to the suspension, and mix for 3–4 h. Thiophenol can also be used in place of 2-mercaptoethanol. The solution will turn bright yellow as the Dnp group is liberated from the histidine.
3. Filter over a fritted-glass funnel, and wash the resin several times with DMF followed by three washes of each solvent in the following order: H_2O followed by ETOH and, finally, DCM.

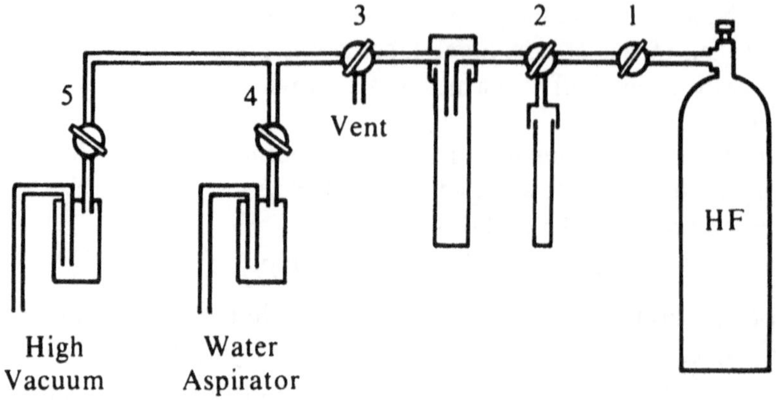

Fig. 1. Diagram of the HF manifold.

3.2. Standard HF Procedure

It is essential that the chemist establish a general protocol from which other variations can be derived. This section describes the cleavage protocol for a "simple peptide." Although the definition of such a peptide is rather ill-defined, one may consider peptides not containing tryptophan, methionine, cysteine, glutamic acid, aspartic acid, and tyrosine as simple (*see* Note 1). Even in cases where a peptide does contain these amino acids, many times the standard cleavage will be successful. An excellent description of the overall HF deprotection procedure may be found in Stewart and Young *(19)*.

1. A portion of the final deblocked peptide resin is placed into a dry Kel-F reaction vessel with a dry Teflon™-coated magnetic stir bar. Unless resin quantity is limited, no more than half of the resin should be cleaved until a successful cleavage of the peptide has been performed.
2. Anisole is pipeted into the reaction vessel at a ratio of 1 mL/g of resin. The anisole serves as a scavenger and also swells the resin support.
3. The reaction vessel is placed into the HF apparatus manifold. In an older HF apparatus, the user will use an aspirator to evacuate the entire system (*see* Fig. 1 for simple diagram). More recent instruments use nitrogen to purge the system.
4. The reaction vessel is cooled with dry ice-ethanol or with liquid nitrogen. For larger quantities of resin (>4 g), it may be advisable to use liquid nitrogen to speed up the overall cooling and distillation steps. Cooling for 5 min in liquid nitrogen or 10–15 min dry ice-ethanol is sufficient.

5. Begin the distillation of HF by closing of the valve to the aspirator. Next, slowly open the valve on the HF cylinder while keeping the reaction vessel cooled in the appropriate bath. HF will slowly begin to condense in the reaction vessel. Although many researchers distill HF to an initial HF reservoir and subsequently distill the HF from this reservoir to the reaction vessel, for larger quantities of resin, this may be impractical because of the time required to accomplish two large distillation steps. We have found that we can use the HF reservoir as a trap to catch the impurities and distill the HF directly into the reaction vessel. With nitrogen-pressured systems, the distillation time is dramatically shortened because of the nitrogen gas driving the HF into the cold vessel.
6. The distillation time required depends on the amount of HF needed. Use of liquid nitrogen will speed up the distillation process and works much better when volumes >100 mL of HF are required. Once the appropriate amount of HF (approx 9 mL/g of resin) has been distilled, the valve to the HF cylinder is closed as well as the valve to the reaction vessel. The remainder of the system can then be evacuated to remove any residual HF vapor from the system by opening the valve to the water aspirator.
7. The reaction vessel is now placed into an ice-water bath and allowed to warm up to 0–4°C. This warming process is slowed because of the low thermal conductivity of Kel-F. As the vessel warms, the frozen scavengers melt allowing the stir bar to be actuated.
8. Stirring of the resin-HF-anisole slurry must be thorough. Depending on the type of resin, the color of the slurry can vary from a warm orange-cream color for Pam (phenylacetamidomethyl) or Merrifield (chloromethyl) resins, or magenta for MBHA resins.
9. Under usual conditions, the slurry is stirred at 0°C for 45–60 min. If an unusually high number of Cys (Meb) or Arg (Tos) is found in the peptide, a longer reaction time may be required (up to 2 h) for complete deprotection of these more resistant protecting groups.
10. At the end of the reaction time, removal of the HF is accomplished by evaporation. With the water aspirator turned on and the system evacuated, the valve to the reaction vessel is slowly opened. This must be done very carefully to avoid the rapid bumping that will carry resin into the manifold and contaminate the system. As the HF begins to boil, stop turning the valve until the bumping subsides. Repeat this until the valve is fully open to the system. Continue magnetic stirring at 0°C until all of the liquid is evaporated.
11. At this point, many researchers proceed with work-up of the peptide. However, others remove any residual HF by pulling a high vacuum on the system with a vacuum pump. If one desires to perform this additional step,

continue on with these instructions. Otherwise, move on to step 12. A trap in between the vacuum pump and the reaction vessel is cooled with the dry ice-ethanol thermos for 10–15 min. Once the trap is cold, close the valve to the water aspirator. Close the valve to the system, and open the valve to the vacuum pump. Activate the vacuum pump, and slowly open the system valve to the vacuum pump. Residual HF can take up to 90 min to be completely removed. Once the HF has been eliminated, close the valve to the vacuum pump and disengage the pump.

12. Slowly open the system valve to the air. Remove the reaction vessel from the manifold and unscrew the top. Use of protective gear is still essential to prevent any possible residual HF burns.

3.3. Peptide Work-Up

Isolation of the crude product is the final step in this process. Depending on the solubility of the product, several different protocols may be used.

1. Add ether (*see* Note 2) directly to the reaction vessel, and stir the resin-scavenger-peptide mixture for 30–60 s. Pour this mixture over a filter flask equipped with a fritted-glass funnel (coarse or medium porosity). Filter the ether solution through the fritted-glass funnel. Repeat this ether extraction-wash step two to three times to remove residual scavengers and organic byproducts (*see* Note 3).
2. Transfer the fritted funnel containing the precipitated peptide to a clean filter flask, and extract the peptide from the resin mixture by stirring the mixture in 30–50 mL of 20% (v/v) acetic acid in water/g of resin. Although most peptides will dissolve in this solution, some more hydrophobic peptides require higher concentrations of acetic acid (*see* Note 4). Additionally, solutions containing chaotropic agents, such as 8*M* urea buffered with Tris acetate or 6*M* guanidinium hydrochloride again buffered with Tris acetate, may also be effective solvents. We use these solvents in cases where a ferricyanide oxidation will follow HF cleavage. For very acidic peptides (rich in Glu or Asp), we extract with aqueous ammonia. For particularly insoluble peptides, neat TFA or formic acid is an effective solvent. When using either neat formic acid or TFA, peptides containing Trp and Met are prone to oxidize. Incorporation of dithiothreitol or β-mercaptoethanol (1 to 2% wt/vol) will help prevent these side reactions. Filter this solution through the funnel. Repeat this two more times to extract all of the peptide. This peptide solution is appropriately called the "crude" peptide.
3. Solutions of acetic acid can be lyophilized directly by diluting the acetic acid concentration to <10% and shell freezing the solution in a lyophilization flask. Lyophilization concentrates the sample, making yield calcula-

tions much simpler, and also eliminates most of the residual volatile scavengers, such as anisole and *p*-cresol.

4. Peptides extracted with neat TFA can be precipitated directly to a solid in 5 vol of diethylether. The solid product is then washed with ether 3 × 30 mL/g of resin. The solid product may be dried overnight in a desiccator attached to a vacuum pump equipped with a dry-ice acetone trap.
5. For peptides containing Ser and Thr, an *N-O* acyl shift can occur under acidic conditions. The acetic acid solution can either be neutralized prior to lyophilization with NH_4OH, or by simply redissolving the peptide in a weak base, such as 5–10% (w/v) NH_4HCO_3 and allowing to stand for 30 min in this buffer prior to purification.

3.4. Strategies to Minimize Side Reactions

Several different strategies exist to minimize side reactions occurring during an HF cleavage. Some of these are simply different scavenger cocktails to eliminate deleterious alkylations or oxidations of amino acid side chains. Table 1 shows several of the scavengers used in Boc/Bzl-based strategies. Combinations of these scavengers have been very useful in minimizing a number of these side reactions. We have found the following combinations to be very effective.

1. As a standard rule, we only cleave a small portion of the completed peptide resin, usually one-third or less, until suitable cleavage conditions have been established.
2. For the standard peptide containing no difficult residues, the 1 part anisole or *p*- or *m*-cresol to 1 part resin to 9 parts HF works quite well. Anisole and *m*-cresol are liquids at room temperature and a little easier to use. *p*-Cresol must be melted at an elevated temperature in order to pipet accurately.
3. Lowering the temperature to between –10 and –5°C by incorporating a salt-ice bath will slow down aspartimide formation and anisylated glutamic acid. Synthesis of the peptide with cyclohexyl-protecting groups on Asp residues will further reduce the formation of aspartimide at reduced temperatures *(20)*. Occasionally, a longer reaction time may be required at these lower temperatures to remove more stable protecting groups, such as the Arg (Tos) and Lys (Cl-Z), completely.
4. For peptides containing Cys (Meb), we routinely use 1 part *p*-cresol to 1 part *p*-thiocresol to 18 parts HF (v/v)/g of resin *(21)*. This cocktail has been very useful for complex peptides that require postcleavage oxidative folding *(22)*.
5. For peptides containing Trp (For), which has not been removed by the abovementioned procedure, incorporation of 1,2 ethanedithiol (*see* Note 6) at

a ratio of 0.6 mL/g of resin in addition to the other appropriate scavengers (anisole or *p*-cresol) will effectively deformylate the Trp *(18)*. Addition of methylindole (skatole) at a ratio of 10 mg/g of resin will help protect the indole ring from deleterious reactions. We do not routinely use methylindole except in cases where destruction of Trp has been identified in an earlier small-scale cleavage.

6. For peptides containing Met, addition of dimethyl sulfide helps prevent alkylations of methionine by competing for carbocations. When necessary, 0.4 mL/g of resin has been a useful ratio of DMS in addition to the other scavengers required.
7. For peptides containing combinations of Met, Trp, and Cys, the standard high cleavage may be attempted with a combination of *p*-cresol, *p*-thiocresol, and DMS at a ratio of 1:1:0.4/g of resin with 18 parts HF for 1 h at 0°C, which usually gives satisfactory results (*see* Notes 7 and 8). However, the low-high procedure described in the next section may give better results in those cases where the standard high procedure fails or gives a very low yield.
8. Use of peroxide-free ether is mandatory for peptides containing oxidation-sensitive residues. Methyl-*tert*-butyl ether is less prone to peroxide formation than diethylether.

3.5. Low-High HF Procedure

In order to minimize or eliminate many of the deleterious side reactions in peptides containing some or all of these problematic amino acids, the low-high HF procedure was developed *(14)*. This procedure eliminates most of the alkylation problems by employing a nucleophilic-type cleavage, which does not generate carbocations. Furthermore, Trp(For) is cleaved, and Met(O) is reduced during the course of cleavage. This is accomplished by performing the cleavage in a low concentration of HF with a high concentration of scavengers. Subsequent to this low step, a high step is then required to cleave the more resistant protecting groups, such as Arg(Tos) and Cys(Meb), as well as some of the more resistant resin linkages (*see* Note 5).

1. Place 1–2 g of peptide resin and a stirring bar in the cleavage vessel. Use the following ratio of scavengers per gram of resin: 6.5 mL of DMS and 1.0 mL of *p*-cresol (for peptides containing Trp[For], use 0.75 mL *p*-cresol and 0.25 mL *p*-thiocresol) (*see* Note 6).
2. Place the reaction vessel into the HF apparatus manifold. Evacuate the system for 1–2 min. Note that the DMS will begin to boil. Close the valve to the reaction vessel, and cool with liquid nitrogen or dry ice-acetone. Following this cooling, the *p*-cresol and *p*-thiocresol will be solid.

3. Distill 2.5 mL of HF/g of resin. Allow the reaction vessel to warm up to 0°C. The *p*-cresol and *p*-thiocresol will slowly melt, freeing the stir bar. Cleave this mixture for 2 h at 0°C. At this point, peptides that do not contain Asp(cHx), Glu(cHx), Cys(Bzl), Cys(Meb), or Arg(Tos) synthesized on a chloromethyl-resin support will be completely cleaved and deprotected by extending the cleavage time to 3 h. Peptides synthesized with the more acid-stable acetamidomethyl (Pam) linkage are cleaved in 4 h. Other more stable resin linkages, such as benzhydrylamine and phenacyl linkages, require the high-HF procedure to give good yields.
4. Using the water aspirator, evaporate the HF and most of the DMS. This usually takes 1–1.5 h. At this point, we have found that removal of all of the DMS and sulfonium salts generated during this low cleavage significantly improves the quality of our product. We accomplish this by removing the reaction vessel from the manifold and extracting the peptide resin with diethyl ether. Perform all of these procedures in the hood since some residual HF may be observed fuming from the vessel. Following three ether washes of the partially cleaved peptide resin, we dry the resin *in vacuo* for 30 min. For peptides not containing any of the protecting groups mentioned in step 3, the high-HF step is not necessary, and the peptide may be solubilized following the ether extractions as described above in Section 3.3., steps 2–5. Otherwise, we replace the resin, *p*-cresol, and *p*-thiocresol in the same ratio as described above, back into the reaction vessel and place it back on the HF apparatus manifold for continuation of the high procedure.
5. The reaction vessel is evacuated and cooled as before, and recharged with HF. We generally charge the vessel with 10 mL of HF/g of resin. The peptide resin is cleaved for 1–1.5 h at –5 to 0°C.
6. Evaporate the HF, and follow the procedures described in the peptide workup in Section 3.3.

3.6. Cleavage Examples

Although no peptide is considered trivial, the cleavage procedure greatly affects the overall yield of the desired product. The following examples are provided to help in the selection of cleavage cocktails for a variety of different peptides.

3.6.1. Amylin (Human) (23)

- Sequence: H-Lys-Cys-Asn-Thr-Ala-Thr-Cys-Ala-Thr-Gln-Arg-Leu-Ala-Asn-Phe-Leu-Val-His-Ser-Ser-Asn-Asn-Phe-Gly-Ala-Ile-Leu-Ser-Ser-Thr-Asn-Val-Gly-Ser-Asn-Thr-Tyr-NH_2.

- Characteristics: 37-mer, basic with amidated C-terminus; contains 1 intramolecular disulfide bond; rather hydrophobic; considered difficult to make.
- Resin: Methylbenzhydrylamine (results in C-terminal amide).
- Cleavage: 4.0 g resin with 4.5 mL anisole and 40 mL HF at 0°C for 60 min.
- Extraction: Peptide dissolved in 400 mL of 8*M* urea containing 0.1*M* Tris-acetate, pH 7.2, followed by a ferricyanide oxidation procedure *(24)*.
- Remarks: Amino acid analysis and FAB/MS (*see* Chapter 7, *PAP*) (both after oxidation and purification) in accordance with theory. *(See* Fig. 2 and Note 7). Overall result, difficult peptide.

3.6.2. *Glu^{3,4,7,10,14}-Conantoxin-G* (25)

- Sequence: H-Gly-Glu-Glu-Glu-Leu-Gln-Glu-Asn-Gln-Glu-Leu-Ile-Arg-Glu-Lys-Ser-Asn-NH_2.
- Characteristics: 17-mer; very acidic; no Cys, Tyr, Trp, Met (*see* Note 8); considered easy to make; possible anisylation problem because of 6 Glu.
- Resin: Methylbenzhydrylamine.
- Cleavage: 2.0 g of resin, 2.0 mL *p*-cresol, and 20 mL HF at –5°C for 60 min.
- Extraction: Peptide dissolved in 20% acetic acid.
- Remarks: Amino acid analysis in accordance with theory. Excellent yield, easy peptide, no anisylation problem (*see* Fig. 3).

3.6.3. *α-MSH Antagonist* (26)

- Sequence: H-His-(D)-Trp-Ala-Trp-(D)-Phe-Lys-NH_2.
- Characteristics: Contains two Trp; basic peptide; rather easy.
- Resin: Methylbenzhydrylamine; Trp not deformylated prior to cleavage.
- Cleavage: 2.0 g resin, 2.0 mL anisole, 1.2 mL EDT, and 30 mL HF for 60 min at 0°C.
- Extraction: 20% Acetic acid.
- Remarks: Amino acid analysis in accordance with theory. Product successfully deformylated during cleavage by incorporation of EDT (absorbance max shifts from 290 to 280 nm). Easy peptide (*see* Fig. 4).

3.6.4. *Anthopleurin A* (27)

- Sequence: H-Gly-Val-Ser-Cys-Leu-Cys-Asp-Ser-Asp-Gly-Pro-Ser-Val-Arg-Gly-Asn-Thr-Leu-Ser-Gly-Thr-Leu-Trp-Leu-Tyr-Pro-Ser-Gly-Cys-Pro-Ser-Gly-Trp-His-Asn-Cys-Lys-Ala-His-Gly-Pro-Thr-Ile-Gly-Trp-Cys-Cys-Lys-Gln-OH.
- Characteristics: 49-mer: extremely difficult peptide contains 3 Trp, 6 Cys, Tyr, and acid-sensitive Asp-Gly bond. Oxidative folding required (*see* Note 9).
- Resin: Pam resin (results in C-terminal acid).
- Cleavage: Low-high HF procedure on 2.0 g resin, 13 mL DMS, 1.5 mL *p*-cresol, 0.5 mL *p*-thiocresol, and 5 mL HF for 2 h at 0°C. Followed by

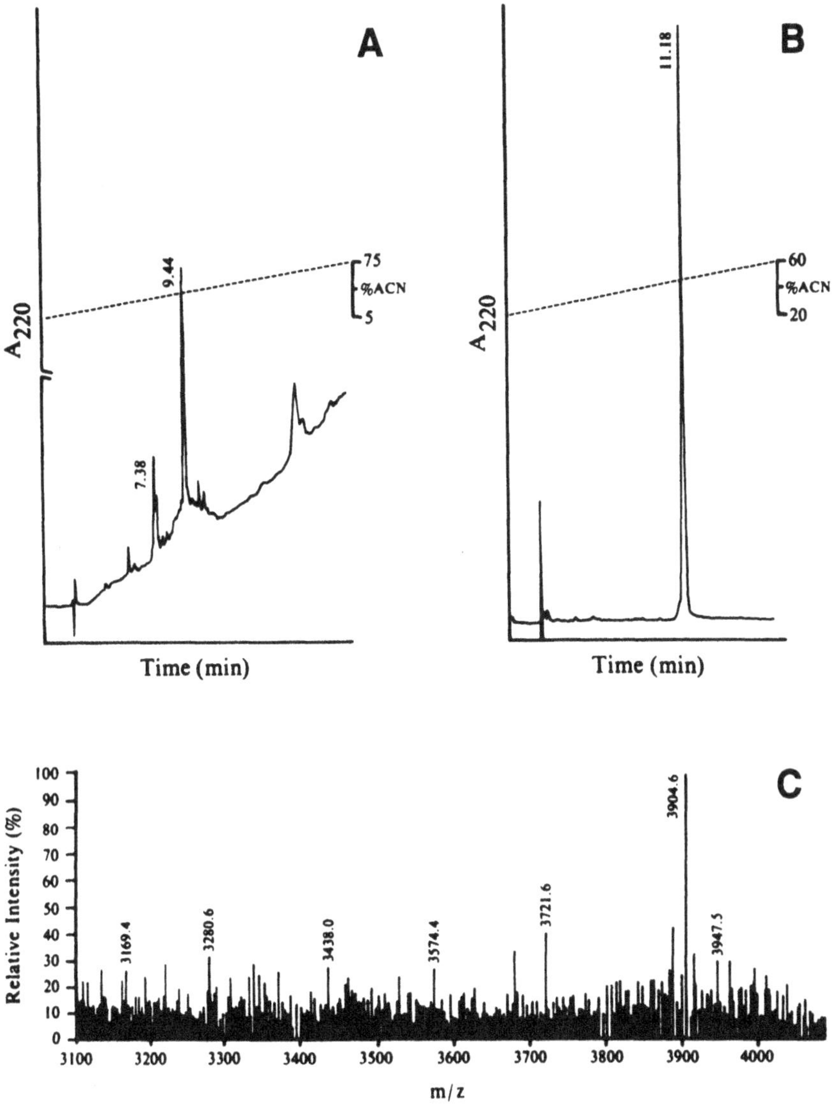

Fig. 2. RP-HPLC gradient elution of amylin. (**A**) Crude profile following ferricyanide oxidation. (**B**) Purified product. (**C**) FAB-MS analysis of purified amylin.

high cleavage with 1.5 mL *p*-cresol, 0.5 mL *p*-thiocresol, and 22 mL HF for 60 min at 0°C (*see* Note 10).

- Extraction: Required Tam-type dialysis oxidative refolding *(28)*. Peptide initially extracted into 8*M* urea containing 90 m*M* DTT and 0.1*M* Tris-acetate, pH 8.0.

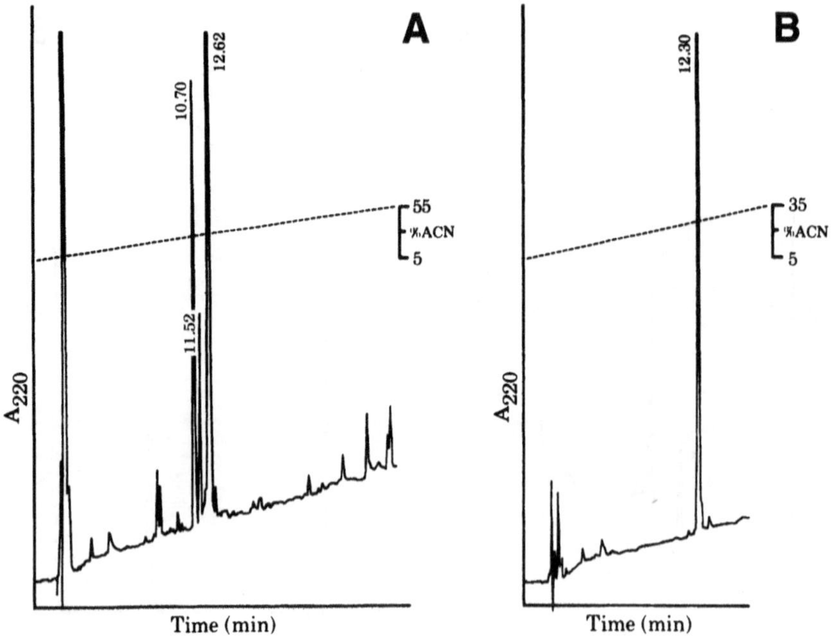

Fig. 3. RP-HPLC gradient elution of Glu-Con-G. (**A**) Crude profile—note residual scavenger (*p*-cresol) at 10.7 min—and (**B**) purified product.

- Remarks: Amino acid analysis and FAB-MS in accordance with theory (*see* Fig. 5). Very difficult peptide with extremely poor yield. Difficulties above expectations.

3.6.5. HIV-Substrate III-B (Native Sequence) (29)

- Sequence: H-His-Lys-Ala-Arg-Val-Leu-Ala-Glu-Ala-Met-Ser-NH_2.
- Characteristics: 11-mer; slightly basic peptide; contains Met; rather simple peptide to make.
- Resin: Methylbenzhydrylamine.
- Cleavage: 2.0 g resin, 2.0 mL anisole, 0.8 mL DMS, and 25 mL HF for 60 min at 0°C.
- Extraction: Easily solubilized in 10% acetic acid.
- Remarks: Amino acid analysis in accordance with theory. No problem with methionine oxidation. Easy peptide (*see* Fig. 6).

3.6.6. Cathepsin G (77–83) (30)

- Sequence: His-Pro-Gln-Tyr-Asn-Gln-Arg-OH.
- Characteristics: heptapeptide; basic; no Cys, Trp, Met; considered an easy peptide.

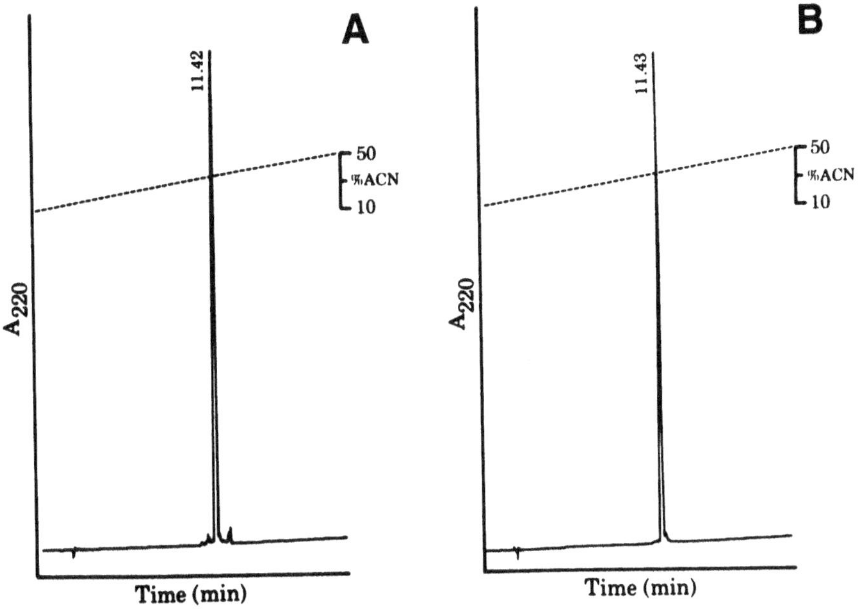

Fig. 4. RP-HPLC gradient elution profile of α-MSH antagonist. (**A**) Crude profile and (**B**) purified product.

- Resin: Merrifield resin.
- Cleavage: 2.0 g resin, 2.0 mL anisole, and 20 mL HF for 60 min at 0°C.
- Extraction: Peptide soluble in H_2O.
- Remarks: Major impurity found in cleavage mixture (*see* Note 11). FAB-MS analysis of this peak determined it to be an alkylation product of the Tyr residue with the 2,6 dichlorobenzyl-protecting group used to protect the Tyr hydroxyl group. The correct product was also isolated. This difficulty was not expected (*see* Fig. 7).

3.6.7. Echistatin (31)

- Sequence: H-Glu-Cys-Glu-Ser-Gly-Pro-Cys-Cys-Arg-Asn-Cys-Lys-Phe-Leu-Lys-Glu-Gly-Thr-Ile-Cys-Lys-Arg-Ala-Arg-Gly-Asp-Asp-Met-Asp-Asp-Tyr-Cys-Asn-Gly-Lys-Thr-Cys-Asp-Cys-Pro-Arg-Asn-Pro-His-Lys-Gly-Pro-Ala-Thr-OH.
- Characteristics: 49-mer; contains 8 Cys in 4 disulfide bonds; contains Met, Tyr, and an Asn-Gly bond; slightly basic peptide; requires postcleavage oxidative folding.
- Resin: Pam resin.
- Cleavage: 4.0 g resin, 2 mL *p*-cresol, 2 mL *p*-thiocresol, and 45 mL HF for 90 min at –5°C *(21,22)*.

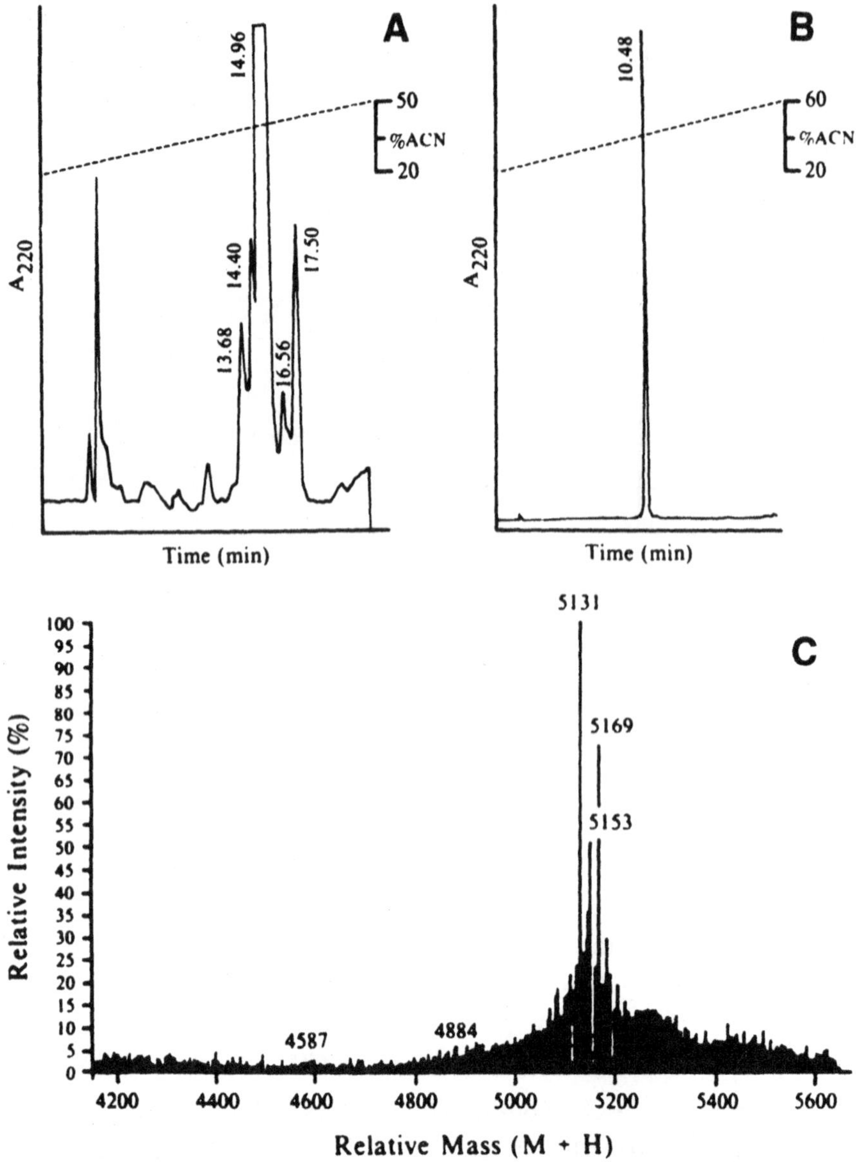

Fig. 5. RP-HPLC gradient elution profile of anthopleurin-A. (**A**) Crude oxidized product, (**B**) purified product, and (**C**) FAB-MS analysis of final purified product.

- Extraction: Dilute aqueous acetic acid.
- Remarks: Amino acid analysis and FAB-MS are in accordance with theory. Peptide reasonably well behaved during work-up. No precipitation problems. Moderately difficult (*see* Fig. 8).

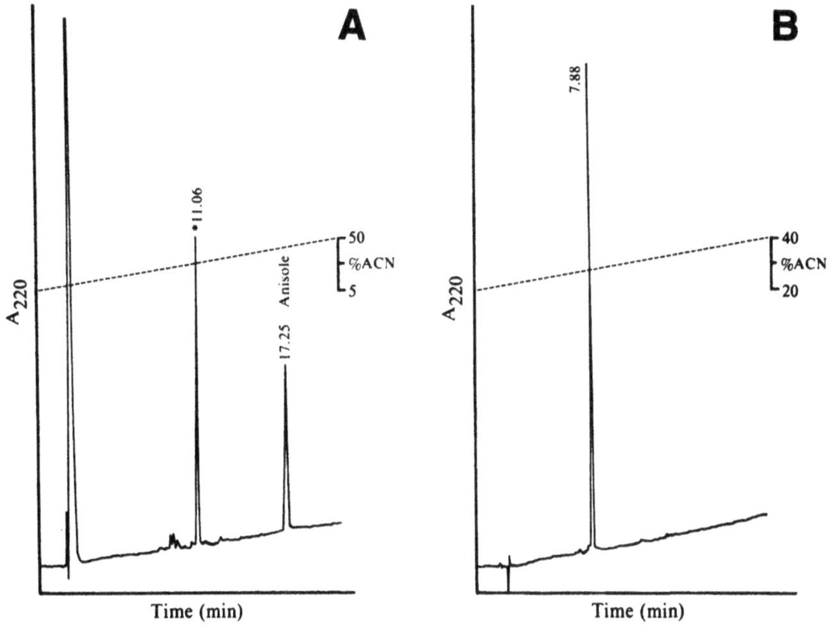

Fig. 6. RP-HPLC gradient elution profile of HIV substrate III-B. (**A**) Crude profile and (**B**) final purified product.

3.6.8. M-15 or Galantide (32)

- Sequence: Gly-Trp-Thr-Leu-Asn-Ser-Ala-Gly-Tyr-Leu-Leu-Gly-Pro-Gln-Gln-Phe-Phe-Gly-Leu-Met-NH_2.
- Characteristics: 20-mer; contains Trp, Met, and Tyr; very hydrophobic peptide; considered a difficult peptide.
- Resin: Methylbenzhydrylamine: Trp deformylated prior to cleavage.
- Cleavage: 4.0 g resin, 4.0 mL anisole, 1.6 mL DAMS, and 45 mL HF for 60 min at 0°C.
- Extraction: Neat TFA containing 2% β-mercaptoethanol. Product fairly insoluble.
- Remarks: Amino acid analysis in accordance with theory. Crude product contaminated with Trp oxidation product and Met(O) product (*see* Fig. 9). A rather difficult peptide where the Met is prone to oxidation.

4. Notes

1. Standard HF deprotection procedures will not remove the Cys(Acm)-protecting group. If this group is used, it is necessary to remove the Acm

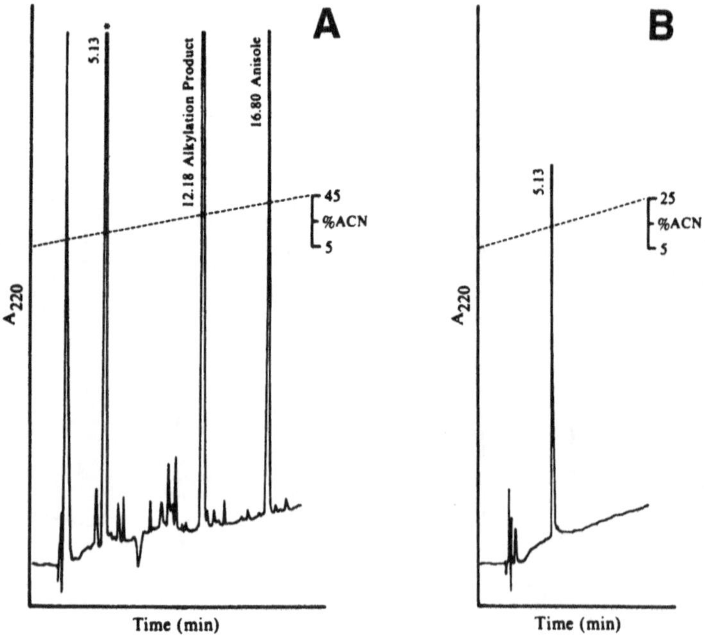

Fig. 7. RP-HPLC gradient elution profile of cathepsin-G 77-85. (**A**) Crude profile prior to lyophilization. Peak A correct product, peak B is the tyrosine alkylation product as determined by FAB-MS analysis, and peak C represents residual scavenger anisole. (**B**) Purified product.

postcleavage by treatment with mercuric acetate or by treatment with iodine *(33,34)*.

2. Peptides containing oxidation-sensitive residues, such as Met or Cys, should always be extracted with peroxide-free ether.
3. Never throw away the ether extract or the spent resin beads until your peptide has been successfully recovered. Many times during a poor extraction step, a great deal of peptide remains adhered to the resin beads and must be extracted with an alternative solvent.
4. Incorporation of N-terminal fluorophores, such as dansyl, dabcyl, 2,4-dinitrophenyl, and anthranilic acid, may reduce solubility. These fluorophores are stable to standard HF deprotection procedures. Isothiocyanate-type fluorophores are not stable to HF treatment.
5. The low-high HF procedure generally takes 5–7 h to perform and should be started at the beginning of the day so as to be able to complete in a reasonable time frame.

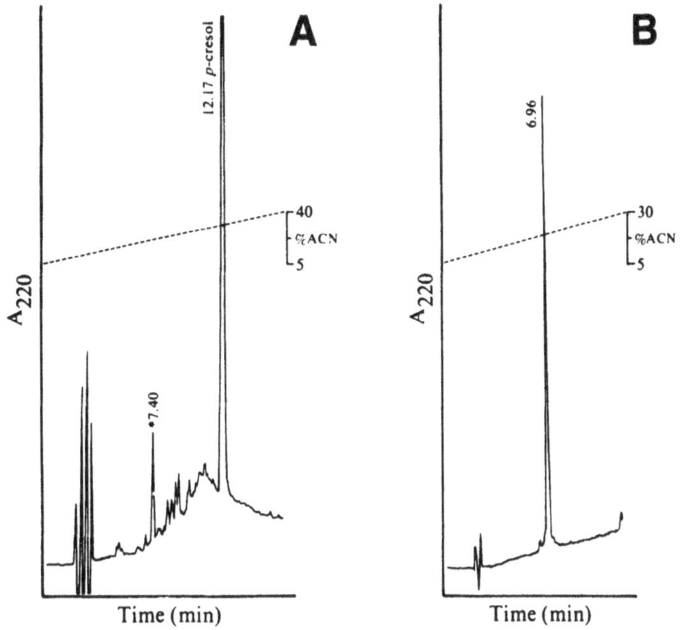

Fig. 8. RP-HPLC gradient profile of echistatin. (**A**) Crude oxidation profile. Note large residual *p*-cresol peak. (**B**) Purified product.

6. Always handle thiol-containing substances in proper ventilation. These compounds have an offensive odor that can be neutralized with bleach.
7. We routinely run an RP-HPLC analysis (*see* Chapter 3, *PAP*) on the crude product with a steep 5–95% acetonitrile gradient at 1.5 mL/min in 45 min (a 2%/min change) so as to identify unambiguously the product. This type of gradient allows the more hydrophobic peptides to be identified that may not elute on a shorter, shallower gradient.
8. As a general rule, Met(O) peptides elute slightly earlier on RP-HPLC than do the corresponding Met peptides. Additionally, Trp(For)-containing peptides elute slightly earlier than the corresponding Trp peptides.
9. Reduced peptides (Cys-SH) generally elute later than the oxidized (Cys-S-S-Cys) peptides.
10. Residual scavenger material may complicate an RP-HPLC analysis. Lyophilization will eliminate this problem. However, when this is not possible, diode array analysis will quickly identify the scavengers from peptide by the differing absorbance maxima.

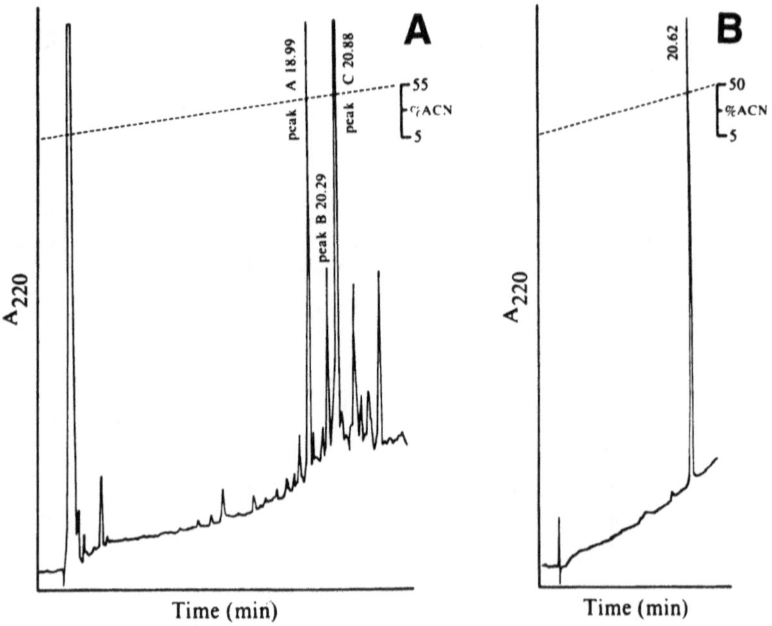

Fig. 9. RP-HPLC gradient elution profile of M-15 peptide. **(A)** Crude cleavage product. Note large side product peak A determined to be $Met(O)_2$, peak B determined to be Met(O), and peak C, which is Met. **(B)** Purified product.

11. Optimization of time parameters and temperature conditions must be determined on small-scale reactions. Never dedicate all of your resin to one cleavage unless you are confident of your procedure.

References

1. Merrifield, R. B. (1963) Solid phase synthesis peptide synthesis: the synthesis of a tetrapeptide. *J. Am. Chem. Soc.* **85,** 2149–2154.
2. Hyman, H. H. and Garber, R. A. (1959) The Hammet acidity function H_0 for trifluoroacetic acid solutions of sulfuric and hydrofluoric acids. *J. Am. Chem. Soc.* **81,** 1847–1849.
3. Sakakibara, S. (1971) The use of hydrogen fluoride in peptide chemistry, in *Chemistry and Biochemistry of Amino Acids, Peptides and Proteins.* vol. 1 (Weinstein, B., ed.), Dekker, New York, pp. 51–85.
4. Erickson, B. W. and Merrifield, R. B. (1973) The acid stability of several benzylic protecting groups used in solid-phase peptide synthesis. The rearrangement of *O*-benzyltyrosine to 3-benzyltyrosine. *J. Am. Chem Soc.* **95,** 3750–3756.
5. Engelhard, M. and Merrifield, R. B. (1978) Tyrosine protecting groups: minimization of rearrangement to 3-alkyltyrosine during acidolysis. *J. Am. Chem. Soc.* **100,** 3559–3563.

6. Yamashiro, D. and Li, C. H. (1973) Protection of tyrosine in solid-phase peptide synthesis. *J. Org. Chem.* **38,** 591,592.
7. Bodanszky, M. and Martinez, J. (1981) Side reactions in peptide synthesis. *Synthesis* 333–356.
8. Lundt, B. F., Johansen, N. C., Voelund, A., and Markussen, J. (1978) Removal of *t*-butyl and *t*-butyloxycarbonyl protecting groups with trifluoroacetic acid. *Int. J. Peptide and Protein Res.* **12,** 258–268.
9. Masui, Y., Chino, N., and Sakakibara, S. (1980) The modification of tryptophan residues during the acidolytic cleavage of Boc-groups. *Bull. Chem. Soc. Jpn.* **53,** 464–468.
10. Tam, J. P., Heath, W. F., and Merrifield, R. B. (1982) S_N1 and S_N2 mechanism for the deprotection of synthetic peptides by hydrogen fluoride. Studies to minimize tyrosine alkylation side reaction. *Int. J. Peptide Protein Res.* **21,** 57–65.
11. Feinberg, R. S. and Merrifield, R. B. (1975) Modification of peptides containing glutamic acid by hydrogen fluoride-anisole mixtures. Gamma-acylation of anisole or the glutamyl nitrogen. *J. Am. Chem. Soc.* **97,** 3485–3496.
12. Baba, T., Sugiyama, H., and Seto, S. (1973) Rearrangement of α and β-aspartyl peptide with anhydrous HF. *Chem. Pharm. Bull. (Tokyo)* **21,** 207-209.
13. Yang, L. L. and Merrifield, R. B. (1976) The α-phenacyl ester as a temporary protecting group to minimize cyclic imine formation during subsequent treatment of asparty peptides with HF. *J. Org. Chem.* **41,** 1032–1041.
14. Tam, J. P., Heath, W. F., and Merrifield, R. B. (1983) S_N2 deprotection of synthetic peptides with a low concentration of HF in dimethylsulfide: evidence and application in peptide synthesis. *J. Am. Chem. Soc.* **103,** 6442,6455.
15. Tam, J. P., Heath, W. F., and Merrifield, R. B. (1986) Mechanism for the removal of benzyl protecting groups in synthetic peptides by trifluoromethanesulfonic acid-trifluoroacetic acid-dimethylsulfide. *J. Am Chem. Soc.* **108,** 5242–5251.
16. Tam, J. P. and Merrifield, R. B. (1987) Strong acid cleavage of synthetic peptides: mechanisms and methods, in *The Peptides: Analysis, Synthesis and Biology,* vol. 9 (Meienhofer, J. and Udenfriend, eds.), Academic, Orlando, FL, pp. 185–248.
17. Yajima, H., Fujii, N., Ogawa, H., and Kawatani, H. (1974) Trifluoromethanesulfonic acid, as a deprotection reagent in peptide chemistry. *J. Chem. Soc. Chem. Commun.* pp. 107,108.
18. Matsueda, G. R. (1982) Deprotection of N^{in}-formyl tryptophan residues using 1,2 ethanedithiol in liquid hydrogen fluoride. *Int. J. Peptide Protein Res.* **20,** 26–34.
19. Stewart, J. M. and Young, J. D. (1984) in *Solid Phase Peptide Synthesis,* Pierce Chemical Co., Rockford IL.
20. Tam, J. P., Wong, T. W., Riemen, M. W., Tjoeng, F. S., and Merrifield, R. B. (1979) Cyclohexyl ester as a new protecting group for aspartyl peptides to minimize aspartimide formation in acidic and basic treatments. *Tet. Lett.* **42,** 4033–4036.
21. Heath, W. F., Tam, J. P., and Merrifield, R. B. (1986) Improved deprotection of cysteine-containing peptides in HF. *Int. J. Peptide Protein Res.* **28,** 498–507.
22. Garsky, V. M., Lumma, P. K., Freidinger, R. M., Pitzenberger, S. M., Randall, W. C., Veber, D. F., Gould, R. J., and Friedman, P. A. (1989) Chemical synthesis of

echistatin, a potentent inhibitor of platelet aggregation from *Echi carinatus*: synthesis and biological activity of selected analogs. *Proc. Natl. Acad. Sci. USA* **86,** 4022–4026.
23. Westerberg, P., Wernstedt, C., Wilander, E., Hayden, D. W., O'Brien, T. D., and Johnson, K. H. (1987) Amyloid fibrils in human insulinoma and islets of Langerhans of the diabetic cat are derived from a neuropeptide-like protein also present in normal islet cells. *Proc. Natl. Acad. Sci. USA* **84,** 3881–3885.
24. Hope, D. B., Murti, V. V. S., and Du Vigneaud, V. (1962) A highly potent analog of oxytocin, desaminooxytocin. *J. Biol. Chem.* **237,** 1563–1566.
25. Sawyer, T. J., Staples, D. J., de Launo Castucci, A. M., and Hadley, M. E. (1989) Discovery and structure–activity relationships of novel α-melanocyte stimulating hormone inhibitors. *Peptide Res.* **2,** 140–146.
26. Chandler, P., Pennington, M. W., Maccecchini, M. L., Nashed, N., and Skolnick, P. (1993) Polyamine-like actions of peptides derived from conantokin-G, an NMDA antagonist. *J. Biol. Chem.* **268,** 17,173–17,178.
27. Tanaka, M., Haniu, M., Yasunobu, K. T., and Norton, T. R. (1977) Amino acid sequence of the *Anthopleura xanthogrammica* heart stimulant anthopleurin-A. *Biochemistry* **16,** 204–208.
28. Tam, J. P. (1987) Synthesis of biologically active transforming growth factor alpha. *Int. J. Peptide Protein Res.* **29,** 421–431.
29. Pennington, M. W., Festin, S. M., Maccecchini, M. L., Dick, F., and Scarborough, P. E. (1991) HIV protease, chromogenic substrate and inhibitor, in *Peptides 1990* (Giralt, E. and Andreu, D. eds.), Escom, Leiden, Netherlands, pp. 787–789.
30. Bangalore, N., Travis, J., Onuka, V. C., Pohl, J., and Shafer, W. M. (1990) Identification of the primary antimicrobial domains in human neutrophil cathepsin-G. *J. Biol. Chem.* **265,** 13,584–13,588.
31. Gan, Z. R., Gould, R. J., Jacobs, J. W., Friedman, P. A., and Polokoff, M. A. (1988) Echistatin, a potent platelet aggregation inhibitor from the venom of the viper, *Echis carinatus*. *J. Biol. Chem.* **263,** 19,827–19,832.
32. Bartfai, T., Bedecs, K., Land, T., Langel, U., Bertorelli, R., Girotti, P., Consolo, S., Xu, X., Hallin, Z., Nilsson, S., Pieribone, V., and Hokfelt, T. (1991) M-15: high affinity chimeric peptide that blocks the neuronal actions of galanin in the hippocampus, locus coeruleus and spinal cord. *Proc. Natl. Acad. Sci. USA* **88,** 10,961–10,965.
33. Veber, D. F., Milkowski, J. D., Varga, S., Denkewalter, R. G., and Hirshmann, R. (1972) Acetamidomethyl: a novel thiol protecting group for cysteine. *J. Am. Chem. Soc.* **94,** 5456–5461.
34. Kamber, B. (1971) Cystin peptide aus (*S*-acetamidomethyl-cystein)-peptiden durch oxydation mit jod: die synthese con cyclo-cystin. *Helv. Chim. Acta* **54,** 927–930.

CHAPTER 5

Acid Cleavage/Deprotection in Fmoc/*t*Bu Solid-Phase Peptide Synthesis

Fritz Dick

1. Introduction

In general, a solid-phase peptide synthesis (SPPS) consists of the assembly of a protected peptide chain on a polymeric support (=synthetic step) and the subsequent cleavage/deprotection to release the crude, deprotected peptide from the solid support (=cleavage step). Usually, these two steps are followed by chromatographic purification of the crude peptide (*see* Chapters 1, 2, 4, and 5, *PAP*). The techniques to synthesize protected peptides or peptide fragments are discussed in Chapter 14. This chapter describes exclusively the cleavage step in the Fmoc/*t*Bu-SPPS with TFA as cleavage reagent leading to fully deprotected peptides (*see* Note 1) *(1)*. All of the cleavage procedures that are exemplified in Section 3. of this chapter hold for peptides synthesized under the following conditions:

1. Resins used:
   a. Sasrin resin™ (2-Methoxy-4-alkoxybenzyl alcohol resin) leads to peptide acids. It is also used to synthesize protected peptide fragments (*see* Note 2).
   b. Wang resin (4-Alkoxybenzyl alcohol resin) leads to peptide acids (*see* Note 2).
   c. Fmoc-amide resin (2,4-Dimethoxy-4'-[carboxymethyloxy]-benzhydrylamine linked to amino methyl resin) leads to peptide amides (*see* Note 3).
   d. Others with similar properties to a–c, specifically relative to the stability of the peptide-resin bond.

From: *Methods in Molecular Biology, Vol. 35: Peptide Synthesis Protocols*
Edited by: M. W. Pennington and B. M. Dunn

2. N-terminal-protecting group: Boc or none.
3. Side-chain-protecting groups: *t*Bu for Asp, Glu, Ser, Thr, and Tyr; Boc for Lys; Pmc for Arg (Note 4); Trt, Mtt *(2)*, or none for Asn and Gln; Trt for His; Trt or Acm for Cys (*see* Note 1); none for Met and Trp.

Thirty-seven years have elapsed since the introduction of the Boc-protecting group by Carpino *(3)* and its application as a TFA-labile amino-protecting group in peptide chemistry *(4)*. During this time, Boc and other *t*Bu-type-protecting groups (esters and ethers) have proven to be very useful *(5)*. Carpino again, together with Han, introduced the Fmoc-protecting group in 1970 *(6,7)*. The combined application of the orthogonal Fmoc/*t*Bu-pair in a new SPPS strategy followed in 1978 by Atherton et al. *(8)* and independently by Chang and Meienhofer *(9)*. Only in recent years, however, has the Fmoc/*t*Bu-SPPS really been established as an extremely valuable alternative to the Boc/Bzl-SPPS *(1)*.

During the cleavage step, highly reactive species (*t*Bu-cations and *t*Bu-trifluoroacetate, among others) are generated that can undergo undesired side reactions with sensitive amino acids, such as Cys, Met, Trp, and Tyr. These reactive species have to be trapped chemically by the addition of appropriate scavengers to the actual cleavage agent TFA. Table 1 shows a choice of possible scavengers widely used in Fmoc/*t*Bu-SPPS. For a broader discussion, *see* refs. *1* and *10*, and literature cited therein.

Whenever an SPPS (including purification) leads to the target peptide without severe problems, the peptide chemist assumes that every single step has proceeded satisfactorily. In the case of "difficult peptides," on the other hand, it is not necessarily very obvious whether the synthetic step or the cleavage step is mainly responsible for the difficulties. A poor synthetic step cannot be saved even by an optimal cleavage step, but in the opposite case, an easy synthesis can be ruined by inappropriate cleavage conditions. It is therefore strongly recommended to perform small-scale trial cleavage runs with 10–50 mg of peptide resin and different cleavage cocktails, as well as variations in reaction time and reaction temperature (with HPLC analysis of the resulting different crude peptides) to determine optimum cleavage conditions prior to the main cleavage run.

2. Materials

1. All reagents required (TFA, scavengers listed in Table 1) are commercially available and used as such. TFA may be distilled under normal pressure if colored.
2. Cleavage cocktails should always be prepared freshly.

Table 1
Scavengers in Fmoc/*t*Bu–SPPS

	Protection* of		
Scavenger	Met	Trp	Tyr
Phenol	o	o	+
p-Cresol	o	o	+
Anisole	–	–	–
Thioanisole	–	o	+
Ethanedithiol	+	o	–
Dithioerythritol	+	o	–
Dimethylsulfide	+	o	o
2–Methylindole	o	+	–
Water**	o	o	o

*+ = good, o = fair, – = poor.
** Indispensable if Arg(Pmc) is present.

3. Solvents, such as dichloromethane, diethylether, diisopropylether, and methyl-*tert*-butylether, are commercially available, too, and used as such. Ethers must be free of peroxides if an oxidizable peptide (containing Met and/or Cys) is to be treated (check, e.g., with test strips Merckoquant 10011, from Merck).
4. Equipment for HPLC analysis (*see* Chapter 3, *PAP*).
5. Further analytical tools to check peptides for purity and identity, as needed (*see* Chapters 6, 7, and 9, *PAP*).

3. Methods

3.1. General Considerations

3.1.1. Pretreatment of the Peptide Resin

After completion of the synthetic step, the peptide resin is usually thoroughly washed with DCM or any of the ethers mentioned above, and subsequently dried in a vacuum desiccator. Such a peptide resin is ready for cleavage. In some cases, however, it can be advisable to preswell the dry peptide resin for 10 min in DCM (the remaining DCM in the resin dilutes the subsequently added cleavage cocktail and can smoothen the cleavage reaction). This means, on the other hand, that drying after a final DCM wash is not strictly necessary.

3.1.2. Cleavage Cocktails

The four cleavage cocktails given below are just a selection. Other scavengers (cf Table 1) or scavenger ratios may give better results in a

Table 2
Peptide Composition and Cleavage Cocktail of Choice

	Cleavage cocktail
The peptide contains	
Any amino acid, except Met, Ser, Thr, Trp, Tyr, or Cys(Trt)	1 or 2
Any amino acid, except Met, Trp, or Cys(Trt)	2
Any amino acid, including Met, Trp, and/or Cys(Trt)	3 or 4

particular case. Trial runs are recommended as mentioned in Chapter 4. Quantities to cleave 1 g of peptide resin (dry wt) follow:

1. 9.5 mL of TFA are mixed with 0.5 mL of water.
2. 9 mL of TFA are mixed with 0.5 g of *p*-cresol and 0.5 mL of water.
3. 7.5 mL of TFA are mixed with 1.5 mL of EDT, 0.5 g of *p*-cresol, and 0.5 mL of water.
4. 8.3 mL of TFA are mixed with 0.25 mL of EDT, 0.5 g of phenol, 0.5 mL of thioanisole, and 0.5 mL of water (Reagent K *[10]*).

3.1.3. How to Choose a Cleavage Cocktail

The choice of an appropriate cleavage cocktail depends on the amino acid composition of the peptide to be cleaved and can be made according to Table 2.

3.1.4. General Cleavage Procedure

The cleavage should be performed under an inert gas.

1. One gram of peptide resin (dry wt) is placed in a round-bottom flask with a magnetic stirring bar and cooled in an ice bath.
2. Ten milliliters of precooled (ice bath), freshly prepared cleavage cocktail are added.
3. The reaction mixture is allowed to warm up to room temperature and stirred for 90 min (*see* Note 5).
4. After this time, the suspension is filtered through a fritted-glass funnel of medium porosity.
5. The resin is washed on the filter with 1–2 mL of neat TFA and 5–10 mL of DCM in several portions (*see* Note 6).
6. The combined filtrates are concentrated quickly to a volume of approx 2 mL on a rotatory evaporator (temperature of the water bath below 40°C) (for a milder alternative, *see* Note 7).
7. The concentrate is added dropwise to 100 mL of ice-cold ether with good stirring (*see* Note 8).

8. The precipitated peptide is collected by filtering the suspension from step 7 through a fritted-glass funnel of fine porosity and subsequent washing (two to three times) with ether (*see* Notes 8–10).
9. Finally, the wet, crude peptide is dried in a vacuum desiccator and weighed (*see* Note 11).
10. The peptide is now ready for HPLC analysis. In case of incomplete cleavage of protecting groups, steps 2, 3, and 6–9 can be repeated accordingly with the crude peptide (*see* Note 12).

3.2. Examples

3.2.1. Sex Pheromone Inhibitor iPD 1 (11)

- Sequence: H-Ala-Leu-Ile-Leu-Thr-Leu-Val-Ser-OH.
- Characteristics: 8-mer; neutral, rather hydrophobic; no Tyr, Cys, Met, Trp; in spite of (anticipated) hydrophobicity, considered to be a peptide that is rather easy to make (because short).
- Resin: Wang (*see* Section 1.).
- Cleavage: 1.93 g of peptide resin, 20 mL of cleavage cocktail 2, 90 min at room temperature.
- Crude: 1.05 g (>100% of theory, contains scavenger) with a purity of 56% (HPLC).
- Remarks: Amino acid analysis and FAB-MS (both after purification) in accordance with theory. The peptide is not an easy one as expected, mainly because of its low solubility.

3.2.2. Angiotensin II (Human)

- Sequence: H-Asp-Arg-Val-Tyr-Ile-His-Pro-Phe-OH.
- Characteristics: 8-mer; slightly basic; no Cys, Met, Trp; considered to be a peptide that is rather easy to make.
- Resin: Wang (*see* Section 1.).
- Cleavage: 863 mg of peptide resin, 8 mL of cleavage cocktail 2, 90 min at room temperature.
- Crude: 463 mg (>100% of theory, contains scavenger) with a purity of 87% (HPLC; *see* Chapter 3, *PAP*).
- Remarks: Amino acid analysis (after purification) in accordance with theory. Easy peptide.

3.2.3. α-MSH Agonist (12)

- Sequence: Ac-Cys-Glu-His-D-Phe-Arg-Trp-Cys-Lys-Pro-Val-NH_2.
- Characteristics: 10-mer; both termini blocked, slightly basic; contains two Cys (anticipated) (bridged) and Trp; could be sensitive.
- Resin: Fmoc-amide resin (*see* Section 1.).

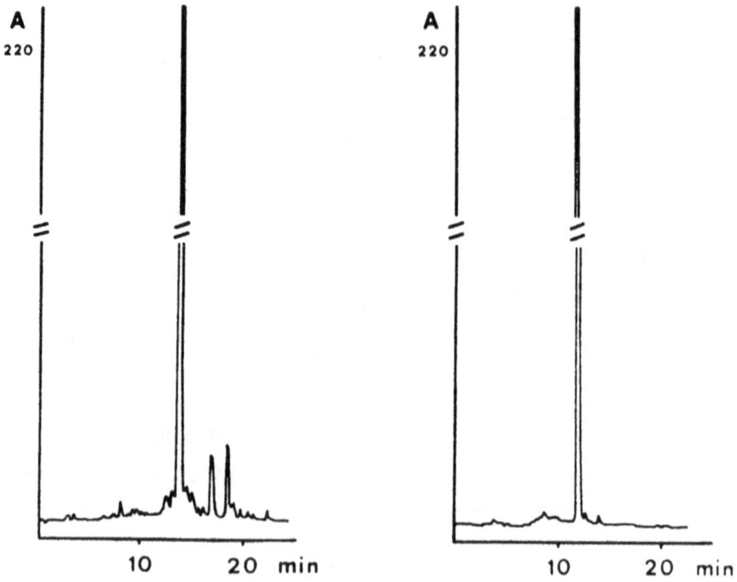

Fig. 1. RPHPLC gradient elution profile of *3*, crude: left, purified: right.

- Cleavage: 1.0 g of peptide resin, 10 mL of cleavage cocktail 3, 90 min at room temperature, resin cleaved for a second time.
- Crude: 365 mg (73% of theory) of peptide in the reduced state, i.e., with free sulfhydryl groups, both crops of equal purity (82%, HPLC; *see* Fig. 1, left).
- Remarks: Amino acid analysis and FAB-MS (both after oxidation and purification) in accordance with theory (HPLC; *see* Fig. 1, right; FAB-MS [*see* Chapter 7, *PAP*]; *see* Fig. 2). Difficulties below expectations.

3.2.4. Calmodulin Antagonist (13)

- Sequence: H-Leu-Lys-Lys-Phe-Asn-Ala-Arg-Arg-Lys-Leu-Lys-Gly-Ala-Ile-Leu-Thr-Thr-Met-Leu-Ala-OH.
- Characteristics: 20-mer; basic, with accumulation of basic amino acids in the first half from the N-terminus; contains Met near the C-terminus; not an easy peptide.
- Resin: Wang (*see* Section 1.).
- Cleavage: 1.0 g of peptide resin, 10 mL of cleavage cocktail 3, 60 min at room temperature. Second cleavage of the crude peptide necessary, same conditions.
- Crude: 545 mg (65% of theory) after two cleavages, purity 31% (HPLC).
- Remarks: Amino acid analysis and FAB-MS (both after purification) in accordance with theory. A rather difficult peptide. Met easily oxidized.

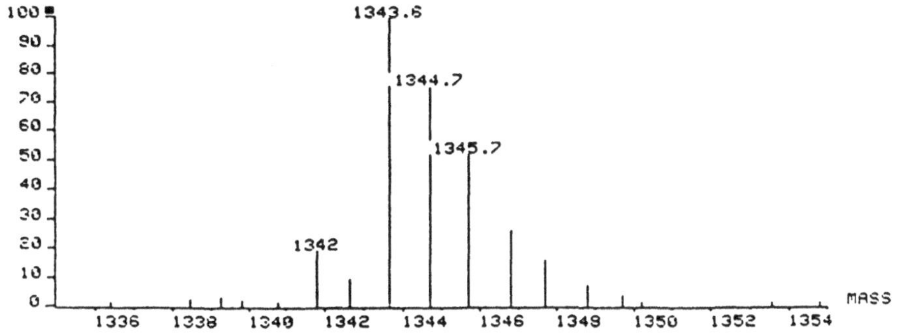

Fig. 2. FAB-MS of purified *3* $[M + H]^+$: calculated = 1343.6, found = 1343.6.

3.2.5. Dermaseptin (14)

- Sequence: H-Ala-Leu-Trp-Lys-Thr-Met-Leu-Lys-Lys-Leu-Gly-Thr-Met-Ala-Leu-His-Ala-Gly-Lys-Ala-Ala-Leu-Gly-Ala-Ala-Ala-Asp-Thr-Ile-Ser-Gln-Gly-Thr-Gln-OH.
- Characteristics: 34-mer; basic, with a marked hydrophobic domain of seven amino acids; contains two Met and one Trp; considered as difficult peptide.
- Resin: Wang (*see* Section 1.).
- Cleavage: 1.0 g of peptide resin, 10 mL of cleavage cocktail 3, 90 min at room temperature.
- Crude: 429 mg (44% of theory) with a purity of 39% (HPLC; *see* Fig. 3, left).
- Remarks: Amino acid analysis and ES-MS (both after purification) in accordance with theory (HPLC; *see* Fig. 3, right; ES-MS; *see* Fig. 4). A difficult and sensitive peptide. Prone to oxidation.

4. Notes

1. Any Cys(Acm) as part of the peptide chain on the resin remains unchanged, i.e., Cys will not be deprotected.
2. Commercially available with the first (C-terminal)-protected amino acid linked to the resin (from BACHEM Feinchemikalien AG, Switzerland).
3. Commercially available with Fmoc protection, i.e., without amino acid loading (e.g., from BACHEM Feinchemikalien AG, Switzerland).
4. Mtr as protecting group for Arg is no longer recommended (except where explicitly required), since it needs more drastic cleavage conditions, which may damage the peptide.
5. It may be necessary to optimize the reaction conditions (reaction time and/or reaction temperature).
6. The resin should not be discarded yet at this stage, but stored at –25°C for a possible second cleavage (*see* Note 11).

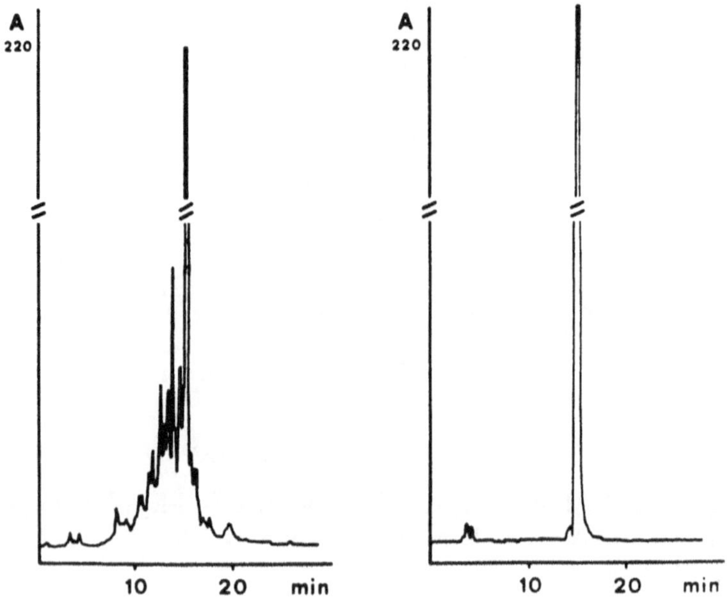

Fig. 3. RPHPLC gradient elution profile of *5,* crude: left, purified: right.

7. Step 6 can be omitted in many cases, i.e., the combined filtrates can be added directly to the cold ether (*see* Note 8) without previous concentration. It is, however, advisable to check if the peptide is precipitable by this alternative treatment by adding a few drops of the filtrate to 5–10 mL of cold ether (*see* Note 8).
8. Methyl-*tert*-butylether is preferred, since it is less prone to peroxide formation than diethylether and diisopropylether.
9. This step can be laborious and time-consuming, because many precipitated peptides are of slimy consistency. The isolation could also be done by centrifugation.
10. Mother liquors must be disposed of properly. To destroy bad-smelling components, such as mercaptans, the mother liquors are carefully neutralized with aqueous NaOH and subsequently treated with commercial laundry bleach *(15)*.
11. If the crude yield is low (<50–70% of theory), the whole procedure can be repeated with the resin (*see* Note 6). The second crop (if any) may be of poorer quality than the first one.
12. In most cases, the RPHPLC-trace shows one clear main peak. The occurrence of peaks with longer retention times than the main peak is a hint for incomplete cleavage.

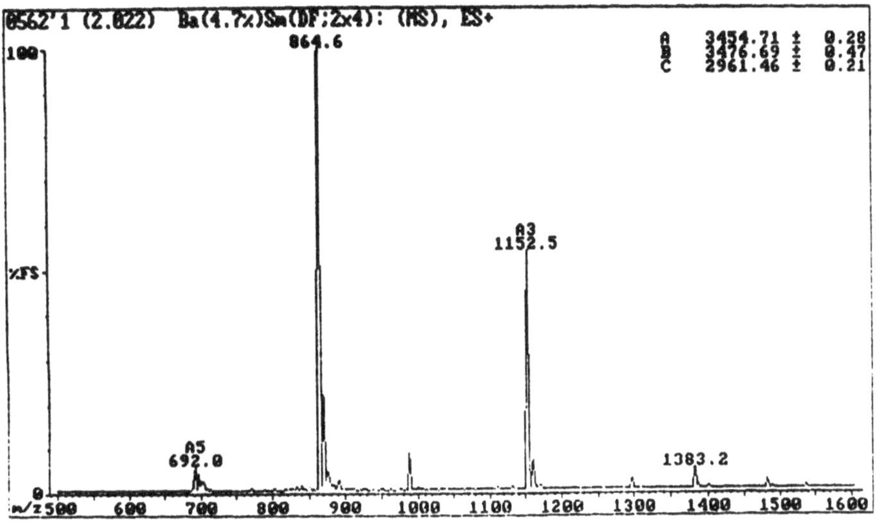

Fig. 4. ES-MS of purified **5**, M_r: calculated = 3455.1, found = 3454.7 (**A**), sodium adduct (**B**), unknown impurity (**C**).

Acknowledgments

I want to thank E. Kulow, O. Hässler, and M. Schwaller for important experimental contributions.

References

1. Fields, G. B. and Noble, R. L. (1990) SPPS utilizing 9-fluorenylmethoxycarbonyl amino acids. *Int. J. Peptide Protein Res.* **35,** 161–214 (an excellent review on the Fmoc/*t*Bu-SPPS with 753 refs.).
2. Sax, B., Dick, F., Tanner, R., and Gosteli, J. (1992) 4-Methyltrityl (Mtt), a new protecting group for the side-chain of Asn and Gln in solid-phase peptide synthesis. *Peptide Res.* **5,** 245,246.
3. Carpino, L. A. (1957) Oxidative reactions of hydrazines. II. Isophthalimides. New protective groups on nitrogen. *J. Am. Chem. Soc.* **79,** 98–101.
4. McKay, F. C. and Albertson, N. F. (1957) New amine-masking groups for peptide synthesis. *J. Am. Chem. Soc.* **79,** 4686–4690.
5. Gross, E. and Meienhofer, J. (eds.) (1981) *The Peptides. Analysis, Synthesis, Biology, vol. 3, Protection of Functional Groups in Peptide Chemistry.* Academic, New York.
6. Carpino, L. A. and Han, G.Y. (1970) The 9-fluorenylmethoxycarbonyl function, a new base-sensitive amino-protecting group. *J. Am. Chem. Soc.* **92,** 5748,5749.
7. Carpino, L. A. and Han, G. Y. (1972) The 9-fluorenylmethoxycarbonyl amino-protecting group. *J. Org. Chem.* **37,** 3404–3409.

8. Atherton, E., Fox, H., Harkiss, D., Logan, C. J., Sheppard, R. C., and Williams, B. J. (1978) A mild procedure for solid phase peptide synthesis: use of fluorenylmethoxy-carbonylamino-acids. *J. Chem. Soc., Chem. Comm.,* 537–539.
9. Chang, C. D. and Meienhofer, J. (1978) Solid-phase peptide synthesis using mild base cleavage of N^{α}-fluorenylmethyloxycarbonylamino acids, exemplified by a synthesis of dihydrosomatostatin. *Int. J. Pept. Prot. Res.* **11,** 246–249.
10. King, D. S., Fields, C. G., and Fields, G. B. (1990) A cleavage method which minimizes side reactions following Fmoc solid phase peptide synthesis. *Int. J. Pept. Prot. Res.* **36,** 255–266.
11. Mori, M., Tanaka, H., Sakagami, Y., Isogai, A., Fujino, M., Kitada, C., Clewell, D. B., and Suzuki, A. (1978) Isolation and structure of the sex pheromone inhibitor, iPD1, excreted by *Streptococcus faecalis* donor strains harboring plasmid pPD1. *J. Bacteriol.* **169,** 1747–1749.
12. Cody, W. L., Mahoney, M., Knittel, J. J., Hruby, V. J., Castrucci, A., and Hadley, M. E. (1985) Cyclic melanotropins. 9. 7-D-Phenylalanine analogues of the active-site sequence. *J. Med. Chem.* **28,** 583–588.
13. Payne, M. E., Fong, Y., Ono, T., Colbran, R. J., Kemp, B. E., Soderling, T. R., and Means, A. R. (1988) Calcium/calmodulin-dependent protein kinase II. *J. Biol. Chem.* **263,** 7190–7195.
14. Mor, A., Van Huong, N., Delfour, A., Migliore-Samour, D., and Nicolas, P. (1991) Isolation, amino acid sequence, and synthesis of dermaseptin, a novel antimicrobial peptide of amphibian skin. *Biochemistry* **30,** 8824–8830.
15. Committee on Hazardous Substances in the Laboratory; Commission on Physical Sciences, Mathematics, and Resources; National Research Council (1983) *Prudent Practices for Disposal of Chemicals from Laboratories.* National Academy Press, Washington, DC.

CHAPTER 6

Bromoacetylated Synthetic Peptides

Starting Materials for Cyclic Peptides, Peptomers, and Peptide Conjugates

Frank A. Robey

1. Introduction

Modern peptide synthesis techniques make it straightforward to synthesize linear peptides that are based on amino acid sequences of parent proteins. It is well known that the linear peptide itself often will not share the same activity as the peptide in its original biological environment provided by the native protein. In addition, the scientific reasons for making a peptide will often not be simply to mimic an activity of a protein—whatever it may be—with a small synthetic substitute, but the goal may be to enhance or decrease an activity found in a native protein. Therefore, it is important to be able to modify the conformations and configurations of a linear peptide by cyclizing, polymerizing, or conjugating the peptide.

The main obstacle to modification approaches is the lack of available techniques for synthesizing these peptide-containing materials. However, methods are being developed to use automated peptide synthesizers for additional chemistry in order to perform controlled modifications on the peptide after synthesis.

Until recently, there were three obvious limitations to the available methods in peptide chemistry for the design and synthesis of cyclic, polymeric, and/or conjugated peptides. First, it was not demonstrated that leaving groups could be incorporated into a peptide backbone prior to

From: *Methods in Molecular Biology, Vol. 35: Peptide Synthesis Protocols*
Edited by: M. W. Pennington and B. M. Dunn Copyright ©1994 Humana Press Inc., Totowa, NJ

deprotection. It was never demonstrated that the placement of a leaving group at a defined position in the peptide backbone could be accomplished, and this is the second limitation. Finally, for peptide conjugates, there were very few accurate quantitation methods for determining the amounts of peptide covalently linked to other molecules, especially to carrier proteins. All of these limitations are overcome by the use of bromoacetyl-derivatized peptides as described in this chapter. The basic finding was that bromoacetyl and chloroacetyl moieties are stable under the reactions conditions used in *t*-butoxycarbonyl (*t*BOC)-based peptide chemistry, including the deprotection steps using anhydrous hydrogen fluoride at –5 to 0°C (*see* Chapter 4). Bromo- and chloroacetyls are expected to be stable also under the conditions of fluorenylmethoxycarbonyl (Fmoc)-based peptide synthesis when the haloacetyl is added at the N^{α}-position.

Examples of haloacetyl-derivatized peptides as starting materials for new peptide-based configurations have appeared from several laboratories. Peptide polymers that were shown to induce T-cell-independent antibody production (*see* Chapters 10–12, *PAP*) in nude mice were synthesized from haloacetyl-modified peptides *(1)*, a chemically well-defined sugar-peptide conjugate from *Neisseria meningitidis* as a synthetic vaccine candidate has been made using a bromoacetylated peptide conjugated to a thiol-derivatized sugar *(2)*, multiple antigenic peptides (MAPS) are based on a chloroacetyl-derivatized backbone *(3)*, and an N^{α}-bromoacetyl peptide has been ligated to a *C*-terminally placed α-thiocarboxylic acid of another peptide to form a backbone-engineered HIV protease *(4)*.

The detailed descriptions that follow are from previously published methods based on *t*BOC chemistry *(5–7)*. The original work introducing haloacetyl-derivatized peptides was developed using an automated peptide synthesizer, and therefore, the assumption is made here that the reader is familiar with the routine uses of automated peptide synthesizers. However, manual syntheses of haloacetyl-derivatized peptides are readily accomplished by anyone with experience in those techniques and are also described below.

2. Materials

1. Bromoacetic acid, anisole, *S*-carboxymethylcysteine (CMC), 5,5'-dithiobis(2-nitrobenzoic acid) (DTNB), bromoacetylbromide, dithiothreitol (DTT), tri-*n*-butylphosphene (Bu_3P), and *N,N'* diisopropylcarbodiimide

were from Aldrich Chemical Co. (Milwaukee, WI). Reagent-grade chemicals and solvents used in the synthesis of BBAL were obtained from Fisher Scientific (Pittsburgh, PA). β-Alanine, *N*-hydroxysuccinimide, and bovine serum albumin (BSA) were from Sigma Chemical Co. (St. Louis, MO). N^{α}-BOC-L-lysine was obtained from Vega Biochemicals (Tucson, AZ). 2-Iminothiolane was from Pierce Chemical Co. (Rockford, IL).

2. All reagents, except bromoacetic acid and BBAL *(vide infra)*, used for the automated syntheses of peptides were purchased from Applied Biosystems, Inc. (Foster City, CA).
3. Sodium dodecylsulfate polyacrylamide gel electrophoresis (SDS-PAGE) was performed with the gel electrophoresis system sold by Novex (Encinitas, CA) (*see also* vol. 32 of this series).
4. Two amino acid analysis systems were used as described by the manufacturers for the analyses of products: Amino Quant® by Hewlett Packard, Inc. (Gaithersburg, MD) and Picotag® by Waters Associates (Millipore Corp., Milford, MA).
5. N^{α}-(*tert*-butoxycarbonyl)-N^{ε}-[*N*-(bromoacetyl)-β-alanyl]-L-lysine (BBAL) was synthesized using the reagents listed above and the procedure as described (*see* Section 3.3.) by Inman et al. *(8)*.

3. Methods

All of the bromoacetyl-derivatized peptides were synthesized using an automated solid-phase peptide synthesizer with the exception of the early work where manual bromoacetylations were performed to optimize the method. Both manual (Section 3.1.) and automated (Section 3.2.) approaches using *t*BOC-based peptide chemistry are presented. In addition, *t*BOC-based or Fmoc-based peptide synthesis (for N^{α}-amines) methods can be used, although the *t*BOC approach is outlined in this chapter.

To obtain synthetic peptides that are bromoacetylated at the N^{α}-amino terminus, it is necessary to add the bromoacetyl moiety to the peptide while the peptide is fully protected and, if solid-phase peptide chemistry is being used, covalently linked to the resin provided for normal peptide synthesis. The general step-wise procedure for solid-state peptide bromoacetylation is as follows:

1. Remove the *t*BOC- (*see* Chapter 4) or Fmoc- (*see* Chapter 5) protecting group from the N^{α}-terminus.
2. Neutralize the N^{α}-amine by washing with 10% TEA or DIEA.
3. Add bromoacetic acid anhydride (twofold excess) to the peptide mixture, and stir for 15 min to 1 h.

4. Filter and wash the resin with DMF and DCM to remove soluble, unwanted byproducts.
5. Deprotect the entire peptide with HF (for *t*BOC, *see* Chapter 4) or TFA (for Fmoc, *see* Chapter 5). It is important to omit any thiol-bearing scavengers that may be used here. Other related protocols for deprotection may use thiol-bearing scavengers.
6. Extract the released protecting groups with an organic solvent, generally a 2/1 mixture of ethyl acetate and ether.
7. Filter through a scintered glass filter, and discard the filtrate.
8. Dissolve the peptide in cold aqueous acetic acid.
9. Filter to separate the soluble peptide from the remaining resin.
10. Lyophilize the filtrate to obtain the crude bromoacetylated peptide.

3.1. Manual N^{α}-Bromoacetylation

The chemistry presented here describes in detail the bromoacetylation of a protected peptide at the amino terminal α-amine. The basic procedure calls for the simple formation of bromoacetic acid anhydride from commercially available bromoacetic acid and reacting the anhydride in *N,N'*-dimethylformamide (DMF) with a primary amine on the protected peptide.

The following procedure can be used when the amount of peptide being synthesized is on the 0.5-mmol scale:

1. Two hundred and ninety-eight milligrams of bromoacetic acid (2 mmol) in 5 mL CH_2Cl_2 are treated with 2 mL of a 0.5*M* solution of dicyclohexylcarbodiimide (DCC).
2. The solution is stirred for 15 min at 25°C, and during this time, a white precipitate of dicyclohexylurea (DCU) forms. The DCU is filtered using a scintered glass filter with vacuum suction, and the filtrate is evaporated to approx 2.5 mL using a stream of N_2.
3. The volume is then adjusted to 6 mL with DMF, and the solution is further evaporated to approx 4 mL by bubbling N_2 gas into the solution. The temperature remains unadjusted throughout the procedure. The solution is filtered a second time to remove trace amounts of DCU, which form during the evaporation process. The bromoacetic acid anhydride remains in the DMF for coupling to the protected peptide as outlined below.
4. Prior to adding the bromoacetic acid anhydride to the protected peptide, the *t*BOC group is removed from the N^{α} amine of the protected peptide, and the amine is then deprotonated with a suitable base, such as diisopropylethylamine. To a solution of 0.5 mmol protected peptide on the resin in 10 mL DMF is added 4 mL DMF that contains 1 mmol bromoacetic

acid anhydride. The reaction is allowed to proceed at 25°C for 60 min with stirring, after which time the reaction is complete.

5. The resin containing the N^{α}-bromoacetylated-protected peptide is filtered, washed five times with 50-mL portions of CH_2Cl_2 and air-dried. Alternatively, when solution-phase peptide synthesis is used, the solvent can be removed by vacuum evaporation using a rotary evaporator equipped with a mechanical vacuum pump and an acetone-dry ice trap.
6. Deprotection of the bromoacetylated peptide and release from the resin are described below in Section 3.4.

3.2. Automated Nα-Bromoacetylation

In addition to manual bromoacetylation, N^{α}-bromoacetylation using an automated solid-phase peptide synthesizer can be accomplished readily. This is based on the original solid-phase peptide synthesis procedures described in 1963 by R. B. Merrifield *(9)*.

The description given here is for use with a Model 430A peptide synthesizer, but should be adaptable for any automated peptide synthesizer. As a final step in the synthesis, bromoacetic acid anhydride is reacted with the amino terminal amino acid to form the N^{α}-bromoacetyl-derivatized fully protected peptide. This is carried out on a 0.5-mmol scale simply by substituting 2.0 mmol of bromoacetic acid (277.9 mg) for 2.0 mmol of glycine in an empty glycine cartridge and using the preprogrammed run file of the automated synthesizer for glycine coupling.

3.3. Selective Placement of a Bromoacetyl Moiety at Any Position in a Peptide Using BBAL

3.3.1. Synthesis of BBAL

Synthesizing BBAL involves coupling *N*-(bromoacetyl)-β-alanine to N^{α}-BOC-L-lysine as outlined below and taken from ref. *8*. This is performed by using the succinimidyl ester of the bromoacetic acid β-alanine to react with the free amine in the epsilon position of the N^{α}-BOC-L-lysine.

3.3.1.1. *N*-(BROMOACETYL)-β-ALANINE

A solution of β-alanine (53.5 g, 0.60 mol) in 600 mL of water was cooled to 5°C with an ice-alcohol bath. Bromoacetyl bromide (60.0 mL, 0.66 mol) was added under efficient stirring at such a rate as to maintain the temperature below 12°C. Concurrently, 5*M* NaOH were added at a rate needed to keep the pH near 7. These conditions were maintained for

45 min after completing addition of bromoacetyl bromide. The pH of the solution was then adjusted to 1.9–2.0 using 48% HBr, and its volume was reduced using a rotary evaporator to ca. 150 mL using a 60°C water bath and aspirator vacuum. The heavy precipitate of NaBr was removed by suction filtration and washed with ca. 15 mL water.

The filtrate was treated with a small amount of water to dissolve the NaBr and then shaken once with hexane-ethyl ether 1:1 v/v (450 mL), once with ethyl ether 1:5 v/v (450 mL each time). The upper phase (rich in bromoacetic acid) and final lower phase were discarded. The next five upper phases were pooled, filtered, and rotary evaporated to remove solvent. The residue was dried under vacuum and crystallized from hot ethyl acetate (81 mL) by addition of hexane (about 12 mL) and cooling at 4°C. The dried product (31.4 g) was similarly recrystallized from ethyl acetate plus hexane and dried under vacuum. The yield was 23.8%.

3.3.1.2. Succinimidyl 3-(bromoacetamido)propionate (SBAP)

To a solution of *N*-(bromoacetyl)-β-alanine (21.00 g, 100 mmol) and *N*-hydroxysuccinimide (13.01 g, 113 mmol) in 2-propanol (280 mL) at room temperature was added 1,3-diisopropylcarbodiimide (16.0 mL, 101 mmol). After 8–12 min, an oily precipitate of the product began to appear, and the walls of the container were scratched to induce crystallization. The mixture was allowed to stand for 1 h at room temperature and overnight at 4°C. The crystals were collected, washed with 2-propanol (30 mL), and redissolved in 2-propanol (200 mL brought to reflux). After an overnight stand at 4°C, the crystals were collected, washed with 2-propanol and then hexane, and dried under vacuum/$CaCl_2$. The yield was 74.6%.

3.3.1.3. N^{α}-(*tert*-butoxycarbonyl)-N^{ε}-[*N*-(bromoacetyl)-β-alanyl]-L-lysine (BBAL)

N^{α}-BOC-L-lysine (17.73 g, 72 mmol) was ground to a fine powder and suspended in DMF (600 mL). SBAP (18.43 g, 60.0 mmol) was added to the suspension in five portions at 10-min intervals. The reaction mixture was stirred for 2 h at room temperature, allowed to stand overnight at 4°C, filtered, and rotary evaporated to remove DMF (bath 30°C, vacuum pump). The residue was shaken with a mixture of ethyl acetate (960 mL), 1-butanol (240 mL), and aqueous 0.2*M* $KHSO_4$ (300 and 150 mL, respectively), filtered (Whatman #1 paper), and rotary evaporated to remove solvent (32°C water bath, pump vacuum). Vacuum was applied for at least 2 h to remove traces of solvent. The oily residue was dissolved in 1,2 dichloroethane (400 mL) by gentle warming and swirling. The solution was slowly cooled to 15–20°C, during which time the product initially precipitated as an oil, but was induced to crystallize by scratching. After an overnight stand (4°C), the product was collected, washed in dichloroethane, and dried in vacuum. The yield was 65.8%.

The chemistry for adding BBAL to a synthetic peptide at any desired position involves the formation of the 1-hydroxybenzotriazole (HOBt) ester (Fig. 1) of BBAL in DMF; BBAL is sparingly soluble in CH_2Cl_2, the solvent used to dissolve many of the other *t*BOC-derivatized amino acids in the formation of symmetric anhydrides. BBAL has been used with *t*BOC chemistry only; theoretically, the nucleophiles that remove Fmoc-protecting groups in Fmoc-based peptide synthesis also might react with the bromoacetyl moiety of BBAL, although we have no absolute proof of this occurring.

On a 0.5-mmol scale:

1. A mixture of BBAL and HOBt is made by dissolving 2.0 mmol BBAL (878 mg) in a solution containing 2.0 mmol of HOBt in 4.0 mL DMF and 0.3 mL CH_2Cl_2.
2. A solution containing 4.0 mL of 0.5*M* DCC in CH_2Cl_2 is added to the BBAL-HOBt mixture.
3. Agitate the mixture by bubbling N_2 gas through it for a period of at least 30 min at 25°C.
4. Filter off the DCU byproduct.
5. React the BBAL-HOBt ester that is in the filtrate with a free amine on the protected peptide or resin to couple the BBAL to the peptide.
6. If continuing with peptide elongation, *t*BOC is removed from the coupled BBAL with TFA in CH_2Cl_2 and coupling the remaining amino acids proceeds as usual.

3.4. Deprotection of Bromoacetylated Peptides

As mentioned above, bromoacetyl groups remain intact during the routine deprotection protocols for *t*BOC-based peptide synthesis. The only precaution is that sulfur-containing scavengers, such as thiocresol, thiophenol, or thioanisole, should be avoided.

When a bromoacetylated peptide contains an SH-bearing amino acid (or other strong nucleophile), extra precautions following HF (*see* Chapter 4) or TFA (*see* Chapter 5) deprotection should be taken to prevent undesirable reactions. These include the use of ice-cold extraction solu-

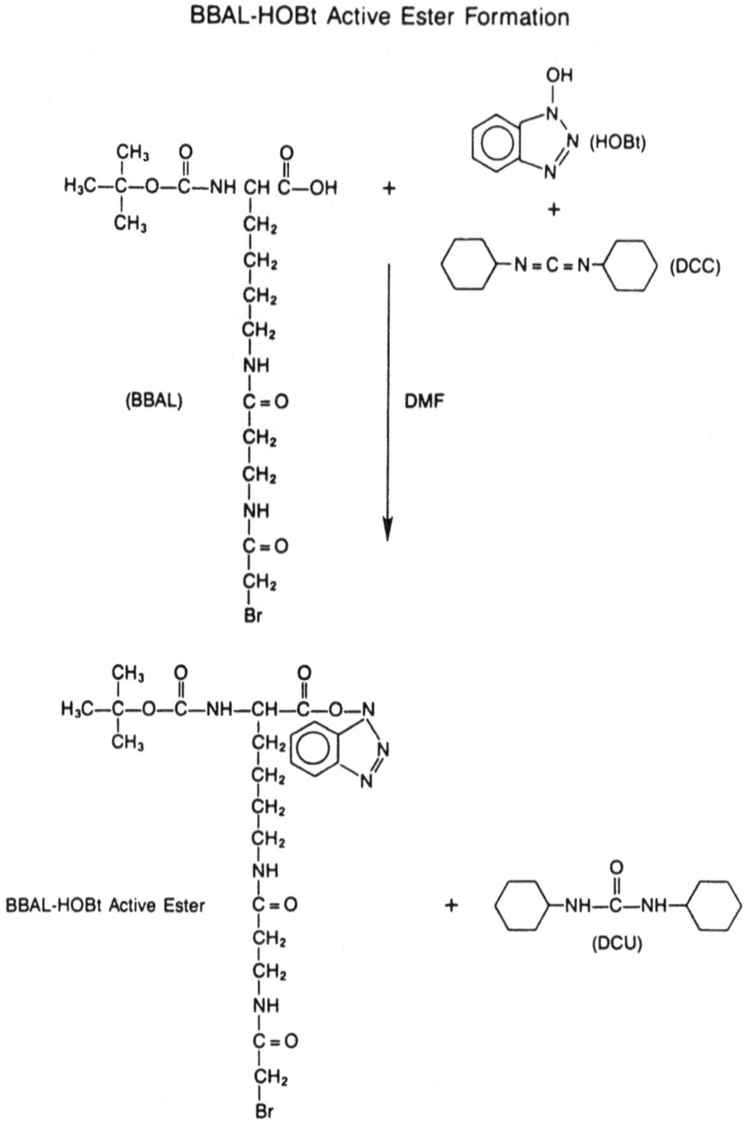

Fig. 1. Reaction scheme for the formation of the HOBt ester of BBAL. The active ester is formed as a result of DCC-mediated coupling of BBAL to HOBt. The ester is then used to couple BBAL to the protected peptide as outlined in Section 3.3.

tions and ice-cold, aqueous acid solvents (such as 10% aq. acetic acid) in which the peptide is dissolved prior to further purification. Bromoacetylated synthetic peptides should be stored dry and frozen in the dark to prevent decomposition, although storage at room temperature in a closed container appears to be suitable for up to 1 wk. Longer periods of time have not been investigated (*see* Note 1).

Finally, it may be necessary to use chloroacetyl moieties instead of bromoacetyl moieties if the bromoacetyl appears to be unacceptably reactive. We have notice that, unpredictably, certain bromoacetylated peptides react with the thiol in a cysteine even in cold solutions of anhydrous hydrogen fluoride (HF) (*see* Chapter 4).

3.5. Synthesis of Cyclic Peptides from Bromoacetylated Peptides

At neutral pH and room temperature, bromoacetyl moieties in buffered aqueous solutions are very reactive toward SH-containing materials, such as the thiol group in cysteine. Thus, if a cysteine is present in a bromoacetyl-containing peptide, it is very likely that the SH will attack the bromoacetyl to form inter- and/or intramolecular thioether bridges. Intramolecular crosslinking of peptides results in a cyclic peptide that will often have activities different from those found in the linear analog. A general reaction scheme to summarize the formation of cyclic peptides starting with a bromoacetyl, cysteine-containing peptide is given in Fig. 2. The ability of a peptide to cyclize cannot be predicted at the present time, but it is strictly dependent on the sequence of amino acids that are present between the bromoacetyl and thiol moieties and to a lesser extent on those outside of these boundaries. Thus, a few pilot experiments suggested below should be performed with a bromoacetylated, thiol-containing peptide to evaluate its ability to cyclize.

A major property of a peptide that is simple to evaluate is its elution time from a reverse-phase (C18, C8, or C4) column. As a general rule, a cyclic peptide will appear in the eluant from a reverse-phase column (*see* Chapter 3, *PAP*) earlier than the noncyclized peptide when the column is eluted with a linear gradient of increasing amounts of an organic solvent, such as CH_3CN. The reason is most likely that the cyclic peptide has less available hydrophobic surface area to bind to the resin than the linear, unfolded peptide.

To determine by reverse-phase HPLC if a bromoacetylated, cysteine-containing peptide can be cyclized:

1. The pure peptide is dissolved at 1 mg/mL in an ice-cold aqueous acidic buffer (such as 0.1% TFA), and an aliquot is immediately analyzed by HPLC. Evaluate the elution time of the peptide from a reverse-phase HPLC

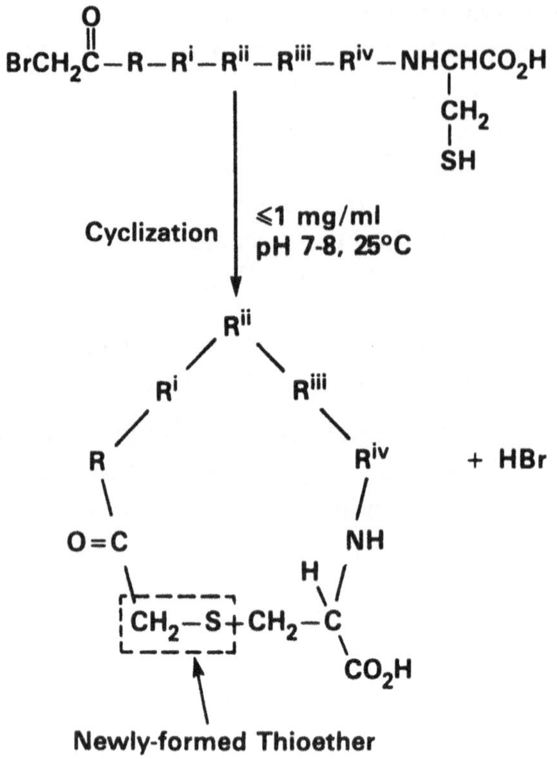

Fig. 2. Reaction scheme for the formation of cyclic peptides using a theoretical bromoacetylated, cysteine-containing peptide as the starting material. When this reaction is buffered with a sodium-containing buffer, such as $NaHCO_3$, the only byproduct of the cyclization reaction is NaBr, an innocuous salt.

column using a linear gradient of 0–70% CH_3CN over 20 min. Use 100 µL/run of peptide-containing solutions having a concentration of 1 mg/mL with detection at 210 nm.

2. Place some of the dry peptide into a buffer at pH 7–8 at a concentration of ≤1 mg/mL.
3. After 1 h at 25°C, run an analytical HPLC of the peptide incubated at a neutral pH, and compare this analysis to that from step 1 above. If there is a peak in front of the original linear peptide peak in the HPLC chromatograph, this is most likely the cyclized peptide.
4. Continue the incubation as in step 2 above until the HPLC profile is no longer changing.

In addition to the elution time by reverse-phase HPLC, further proof for the existence of a cyclic peptide can be obtained by testing the purified materials with DTNB for the presence of detectable SH groups *(10)*, and amino acid analysis for the presence of CMC and the other expected amino acids that would be present in acid hydrolysates of cyclic peptides containing the thioether formed as outlined above *(5–7)*. The most conclusive proof is mass spectrometry to verify an expected mass from the parent molecular ion of the cyclized peptide. Following the proteolysis of a cyclic peptide by enzymatic digestion, products that could only be formed by having the cyclic materials present at the start could be isolated and characterized (*see* Note 2).

3.6. Synthesis of Peptomers from Bromoacetylated Peptides

"Peptomer" is a new term we are using for a peptide polymer formed by the crosslinking of a peptide to itself in a specific fashion *(11)*. The biological characteristics of peptomers are generally not known, but they are being studied very closely as immunogenic materials for vaccine candidates, receptor crosslinking agents, enzyme substrates, or as research tools for the understanding of the chemical and conformational effects of specific amino acid sequences.

As mentioned above, a bromoacetylated, cysteine-containing peptide either will cyclize, polymerize, or do both. The chances of favoring a peptomer-forming reaction appear to be best at high concentrations (≥10 mg/mL) of the bromoacetylated, cysteine-containing peptide dissolved or suspended in an aqueous buffer at pH 6–8. At a pH >9, the thioate anion may compete with other nucleophiles, such as the ε-amine of lysine, and when this occurs, selective control of the peptide polymerization reaction is lost. A scheme for polymerizing peptides starting with a bromoacetylated, cysteine-containing peptide is shown in Fig. 3.

As noted above for cyclic peptides, peptomer formation can be followed by reverse-phase HPLC. In contrast to the formation of cyclic peptides where the cyclized form elutes from the column before the noncyclized starting material, peptomers will elute later than the starting material. An example of this is given in Fig. 4, where the reverse-phase HPLC of an incomplete polymerization reaction is shown. The A panel of Fig. 4 shows the purified bromoacetylated, cysteine-containing starting material. The multiple peaks of the peptomer appear in highest quantity along with small amounts of the cyclic peptide and the starting material as shown in panel B. The C panel shows the HPLC of the

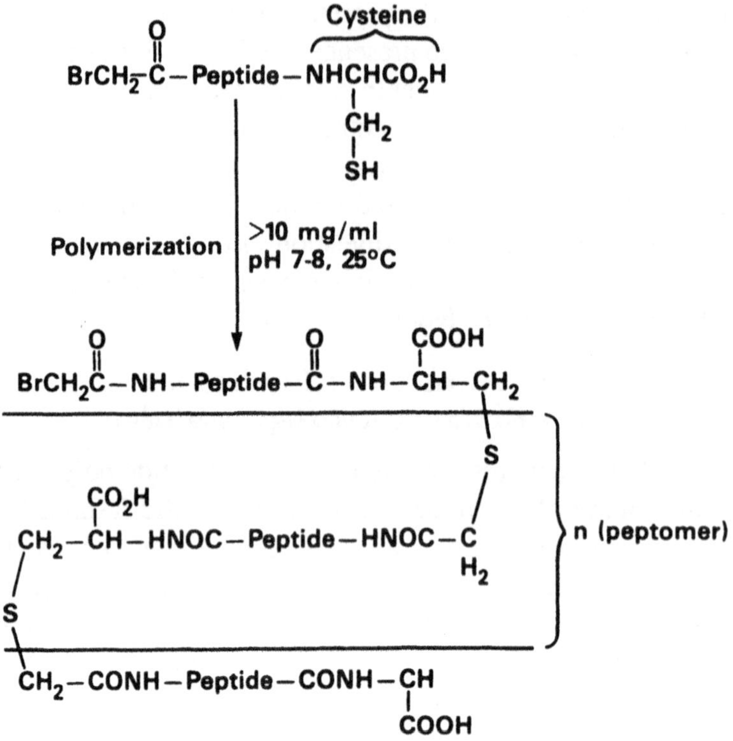

Fig. 3. Reaction scheme for the formation of a peptomer using a hypothetical bromoacetylated, cysteine-containing peptide as detailed in Section 3.6. As with the cyclization reaction shown schematically in Fig. 2, the only byproduct of the polymerization reaction is NaBr.

peptomer after dialysis to remove the low-mol-wt components from the peptomer mixture.

Peptomer formation may be followed using DTNB for the disappearance of the free thiols *(10)* and by SDS-PAGE. In contrast to cyclic peptides, mass spectrometric analyses (*see* Chapter 7, *PAP*) of peptomers may be uninterpretable. The lot-to-lot quality control of peptomers, therefore, would consist of the reverse-phase HPLC and mass spectrometric analysis of the starting peptide, and the reverse-phase HPLC and SDS-PAGE patterns of the peptomer formed. (The key considerations for identity and manufacturing would reside in the lot-to-lot consistency of the starting monomeric peptide and the peptomer formed, but not in the purity of the peptomer itself.) An example of the SDS-PAGE of a sample peptomer (the same as analyzed in Fig. 4) is shown in Fig. 5.

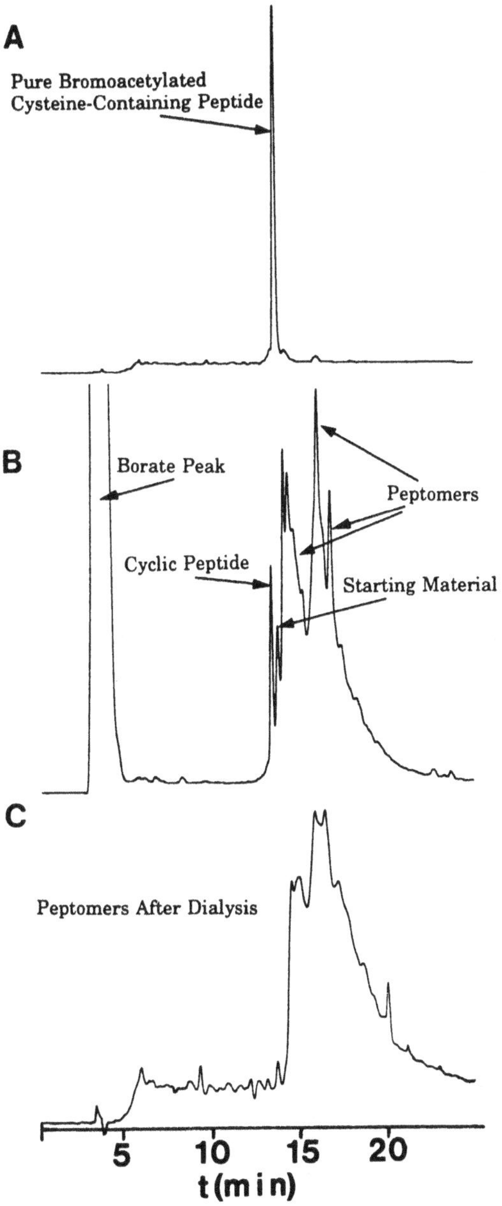

Fig. 4. Reverse-phase HPLC on a Vydac C18 column of: (**A**) pure $BrCH_2CO$-K-R-K-R-I-H-I-G-P-G-R-A-F-C-OH. (**B**) HPLC analysis of the above peptide after stirring at room temperature for 3 h in 0.1*M* $NaHBO_3$, pH 8.5, buffer. The cyclic peptide, starting peptide, and peptomer peaks are noted by the arrows in the figure. (**C**) Chromatogram of the reverse-phase HPLC after dialysis to remove the starting materials and dialyzable byproducts from the peptomer solution.

Fig. 5. SDS-PAGE of a representative peptomer from the same peptide given in the legend to Fig. 4. Mol-wt standard values are given to the right of the figure.

3.7. Synthesis of Peptide Conjugates from Bromoacetylated Peptides

There are three very important advantages of synthesizing protein conjugates with bromoacetylated peptides compared with conventional conjugation chemistry (*see* Note 3). First, the position of the bromoacetyl moiety in a peptide can be controlled. Second, by assaying for the amount of thioether present in acid hydrolysates of a protein-peptide conjugate, reliable values for the amount of covalently linked peptide can be determined. Finally, where in vivo uses of the conjugate are intended, it is reassuring to know that the thioether bonds or the long chain of BBAL (cf Fig. 1) appear to be nonimmunogenic (F. A. Robey, unpublished data).

A general strategy for covalently linking bromoacetylated peptides to sulfhydryl-containing proteins has been given in detail *(7)*. The primary starting point is to introduce the free thiol groups into the carrier for subsequent reactions with bromoacetyl moieties. For proteins, there are three ways to achieve this goal. First, the protein naturally contains free sulfhydryls without any modification. Because of the susceptibility of thiols to oxidize on storage, commercial sources of these proteins are rare. In addition, some proteins, such as papain, are stored in a solution or as a suspension containing excessive amounts of thiol-containing preservatives, and these would certainly destroy the bromoacetyl before it could react with the protein's thiol groups. Removal of the preserva-

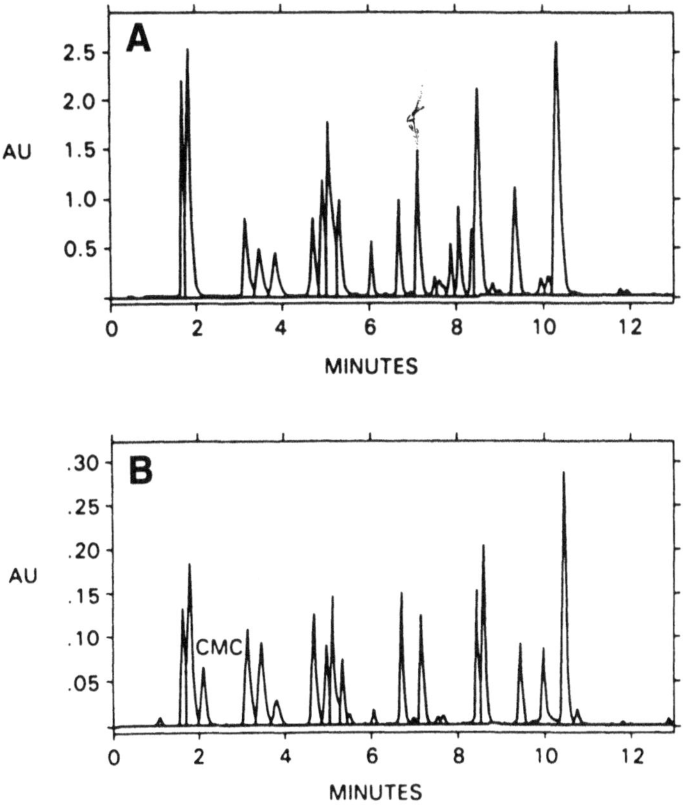

Fig. 6. Amino acid analysis chromatogram of (**A**) BSA and (**B**) the YIGSR conjugated to BSA. Bromoacetyl-YIGSR was reacted with Bu_3P-reduced BSA to form the conjugate. The procedure is outlined in Section 3.7., and the details of the conjugation are taken from ref. *7*. The value taken from the integrated area for the CMC is representative of the minimum amount of YIGSR covalently linked to BSA. With permission, Academic Press, Inc.

tives would probably sacrifice the protein's thiols because of auto-oxidation reactions.

Second, disulfide-rich proteins can be reduced to offer free thiols. Examples are serum albumin and immunoglobulin. Prior to the reaction with the bromoacetylated peptides, the disulfides are reduced with DTT or Bu_3P *(7)*. An amino acid analysis chromatogram for the YIGSR peptide conjugated to BSA is given in Fig. 6. To make the YIGSR-BSA conjugate, the bromoacetylated YIGSR was reacted with Bu_3P-reduced BSA as described in *(7)*. Finally, new thiols may be introduced into a protein by the reaction of a functional group on a protein with certain compounds, such as 2-iminothiolane *(12)*, *N*-acetylhomocysteine thiolactone *(13)*, or the *N*- hydroxysuccinimide ester of *S*-acetylthioacetic acid (SATA) *(14)*. These compounds covalently attach to the α-amine of proteins and the ε-amines of lysine residues in proteins. Although the chemistry of these thiol-producing compounds is very efficient and reproducible, they do not allow for the convenient quantitative analysis of covalently coupled peptide as can be done when a peptide's own cysteines are used for the conjugation *(7)*.

An example of introducing free thiols into bovine serum albumin using 2-iminothiolane and conjugating to a haloacetyl-derivatized peptide is outlined below as reported previously *(5)*.

1. Thirty milligrams of bovine serum albumin is dissolved in 2 mL 0.1*M* $NaHCO_3$ (pH 8.3). To this solution are added 4 mg 2-iminothiolane, and the reaction mixture is stirred for 15 min at room temperature.
2. The protein is separated from the unreacted reagents and byproducts by gel filtration chromatography (Sephadex G-25 (fine); 1.5 × 10 cm column, 0.1*M* $NaHCO_3$ buffer).
3. Fractions containing the modified protein as judged by the 280-nm absorbance and a positive DTNB reaction *(10)* for free thiols are pooled.
4. Protein in solution is reacted with a haloacetyl-containing peptide by adding the solid haloacetyl-containing peptide (20–50 mg) to the solution and stirring the mixture for 3 h at room temperature or until there are no detectable SH groups as tested with DTNB *(10)*.

For the above example, the degree of reaction can be estimated by following the amount of sulfhydryl that is consumed by the haloacetyl-containing peptide. However, it must again be stressed that this does not give a value for the covalently bound peptide.

4. Notes

1. The stability of bromoacetyl peptides is not known, but lyophilized peptides that are stored desiccated should be stable for 1 wk. Because of the lack of information available concerning storage conditions suitable for longer than 1 wk, it is recommended that storage in a desiccator at –20°C be used.
2. Thioether-containing peptides that are formed by reacting bromoacetyl- or chloroacetyl-containing peptides with free SH-containing materials are very stable because of the stability of the thioether bond. However, the sulfur atom can be oxidized, and the peptide material, which is often

hygroscopic, could act as a source of nutrient for microorganisms. Therefore, the storage of the cyclic peptides, peptomers, and peptide conjugates should be under a dry, oxygen-free atmosphere.

3. This chapter emphasized the reactivity of free thiols with haloacetyl moieties. However, there are other nucleophiles, such as histidine, methionine, homocysteine, and thiocarboxylic acid, that could be present on a synthetic peptide and that have the ability to react with the haloacetyl group. During the work-up of haloacetyl-containing peptides that contain various strong nucleophiles, one should be aware of possible side reactions of the nucleophiles with the haloacetyl moiety. This could occur even if they do not occur with cysteine as described here.

References

1. Hillman, K., Shapira-Nahor, O., Blackburn, R., Hernandez, D., and Golding, H. (1991) A polymer containing a repeating peptide sequence can stimulate T-cell-independent IgG antibody production *in vivo*. *Cellular Immunology* **134,** 1–13.
2. Boons, G. J. P. H., Hoogerhout, P., Poolman, J. T., van der Marel, G. A., and van Boom, J. H. (1991) Preparation of a well-defined sugar-peptide conjugate: a possible approach to a synthetic vaccine against neisseria meningitidis. *Bioorg. Med. Chem. Lett.* **1(6),** 303–308.
3. Lu, Y.-A., Clavijo, P., Galantino, M., Shen, Z.-Y., Liu, W., and Tam, J. P. (1991) Chemically unambiguous peptide immunogen: preparation, orientation and antigenicity of purified peptide conjugated to the multiple antigen peptide system. *Molecular Immunol.* **28(6),** 623–630.
4. Schnölzer, M. and Kent, S. B. H. (1992) Constructing proteins by dovetailing unprotected synthetic peptides: backbone-engineered HIV protease. *Science* **256,** 221–225.
5. Lindner, W. and Robey, F. A. (1987) Automated synthesis and use of *N*-chloroacetyl-modified peptides for the preparation of synthetic peptide polymers and peptide-protein immunogens. *Int. J. Pept. Protein Res.* **30,** 794–800.
6. Robey, F. A. and Fields, R. L. (1989) Automated synthesis of *N*-bromoacetyl-modified peptides for the preparation of synthetic peptide polymers, peptide-protein conjugates and cyclic peptides. *Anal. Biochem.* **177,** 373–377.
7. Kolodny, N. and Robey, F. A. (1990) Conjugation of synthetic peptides to proteins: quantitation from *S*-carboxymethylcysteine released upon acid hydrolysis. *Anal. Biochem.* **187,** 136–140.
8. Inman, J. K., Highet, P. F., Kolodny, N., and Robey, F. A. (1991) Synthesis of *N*α-(*tert*-butoxycarbonyl)-*N*ε-[*N*-(bromoacetyl)-β-alanyl]-L-lysine: its use in peptide synthesis for placing a bromoacetyl cross-linking function at any desired sequence position. *Bioconjugate Chem.* **2(6),** 458–463.
9. Merrifield, R. B. (1963) Solid phase peptide synthesis. I. The synthesis of a tetrapeptide. *J. Amer. Chem. Soc.* **85,** 2149–2154.
10. Ellman, G. L. (1959) Tissue sulfhydryl groups. *Arch. Biochem. Biophys.* **82,** 70–77.
11. Robey, F. A., Harris, T. A., Heegaard, N. H. H., Nguyen, A. K., and Batinic, D. (1992) Syntheses, analyses and uses of site-specific bromoacetyl-derivatized syn-

thetic peptides: starting materials for countless new cyclic peptides, peptomers and peptide conjugates. *Chemica Oggi* **Jan/Feb.,** 27–31.
12. Jue, R., Lambert, J. M., Pierce, L. R., and Traut, R. R. (1978) Addition of sulfhydryl groups to *Escherichia Coli* ribosomes by protein modification with 2-iminothiolane (methyl 4-mercaptobutyrimidate). *Biochemistry* **17,** 5399–5406.
13. White, F. H. (1972) Thiolation, in *Methods in Enzymology*, vol. 25 (Hirs, C. H. W. and Timasheff, S. N., eds.), Academic, New York, pp. 541–546.
14. Duncan, R. J., Weston, P. D., and Wrigglesworth, R. (1983) A new reagent which may be used to introduce sulfhydryl groups into proteins, and its use in the preparation of conjugates for immunoassay. *Anal. Biochem.* **132,** 68–73.

CHAPTER 7

Formation of Disulfide Bonds in Synthetic Peptides and Proteins

David Andreu, Fernando Albericio, Núria A. Solé, Mark C. Munson, Marc Ferrer, and George Barany

1. Introduction

Disulfide bridges play a crucial role in the folding and structural stabilization of many important extracellular peptide and protein molecules, including hormones, enzymes, growth factors, toxins, and immunoglobulins *(1–10)*. In addition, the artificial introduction of extra disulfide bridges into peptides or proteins allows the creation of conformational constraints that can improve biological activity *(11–15)* or confer thermostability *(5,16–19)*. Given this intrinsic biological interest, disulfide-containing peptides have long been attractive targets for chemical synthesis. Starting with the pioneering work of du Vigneaud on oxytocin *(20)*, the challenge to reproduce and engineer increasingly complex arrays of disulfide bridges as are found in natural peptides and proteins *(7,10,21–23)* has stimulated the efforts and ingenuities of many peptide chemists. Table 1 provides a representative, but by no means exhaustive, listing of noteworthy syntheses of peptides or small proteins with one or more disulfides. The methods can be readily generalized to analogs in which cysteine residues are replaced by homologs, such as homocysteine, or by sterically restricted derivatives, such as penicillamine (β,β-dimethylcysteine).

Both conformational and chemical considerations determine the ease of disulfide bond formation in synthetic peptides. The conformational aspect becomes especially important with small intramolecular disul-

From: *Methods in Molecular Biology, Vol. 35: Peptide Synthesis Protocols*
Edited by: M. W. Pennington and B. M. Dunn

Table 1
Representative Examples of Synthetic Disulfide-Containing Peptides and Proteins

Name	Residues	*SS* bonds	References
Oxytocin, vasopressin, and analogs	9	1	*20,24–29*
Somatostatin and analogs	13	1	*30–35*
α-Atrial natriuretic factors/peptides (ANF/ANP)	26–33	1	*36–39*
Calcitonin	32	1	*40*
Conotoxins and analogs	13–15	2	*41–49*
Apamin and analogs	18	2	*50–57*
Endothelin and analogs	21	2	*58–63*
Posterior pituitary peptide	21	2	*64*
Sarafatoxin	21	2	*65*
β-hANP (antiparallel dimer of α-ANP)	56	2	*38,39*
E. coli enterotoxin active fragment and analogs	13	2–3	*66–68*
μ-Conotoxins (geographutoxins)	22	3	*69–72*
ω-Conotoxin	27	3	*73–77*
E. elaterium trypsin inhibitor II	28	3	*78*
C. maxima trypsin inhibitor	29	3	*79*
Defensins	29–34	3	*80–83*
Charybdotoxin and analogs	37	3	*84–86*
Bombyxins and analogs	48	3	*87–89*
Sea anemone neurotoxin and analogs	48	3	*90,91*
Human insulin, relaxin, and analogs	51–53	3	*92–97*
Transforming and epidermal growth factors	50–55	3	*98–103*
Bovine pancreatic trypsin inhibitor and analogs	58	3	*104–106*
CCK-releasing peptide	61	3	*107*
C5a anaphylatoxin	74	3	*108*
Angiogenin	123	3	*109*
ω-Agatoxin IVA	48	4	*110*
Echistatin	49	4	*111*
Na^+, K^+-ATPase inhibitor-1 (SPAI-1)	49	4	*112*
Elafin	57	4	*113*
Ribonuclease A	124	4	*114,115*

fides. For example, the minimally sized eight-membered ring formed between two adjacent half-cystinyl residues is only rarely found in nature *(116–119)*, and can be somewhat difficult to form in the laboratory (*see* refs. *120–122* for elegant applications of strained peptide disulfides). With additional intervening residues *n* between a pair of half-cystines, the resultant ($3n + 8$) size rings are relatively unstrained and have considerable conformational flexibility *(3,6,8,10,12,13)*. Still, the exact sequence can make a difference; for example, small "local" loops must

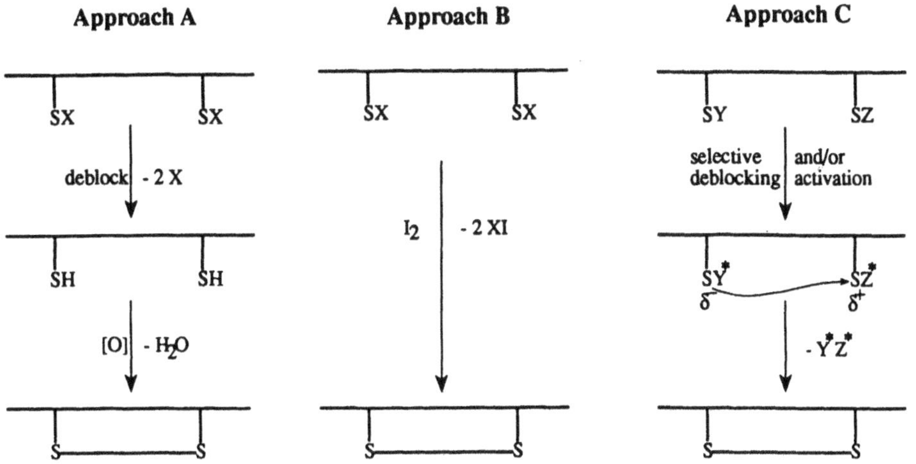

Scheme 1. Synthetic approaches to intramolecular disulfide bridges. X, Y, and Z designate sulfhydryl-protecting groups that are stable to chain assembly. The activation and/or deprotection steps in approach C provide a nucleophilic SY* and an electrophilic SZ* (in the simplest case, Y* = H and Z* = Z). As discussed in the text, these approaches can be generalized to varying extents for the syntheses of molecules with multiple disulfide bridges.

include residues compatible with β- or γ-turns *(47,123–125)*, and in larger structures, the steric environment about cysteine residues affects their accessibility and pK_a, and hence, the ease of oxidation *(126–129)*.

Even for sequences in which the conformations corresponding to disulfide bridging are readily accessible, the success of synthetic efforts may still hinge on the appropriate choice of cysteine-protecting groups and corresponding deblocking conditions (*vide infra*). The present chapter covers some current alternatives for cysteine protection and then turns to procedures for efficient disulfide bond formation. Both solution and solid-phase chemistries are discussed. Three principal approaches to *intramolecular* disulfide formation in synthetic peptides can be envisaged (Scheme 1).

In approach A, both of the cysteine residues to be paired have the same protecting group, which is removed to give the dithiol form of the peptide. Oxidation by molecular oxygen or other appropriate reagents provides the intramolecular disulfide. The disulfide can also be established from the dithiol in the presence of redox mixtures, which catalyze thiol-disulfide exchange reactions. In approach B, the cysteine-protecting groups are removed oxidatively by iodine as shown, or by alternative electrophilic reagents, to furnish the disulfide directly. Finally, the chemically more demanding directed approach C requires two cysteine-protecting groups that can be selectively cleaved and differentially activated; a subsequent displacement reaction gives the disulfide bond. Both of the symmetrical approaches A and B, as well as the unsymmetrical approach C, can suffer from intermolecular dimerization and oligomerization side reactions. These can be mitigated to various extents, depending on the sequence, by the use of *dilute* conditions in solution, or by the *pseudo-dilution* phenomenon that applies to polymer-supported reactions *(124,130–132)*.

The indicated approaches can be extended to the formation of peptides or proteins containing multiple disulfide bridges. Despite the statistical obstacles (3, 15, and 105 intramolecular disulfide isomers theoretically possible for molecules with 4, 6, and 8 half-cystine residues, respectively), approach A is often practical. Scrupulous attention to experimental details (pH, ionic strength, temperature, time, concentration) is critical, and in the cases when such random oxidations are successful, it is because the polypeptide chains in their reduced polythiol form can apparently fold into native-like conformations that favor the proper alignments of disulfides *(5,9)*. However, nonnatural isomers may form inadvertently because of misfolding (or even as obligatory intermediates in the folding pathway!), and intractable oligomers and polymers are often observed as well *(41,99,133,134)*.

Chemical control over the specificity of half-cystine pairings is clearly a desirable goal. For example, the native alignments of multiple bridges are sometimes not known with certainty from biochemical and/or instrumental methods of peptide or protein analysis, and synthetic methods for regioselective construction of the bridges in several alternative ways can become an invaluable tool for deciding the correct structure *(41,43,58, 67,68,77,94,135,136)*.

Typically, approaches A and/or B are applied in series using two or more orthogonally removable sets of pairwise cysteine-protecting groups. Although in principle such strategies generate unambiguous disulfide connectivities, the experimental procedures must be designed and executed with great care in order to prevent or minimize *scrambling* of disulfides through any of the consecutive chemical manipulations.

It is often of interest to pair half-cystine residues that are on separate linear chains. Relevant applications include conjugation of peptides to carriers for immunological studies, preparation of standards corresponding to proteolytic fragments isolated during structural elucidation work on large proteins, development of discontinuous epitopes, and generation of active site models *(36,135–149)*. Formation of the required heterodisulfides cannot be carried out efficiently by approaches A or B, because the statistically predicted distribution that includes both symmetrical homodimers will always form upon random symmetrical oxidation. Therefore, *intermolecular* disulfides are created, whenever possible, by approach C. Because of the ready propensity of unsymmetrical disulfides to disproportionate to the symmetrical species, exposure of the desired products to strong acid or weak base should be minimized.

Although formation of disulfide bridges is usually carried out toward the end of a synthetic plan, it is sometimes advantageous to couple and/or elongate chains that include a preformed disulfide *(93,131,138,150,151)*. This can be done either in solution or on the solid phase, and requires careful choice of deprotection and cleavage conditions (*see* Chapters 4 and 5) that do not affect disulfide bonds. An interesting artifice has been devised to prepare multiply disulfide-linked chains corresponding to putative partial structures encountered in structural work *(72,73)*. A single parent chain is designed that includes a site for proteolysis; intramolecular disulfide formation (by regioselective methods, if the case warrants) followed by cleavage gives the desired product chains.

Several of the themes of the present chapter have been covered within earlier reviews *(131,152–158)*; the fundamental organosulfur chemistry literature can also be a rich source of inspiration in this field *(159,160)*. We assume herein that the required linear peptide sequences can be assembled adequately and with minimal racemization, by stepwise and/or convergent solution *(161–163)* or solid-phase methodology *(131,164–166)*; hence, our focus is on the management of cysteine residues. The subject matter of this chapter is organized parallel to the present introduction, and additionally, modern methods for determining disulfide connectivities are covered. Although the success of any synthetic effort directed at properly folded disulfide-bridged peptides and small proteins often depends critically on the purification and analytical characterization procedures that are applied, we do not address these areas because they are well described in the original literature and in more general reviews and monographs. Similarly, the elegant thiol capture method of Kemp *(167)* demonstrates some interesting disulfide chemistry, but is outside the scope of the present chapter. In accordance with the editorial policy of this volume, emphasis is placed on practical experimental considerations, as filtered through the authors' experiences, and citations to the literature are weighted toward articles providing full details.

2. Cysteine Protection

Side-chain protection for the β-thiol of cysteine is selected foremost in the context of compatibility with the temporary N^{α}-amino protecting group used in stepwise assembly of the linear peptide sequence, and permanent groups protecting other side-chain functions. As an added complication, hydrogenolyzable protecting groups, such as N^{α}-benzyl–oxycarbonyl (Z), are generally incompatible because sulfur-containing residues poison the catalyst. For solution syntheses, strategies depend on graduated acid lability with N^{α}-4-methoxybenzyloxycarbonyl (Moz), N^{α}-*tert*-butyloxycarbonyl (Boc), N^{α}-triphenylmethyl (trityl or Trt), N^{α}-2-(4-biphenylyl)propyl(2) oxycarbonyl (Bpoc), or N^{α}-*ortho*-nitrophenylsulfenyl (Nps), together with side-chain benzyl (Bzl) or *tert*-butyl (*t*Bu) and related permanent groups, whereas current solid-phase syntheses invariably apply either "classical" Boc/Bzl strategies or orthogonal schemes with base-labile N^{α}-9-fluorenylmethyloxycarbonyl (Fmoc) and acidolysable *t*Bu and related permanent groups *(131,161–166)*.

The most widely used or promising cysteine-protecting groups are listed in Table 2, together with indications of stabilities and removal conditions. Note that this tabulation is but a subset of the many ingenious cysteine-protection chemistries that have been proposed and evaluated for peptide chemistry *(131,152,153,156,157,166,167,209–212)*. Deprotection of cysteine can be concurrent with removal of all other side-chain-protecting groups (and simultaneous cleavage from the support in the case of solid-phase peptide synthesis), or, under appropriate circumstances, protected cysteine residues can survive the final cleavage/deprotection step. In further variations, orthogonal deblocking of cysteine can be carried out selectively while retaining other side-chain protection; for solid-phase procedures, this can be done while the peptide chain remains anchored to the support. Orthogonality issues become particularly critical in designing regioselective schemes for syntheses of peptides with multiple disulfides.

Unique to cysteine, protecting group removal can be conducted either to generate the free thiol or to provide directly a disulfide bond. In the former case, the deblocking modes are acid (care must be taken to prevent realkylation of deblocked cysteine residues by carbocations generated under the cleavage conditions), base, reducing agents, or metals, such as Ag(I) or Hg(II). Metal mercaptides need to be treated in a separate step, e.g., with excess β-mercaptoethanol or hydrogen sulfide, to release the free thiol. However, sometimes the metal is very tightly bound and difficult to remove completely. Deprotection procedures should be carried out under an inert atmosphere, to the extent possible, in order to minimize inadvertent or premature oxidation. Intentional oxidative methods to give disulfides directly are covered subsequently, as are treatments with sulfenyl chlorides or other electrophilic reagents to give mixed disulfide intermediates.

The classical *S*-benzyl (Bzl) group developed by du Vigneaud and coworkers has had a distinguished history for the preparation of peptides in the oxytocin family, and was also popular in the early days of the solid-phase methodology *(20,164,168)*. However, *S*-Bzl is applied relatively infrequently in current practice because of the relative harshness of deblocking conditions, e.g., sodium in liquid ammonia or anhydrous hydrogen fluoride (HF) at 25°C (*see* ref. *213* for description of dehydroalanine residue formation upon HF-promoted elimination of benzyl mercaptan from Cys[Bzl]-containing peptide). Introduction of slightly electron-donating substituents on the aromatic ring has led to protecting groups, such as *S*-4-methylbenzyl (Meb; *169*) and *S*-4-methoxybenzyl (Mob; *171*), which have the proper balance of acid lability to allow their removal concurrent to other Bzl-type side-chain-protecting groups by use of HF–anisole (9:1) at 0°C, or acid/scavenger combinations of comparable strength *(170)*. Metal-assisted or oxidative cleavages of *S*-Meb or *S*-Mob appear to be quite difficult when the goal is preparative removal. Nevertheless, cleavage or oxidation of *S*-Acm and related groups (*see* Section 4.) is rarely carried out in the presence of intact *S*-Bzl, *S*-Meb, or *S*-Mob *(155)*.

Given the well-known lability of most *tert*-butyl (*t*Bu) derivatives to acids, it is interesting that the *S*-*t*Bu thioether survives neat HF at 0°C (although it is cleaved at 20°C; *see* ref. *175*). Relatively recently, conditions have been devised for selective acid cleavage of *S*-*t*Bu with HF, in the presence of *certain* efficient scavengers, such as anisole, *m*-cresol, or thioanisole *(39,178)*. Otherwise, adequate removal of *S*-*t*Bu can be achieved with metals or strong electrophilic reagents *(172)*. The conversion of *S*-*t*Bu to

Table 2
Cysteine-Protecting Groups Currently Used in Peptide Synthesis[a,b]

Protecting group	Abbrev.	Partial structure[a]	Stability	Removal	Compatibility[c]	References[a]
S-benzyl	Bz1	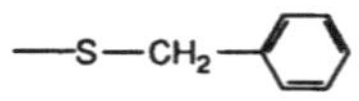	HF (0°C) Tl(III) Base RSCl	Na/liq. NH_3 HF (25°C)	Boc	*164,168*
S-4-methylbenzyl	Meb	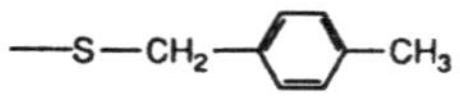	TFA Ag(I) Base RSCl	HF (0°C) Tl(III) (partial)	Boc	*131,155, 169,170*
S-4-methoxybenzyl	Mob	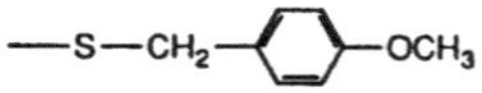	TFA Base RSCl I_2	HF (0°C) Hg(II), Ag(I) Tl(III) $(BrC_6H_4)_3N^{+\cdot}$	Boc	*131,155 171–174*
S-tert-butyl	*t*Bu	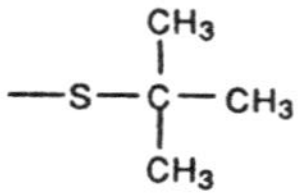	HF (0°C) Base I_2 Ag(I)	HF (20°) HF/scavengers (0°C) Hg(II) NpsC1	Boc, Fmoc	*34,39,172, 175–179*

S-triphenylmethyl	Trt	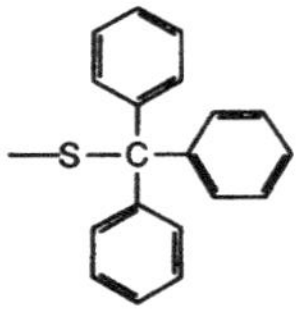	Base Nucleophiles	TFA/scavengers Hg(II), Ag(I) I_2,Tl(III) RSCl	Fmoc, Trt	*92,152,153, 155,179–184*
S-2,4,6-trimethoxy-benzyl	Tmob	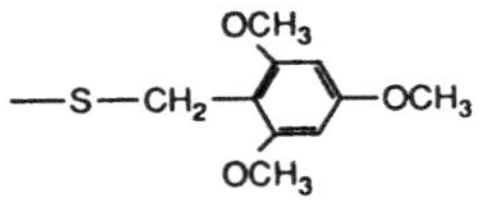	Base Nucleophiles	Dilute TFA/ scavengers I_2, Tl(III)	Fmoc	*49,185,186*
S-4,4',4"-trimethoxy-triphenylmethyl	TMTr	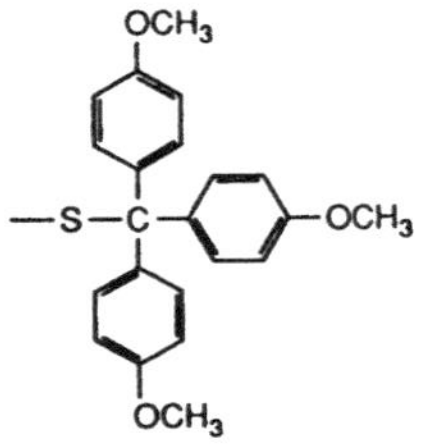	Base Nucleophiles	Very dilute TFA	Fmoc	*186*
S-acetamidomethyl	Acm	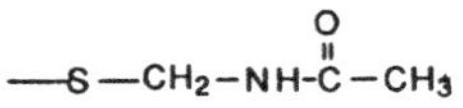	HF (0°C) Base	Ag(I), Hg(II) I_2, Tl(III) RSC1	Boc, Fmoc	*48,146,153 155,174,177, 179,182,183, 187–189*
S-trimethylaceta-midomethyl	Tacm	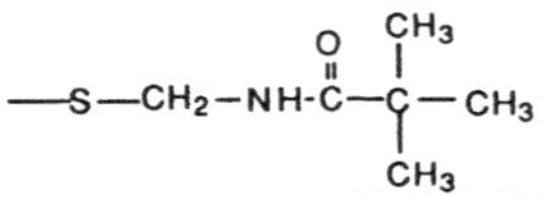	HF Base	Hg(II), Ag(I), I_2	Boc, Fmoc	*190*

(continued)

Table 2 *(continued)*

Protecting group	Abbrev.	Partial structure[a]	Stability	Removal	Compatibility[c]	References[a]
S-phenylacetamido-methyl	Phacm	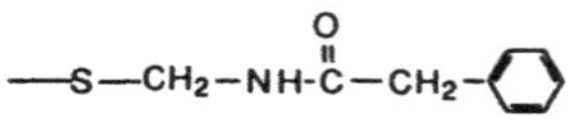	HF Base	Hg(II), I_2, Tl(III) Penicillin amidohydrolase	Boc, Fmoc	*48,191–193*
S-9-fluorenylmethyl	Fm	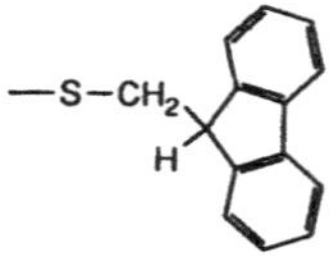	HF I_2	NH_3 in CH_3OH Piperidine Dilute DBU	Boc	*124, 194–196*
S-2-(2,4,dinitro-phenyl)ethyl	Dnpe	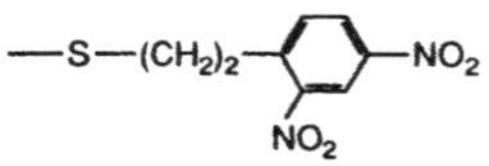	HF	Piperidine, Dilute DBU	Boc	*48,197*
S-*tert*-butylmercapto	S*t*Bu	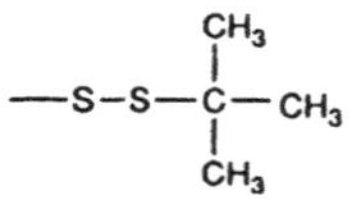	TFA HF (partial) Base RSCl	RSH, Bu_3P, or other reducing agents	Boc, Fmoc, Nps	*35,50,138, 161,177, 198–200*

S-3-nitro-2-pyridine-sulfenyl	Npys	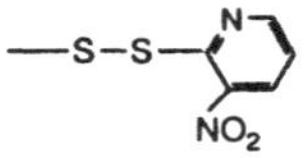	HF (“high”) HOBt	RSH, Bu_3P; *see also* Scheme 3B	Boc	*137,143, 148,178, 202–204*
S-alkoxycarbonyl-sulfenyl	Scm	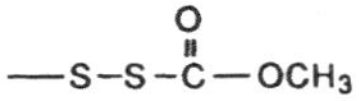	HF	RSH or other reducing agents; *see* Scheme 3A	Note *c*	*93,132, 205–207*
S-[(*N*′-methyl-*N*′-phenylcarbamoyl)-sulfenyl]	Snm	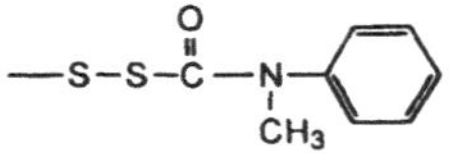	HF	RSH, specifically DTT and 2-mercaptopyridine	Boc	*207*

[a]The protecting group structure is drawn to include the sulfur of the cysteine that is being protected. Groups are listed in the same order as text discussion. Unless indicated otherwise, conditions or reagents outlined under “Removal” are intended for quantitative deprotection and/or oxidative cleavage; the products from these reactions may be free thiols, mercaptides, or disulfides (details in text). For metal-mediated removals, the counterion and solvent are sometimes critical; for acidolytic cleavages with HF or similar strong acids, the nature and amounts of scavengers added may affect significantly the stability and/or lability of the protecting group. References listed in the far right column cover discovery of the protecting group, its introduction onto cysteine, stability properties, and deprotection procedures, and may include relevant review articles in addition to or in place of the primary literature.

[b]Most of the protecting groups listed in this table have been applied to significant targets, as described throughout this chapter. Atherton and coworkers *(208)* have documented the risk of racemization for *C*-terminal-protected cysteinyl *esters* during Fmoc solid-phase synthesis, most serious for S*t*Bu, and less so for Trt, Acm, and *t*Bu. The problem is ***not*** serious for internal protected Cys residues involved in *amide* linkages.

[c]This column provides the compatible repetitively removable N^{α}-amino-protecting group(s) for introduction of the appropriate protected cysteine derivative, and subsequent stepwise chain elongation. There is no entry for Scm, since this protecting group is always introduced indirectly once a cysteine residue has already been incorporated in the peptide chain.

an aromatic mixed disulfide *(176)* by treatment with 2-nitrophenylsulfenyl chloride (NpsCl) is a prototype for a general two-step deprotection method, since further reductive transformations of mixed disulfides are easily achieved (discussed subsequently, with Scheme 3B in Section 5.).

The level of acid lability required to be compatible with Fmoc chemistry is provided by *S*-triphenylmethyl (trityl or Trt; *152,181*), *S*-2,4,6-trimethoxybenzyl (Tmob; *185*), and *S*-4,4',4"-trimethoxytriphenylmethyl (TMTr; *186*) groups (*see also* ref. *214* for description of *S*-9-phenylxanthen-9-yl [pixyl] group, which has properties similar to *S*-Trt]. The *S*-Trt, *S*-Tmob, and *S*-TMTr groups are released by treatment with varying levels of trifluoroacetic acid (TFA) in the presence of appropriate scavengers (relative lability to acid is TMTr > Tmob > Trt). The relatively stable carbocations due to these protecting groups can realkylate cysteine (an acid-mediated equilibrium process), as well as irreversibly modify sensitive tryptophan side chains. Trityl groups are removed generally with cocktails of concentrated TFA that include a mixture of aromatic and/or sulfur-containing scavengers, e.g., Reagent K *(215)*, TFA–phenol–water–thioanisole–1,2-ethanedithiol (82.5:5:5:5:2.5), or Reagent R *(216)*, TFA–thioanisole–1,2-ethanedithiol–anisole (90:5:3:2), 1–4 h, 25°C. Efficient *S*-detritylation can also be achieved by using trialkylsilane scavengers, which quench the trityl carbocation to form triphenylmethane *(217)* and have the added virtue of being less likely to interfere with subsequent oxidation/refolding steps. Reagent B *(218)*, TFA–phenol–water–triisopropylsilane (88:5:5:2), represents an effective and convenient cocktail for removal of *S*-Trt and related acid-labile side-chain-protecting groups used in Fmoc chemistry. The *S*-Tmob and *S*-TMTr groups can be cleaved in the presence of the same scavengers, but at substantially reduced TFA concentrations, e.g., 7 and 1% TFA, respectively, for ~15 min at 25°C *(185,186)*. Hence, it becomes possible to release free sulfhydryls in the presence of other side-chain-protecting groups, or for some solid-phase synthesis applications, while the peptide chain remains anchored to the polymeric support. *S*-Trt, *S*-Tmob, and *S*-TMTr can be used in orthogonal combination with *S*-Acm (discussed subsequently), and can also be removed selectively in the presence of more acid-sensitive groups, such as *S*-Bzl and *S*-Meb. Electrophilic removal conditions to generate disulfides directly are covered separately.

The *S*-acetamidomethyl (Acm; *187*) and related groups (Table 2) are particularly valuable, because they are essentially stable to both acid and base and hence reasonably compatible with both Boc and Fmoc chemistries (*see* refs. *37,38,51,* and *219* for description of partial loss of Acm groups during repetitive chain assembly and final HF cleavage steps of Boc chemistry; the Tacm group described in ref. *190* might be more stable). The *S*-Acm group survives the conditions for acid removal of *S*-Meb, *S*-Mob, *S*-Trt, and *S*-Tmob; conversely, metal-assisted conditions or electrophilic reagents in properly chosen solvents can serve to remove Acm selectively in the presence of relatively acid-stable groups as *S*-Bzl or *S*-*t*Bu. Recent reports indicate that under a variety of N^{α}-Boc or *S*-Acm deblocking conditions, $S \rightarrow N$ or $S \rightarrow O$ transfer of Acm onto side-chain carboxamide or hydroxyl functions can occur; glycerol was suggested as a helpful scavenger to suppress such side reactions *(220,221)*.

With the goal of obtaining a free sulfhydryl, *S*-Acm is removed typically by treatment with mercuric salts in *acidic* aqueous solutions, e.g., mercuric acetate ($Hg[OAc]_2$, 1 Eq; ~0.05*M*]) in pH 4.0 buffer for 1 h at 25°C *(187)*. Alternatively, it is possible to treat *S*-Acm-blocked peptides with silver trifluoromethanesulfonate (10 Eq) in the presence of anisole (10 Eq), using TFA as solvent for 1 h at 0°C *(174)*. In either case, the initial mercaptide is treated in the usual ways, e.g., excess hydrogen sulfide, β-mercaptoethanol, or dithiothreitol, to remove the metal. *S*-Acm removal can also be carried out in solid-phase synthesis while the peptide remains anchored to the support, by treatment with $Hg(OAc)_2$ (0.06*M*) in DMF for 3 h at 20°C (in the dark), followed by washes with DMF and DMF–β-mercaptoethanol (9:1, v/v) to remove Hg^{2+} from the deblocked peptide-resin *(48,146,189)*.

The *S*-9-fluorenylmethyl (Fm; *194,195*) and *S*-2-(2,4-dinitrophenyl) ethyl (Dnpe; *197*) groups are compatible with Boc chemistry and fully orthogonal with *S*-Meb and *S*-Acm groups. *S*-Fm and *S*-Dnpe are stable to strong acids, such as HF, and are removed under basic conditions by β-elimination reactions, generally as promoted by piperidine (10–50%, v/v) or 1,8-diazabicyclo[5.4.0]undec-7-ene (DBU) (2% v/v) in DMF. In solid-phase synthesis, it is possible either to remove these protecting groups while the peptide chain remains anchored to the support, prior to further transformations, or else first to cleave, isolate, and purify *S*-Fm or *S*-Dnpe-protected peptides. Although the chemical properties of *S*-Fm and *S*-Dnpe are similar, the latter group is more polar and less hindered, thereby conferring improved solubility properties to the resultant protected peptides. Also, HF cleavage yields are better with *C*-terminal

Cys(Dnpe), by comparison to Cys(Fm). Base-catalyzed removal of *S*-Fm or *S*-Dnpe in the presence of β-mercaptoethanol provides free thiol functions *(48,195)*; otherwise disulfides are obtained directly (*see* Sections 3., 6., and 8.7.).

An important class of cysteine-protecting groups is mixed alkyl, aryl, and acyl disulfides, such as *S-tert*-butylmercapto (*St*Bu; refs. in Table 2; *see* refs. *50,198,199,222,223* for information about the more acid-stable, base/nucleophile-sensitive ethyl analog), *S*-3-nitro-2-pyridinesulfenyl (Npys; refs. in Table 2), *S*-alkoxycarbonylsulfenyl (Scm shown in Table 2; *see* refs. *150* and *224* for ethyl [Sce], benzyl [Sz], and *tert*-butyl [Scb] homologs), and *S*-([*N'*-methyl-*N'*-phenylcarbamoyl]sulfenyl) (Snm, *207*) that are cleaved under orthogonal reductive conditions. The *S*-S*t*Bu group is compatible with all aspects of Boc or Fmoc chemistries, whereas *S*-Npys is compatible only with Boc, and *S*-Scm and related groups require modified strategies because of the risk of *S* → *N* acyl shifts *(207)*. Since the disulfide-containing protecting groups are reasonably stable to acid, their removal can be accomplished either selectively while a protected peptide remains anchored to a solid-phase synthesis support, or subsequent to acidolytic release from the support and cleavage of all other side-chain-protecting groups (*see* refs. *51* and *201* for discussion of problems with *S*-S*t*Bu in HF with various scavengers). Cleavage of *S*-S*t*Bu in aqueous media occurs with the usual reagents for disulfide bond scission, e.g., β-mercaptoethanol, dithiothreitol, or tri-*n*-butylphosphine (*see* Section 3. for specific examples of conditions for solution deprotection); typical conditions for solid-phase deprotection include treatment with β-mercaptoethanol–DMF (1:1) for 5 h at 25°C *(200)*. Note that these reductions are all designed to provide cysteine residues in the reduced form. In contrast, *S*-Npys, *S*-Scm, and similar groups are set up for regiospecific displacement by free sulfhydryl groups to form unsymmetrical disulfides, as covered in more detail in Section 5., Schemes 3A and B.

An additional level of orthogonality is provided by the use of protecting groups that can be removed enzymatically *(192,225)*. Optimal enzymes for this purpose should operate at near neutral pH values in aqueous media, and should have high selectivity for the bond-cleaving reactions that they catalyze as well as broad substrate specificity with regard to recognizable structures. Hermann and coworkers have suggested that acylamidomethyl cysteine-protecting groups could be deblocked in two stages: (1) enzymatic cleavage of the acylamido bond; followed by (2) spontaneous hydrolysis of the aminomethylmercapto intermediate (for *N*-terminal Cys, this cyclizes further to a thiazolidine-4-carboxyl residue). These concepts were tested initially on Cys(Acm) using an ω-aminoacylase from chicken kidney. The most promising results were obtained using *S*-phenylacetamidomethyl (Phacm) *(48,191–193)* for cysteine protection, and penicillin G acylase from *E. coli*, an enzyme with a P_1 specificity for phenylacetyl moieties. Deprotection is carried out typically for ~24 h at 35°C, using enzyme that is immobilized on acrylic beads, and with the Cys(Phacm) peptide dissolved in pH 7.8 phosphate buffer. When the enzymatic hydrolysis buffer includes β-mercaptoethanol (2% v/v), the product peptide has a free sulfhydryl, whereas in the absence of a reducing agent, the symmetrical homodimer forms. The *S*-Phacm group is compatible with both Boc/Bzl and Fmoc/*t*Bu synthetic strategies, and is fully orthogonal with base-labile *S*-Fm and *S*-Dnpe, and acid-labile *S*-Meb, *S*-Trt, and *S*-Tmob groups. *S*-Phacm can also be applied with *S*-Acm, which survives the initial enzymatic removal of *S*-Phacm *(48,193)*. Thus, the enzymatic approach has the potential to extend substantially the range of options in regiospecific synthesis of disulfide-containing peptides. Possible limitations include relatively long reaction times, and solubility restrictions given the narrow pH optimum of the enzyme.

3. Formation of Disulfides from Free Thiol Precursors

The conceptually simplest approach to prepare disulfide-containing peptides involves complete deprotection, purification of the linear precursor in the free poly(thiol) form, and then careful oxidation to the properly folded product with the correct pairings (Scheme 1, Approach A). This often successful strategy offers the clear advantage of requiring only one type of *S*-protecting group for all cysteine residues, thus minimizing the number of chemical steps that must be conducted following completion of chain assembly.

The crude cleaved material generally needs to be treated with suitable reducing agents to ensure that the chain is a linear monomeric species, free of intra- or intermolecular crosslinks. Toward this end, treatments are generally carried out with: (1) 10–100 m*M* dithiothreitol or (2) 0.1–0.3*M* β-mercaptoethanol, either reducing agent in an appropriate buffer, e.g., 0.1*M* Tris-HCl, 1 m*M* EDTA, at pH 8–9, usually for 2–20 h at 25°C

(226) (*see* ref. *227* for example where heating to 37°C was necessary; *see* refs. *228–230* for alternative dithiols that serve as reducing agents); or (3) tri-*n*-butylphosphine (5- to 10-fold excess over disulfide) in 0.5*M* aqueous sodium bicarbonate–*n*-propanol (1:1), pH 8.0–8.3, at 20–25°C for 1 h (ref. *231*; alcohol needed to dissolve phosphine reagent; however, *see* ref. *232* for an alternative, water-soluble, phosphine: tris[2-carboxyethyl]phosphine [TCEP]). Sometimes, reduction is carried out in the presence of denaturants, such as 6*M* guanidium hydrochloride or 8*M* urea (the latter is less preferred because of partial decomposition to cyanate at basic pH).

Once reduction of crude synthetic material is complete, purification should be carried out at acidic pH, to minimize oxidation and/or disulfide interchange. Low-mol-wt species, e.g., denaturants, salts, and excess reducing agents, should then be removed by dialysis or gel filtration, setting the stage for the folding/renaturation/reoxidation steps. Reversed-phase HPLC (*see* Chapter 3, *PAP*) and/or ion-exchange steps (*see* Chapters 2 and 5, *PAP*) can also separate the desired reduced material efficiently from byproducts. In our experiences, disulfide formation procedures are more likely to be successful if the original reduced polypeptide has undergone initial purification step(s) (*see* ref. *233* for an example where an oxidized product reverted to unwanted polymers on lyophilization, because of concentration of trace thiol scavengers remaining from a peptide-resin cleavage step). Also, some workers prefer to carry out purification at the poly(*S*-sulfonate) level; the required intermediates are generated by oxidative sulfitolysis and subsequently reduced to the poly(thiol) by procedures similar to those already described (*see* refs. *114,131,234* for examples and leading references).

Oxidation in solution needs to be carried out at high dilution, typically at 10–100 μ*M*, in order to avoid aggregation and intermolecular side reactions. Nevertheless, solubility can often be a problem. Commonly, molecular oxygen serves to promote disulfide formation under slightly alkaline conditions, e.g., pH 7.5–8.5; this is done by simple aeration, under gentle stirring, or with slow bubbling, through the dilute peptide solutions. Most likely, a radical mechanism applies *(159,160,235,236)*. Occasionally, moderate levels of denaturants, e.g., 0.5–3*M* urea or 0.1–1.5*M* guanidine hydrochloride, are added to avoid aggregation; a possible added benefit may be to cause a partial unfolding and expedite the renaturation process (*see* ref. *237* for a recent report of intramolecular oxidation of peptide bis(thiols) carried out in 8*M* Gu-HCl). Reactions are conducted at 5–25°C for 2 h to 4 d, and monitored by HPLC, capillary electrophoresis (*see* Chapter 6, *PAP*), and/or other analytical techniques, including Ellman or related tests *(238–240)* to determine disappearance of free thiols. Considerable experimentation is required to devise optimal conditions, and it is not uncommon for misfolded species or oligomers to accumulate in solution and/or precipitate out. Examples of the overall approach and some of the pitfalls, together with experimental documentation, are provided in the syntheses of a variety of toxins, growth factors, bovine pancreatic trypsin inhibitor, and ribonuclease, already listed with references in Table 1.

Air oxidation, under conditions already discussed, is sometimes too slow to achieve useful yields of monomeric products. The organic and biochemical literature includes numerous examples of relatively mild oxidizing agents that can serve to convert thiols to disulfides *(159,160, 236,241)*. Potassium ferricyanide *(24,26,33,242)* has been particularly useful in the preparation of small-size, intramolecular single-disulfide peptides in the oxytocin or somatostatin families. The procedure has also been used recently in the orthogonal formation of the first disulfide bond of several peptides containing two disulfide bonds *(44,46)*. Dimethyl sulfoxide (DMSO)-promoted oxidation of thiols to disulfides was described first by Wallace *(243)*, and the use of DMSO to circumvent some of the difficulties that are encountered occasionally by classical oxidation approaches was recently brought to the fore by the independent studies of Tam, Fujii, and their respective coworkers *(81,146,244)*. Advantages of DMSO oxidation include applicability over the extended pH range of 3–8, faster reaction rates, the effect of DMSO as a denaturing cosolvent, and improved solubility characteristics for the materials being oxidized (*see* Note 2). Further applications of the method have been demonstrated in the conotoxin and trypsin inhibitor families; these experiments were encouraging, although difficulties in removing DMSO from the final products were also noted *(49,106)*.

An interesting way *(245)* to achieve rapid intramolecular cyclization of dithiol peptides involves use of ethoxycarbonylsulfenyl chloride (SceCl). Crude peptides, obtained directly after lyophilization of HF cleavage products, were taken up in mildly acidic (pH 4–7) aqueous media, and combined with the reagent in a 2:1 SH:SceCl ratio (in practice, the amount of SceCl needed was greater). In this method, SceCl acts nominally as an oxidizing agent, although the actual mechanism presumably involves formation of an *S*-Scm derivative of one thiol, which then becomes displaced by a second thiol (compare to Scheme 3A in Section 5.). Because some closely related impurities were noted in several cases tested, the SceCl methodology has been advocated for cases where other methods proved inadequate *(78,245)*. Azo-oxidizing agents, e.g., diethyl azodicarboxylate (DEAD) also react by a net two-step mechanism (compare to Scheme 3C and accompanying discussion in Section 5.).

Another significant strategy for renaturing peptides and small proteins with multiple disulfide bonds involves the use of redox buffers that mimic physiological conditions *(246,247)*. These procedures are carried out with dilute poly(thiol) peptide at pH 7.3–8.7, and proceed by thiol-disulfide exchange mechanisms *(4,9)*. Thus, even should nonnative disulfide bridges be favored kinetically, they can equilibrate ultimately to pair in the thermodynamically preferred arrangements found in the correctly folded structure. Most commonly, reduced and oxidized glutathione are added at 3 and 0.3 m*M*, respectively *(247)*, although in some cases, oxidized glutathione alone has been adequate *(106)*. Other redox systems are based on dithiothreitol *(105,248)* or on the protein thioredoxin *(249)*. Methods involving protein disulfide isomerase *(250–252)* have not been applied as yet to synthetic peptides, but may be important for future work.

The previous section included a description of orthogonal deprotection modes in solid-phase synthesis that allow retention of free cysteine-containing peptides on the polymeric support. This opens up possibilities for carrying out in the solid-phase mode some of the oxidation chemistries covered in the present section. The pseudo-dilution phenomenon will then favor intramolecular cyclization reaction to provide monomeric disulfides as the major (>60%) products, although dimers and higher oligomers also form (all peptide products identified and quantitated *after* cleavage from the support). Other advantages of solid-phase cyclizations include straightforward removal of oxidizing agents and/or solvents by filtration and washings, and the fact that reaction conditions are independent of the solubility characteristics of the sequence to be oxidized.

The principles of resin-bound oxidation of peptide dithiols have been demonstrated in several laboratories. Early examples used air or 1,2-diiodoethane as oxidizing agents *(44,223,253,254)*. An instructive example with Fmoc chemistry used TFA-labile *p*-alkoxybenzyl ester anchoring as a starting point for synthesis of a decapeptide with *S*-S*t*Bu cysteine protection; orthogonal reduction was followed by overnight oxidation with 1*M* aqueous potassium ferricyanide–DMF (1:10 v/v), at 25°C, to form a 26-membered cyclic disulfide *(200)*. An optimal example from Boc chemistry involves oxytocin, which was prepared on an HF-labile MBHA-resin using *S*-Fm protection for the cysteine residues *(124)*. Fm removal under an argon atmosphere with piperidine–DMF–β-mercaptoethanol (10:10:0.7), at 25°C for 3 h, gave the resin-bound dithiol which was oxidized for 1 h with air or with 5,5'-dithiobis(2-nitrobenzoic acid) (DTNB, 0.5 Eq) while the resin was suspended in "buffered" DMF (containing 0.1*M* each of acetic acid [HOAc] and *N*-methylmorpholine [NMM], with further NMM added to achieve a nominal pH of 7.5 at a glass electrode). The overall solid-phase yield was considerably better than when corresponding chemistry was carried out in solution, and marginally improved over a procedure where solid-phase Fm removal with piperidine–DMF (1:1), 25°C, 3 h, gave *directly* the 20-membered cyclic disulfide. Results with *S*-Dnpe in place of *S*-Fm were similar *(197)*. Lastly, resin-bound dithiols from several sequences could be generated by selective removal in dilute TFA of *S*-Tmob from peptides prepared by Fmoc synthesis on tris(alkoxy)benzylamide (PAL) supports *(185,186)*. Subsequent oxidation occurred under novel conditions adapted from Wenschuh and coworkers *(255)*: ~20 m*M* carbon tetrachloride–triethylamine (2 Eq each) in *N*-methylpyrrolidinone, at 20°C for 4 h (49, 186). These conditions were optimized carefully to maximize intramolecular cyclization, and represent a balance between negligible reaction and excessive oligomerization. Interestingly, the same oxidation procedure when applied to resin-bound peptides containing a *single* free cysteine residue represents a key first step in efficient *intermolecular* routes to parallel homodimers (*see* Scheme 5F in Section 6., and ref. *256*).

4. Symmetrical Formation of Disulfides from S-Protected Precursors

A number of oxidizing agents take *S*-protected cysteine derivatives directly to the corresponding disulfides (Scheme 1, Approach B). The prototype reaction in this regard is the conversion with iodine (1–10 Eq = 0.01–0.1*M*, for 1 min to 2 h at 25°C) of *S*-*T*rt, *S*-Tmob, or *S*-Acm groups *(122,182,185,257)*. (On completion of the deprotection, excess iodine is generally destroyed by reaction with solid or aqueous ascorbate or thiosulfate [titration]; with powdered zinc dust; or by dilution with water

Table 3
"Half-times" (t_h) for Iodine Oxidation of (**A**) Boc-Cys(Trt)-Gly-Glu(OtBu)$_2$ and (**B**) Boc-Cys(Acm)-Gly-Glu(OtBu)$_2$ in Various Solvents[a]

	Solvent	t_h for **A**, *S*-Trt	t_h for **B**, *S*-Acm
Group I	MeOH	3–5 s	1 min
t_h(*S*-Trt) < t_h(*S*-Acm)	MeOH–H_2O (4:1)	<1 s	4–6 s
	HOAc	70–80 s	40–45 min
	HOAc-H_2O (4:1)	1–3 s	50–60 s
	Dioxane	1 min	1.5–2 h
	Dioxane–H_2O (4:1)	5–10 s	5–10 min
	MeOH–$CHCl_3$ (1:1)	2–4 s	15 min
Group II	$CHCl_3$, CH_2Cl_2	1–2 s	1.5–2 h
t_h(*S*-Trt) << t_h(*S*-Acm)	HFIP–$CHCl_3$ (1:1)	1–2 s	>2 h
	HFIP–$CHCl_3$ (3:1)	<1 s	>2 h
	TFE–$CHCl_3$ (1:1)	5–6 s	>2 h
	TFE–$CHCl_3$ (3:1)	4–5 s	>2 h
Group III	DMF	25–35 s	2–3 s
t_h(*S*-Trt) > t_h(*S*-Acm)	DMF–H_2O (4:1)	30–40 s	3–5 s

[a]Adapted from Kamber et al. *(182)*. Reactions were carried out at 20–25°C, with 5 m*M* peptide and 15 m*M* iodine. Group II solvents allow selective and quantitative oxidation of *S*-Trt in the presence of *S*-Acm. With Group I solvents, co-oxidation of two different linear peptides, one protected with *S*-Trt and the other with *S*-Acm, showed a surprising preference (as much as 70–80%) for open-chain *asymmetrical* heterodimer formation (*see* refs. *59,144,147,151,182*). The corresponding experiments with Group III solvents (DMF or DMF–H_2O mixtures) gave the expected random statistical mixtures. The recent work of Tesser and coworkers *(258)* illustrates solid-phase variations on this chemistry, taking advantage of cysteine bound as a thioether to an appropriate trityl resin.

followed by extraction with carbon tetrachloride.) Suitable neutral solvents for iodine oxidation include aqueous and neat methanol or trifluoroethanol, dioxane, chloroform, and DMF; in some cases, it is advantageous to carry out reactions with anhydrous or aqueous carboxylic acid solvents or cosolvents such as acetic acid, propionic acid, and trifluoroacetic acid (water often accelerates reaction rates). In fact, the classic studies of Kamber and coworkers *(182)* show that remarkable selectivities (from 10-fold to >1000-fold) between *S*-Trt and *S*-Acm, in either direction, can be achieved by appropriate choices of solvent (Table 3).

As documented throughout this chapter, pairwise iodine oxidations have become the cornerstone of numerous successful applications in the peptide field. Occasionally, sensitive residues, such as tryptophan, tyrosine, or methionine, suffer modification under the conditions used *(62,85,182,221,242)*; overoxidation of cystine bridges has also been noted *(257)*. Such side reactions can be mitigated by appropriate scavengers, control of pH (acid media are preferable), and minimization of reaction times (as shown in ref. *259*, disulfide dimerization can be suppressed, and intramolecular tryptophan-2-thioether formation favored, when the amount of iodine is limiting and "Group II" solvents [*see* Table 3] are used). Iodine oxidations of Cys(Acm) residues at or near the *N*-terminus of a peptide are said to proceed better when the endgroup is blocked *(136)*. Other sources of iodonium ion (I^+), such as cyanogen iodide *(260)* or *N*-iodosuccinimide *(57)*, may be advantageous alternatives to iodine.

Fujii and coworkers *(183)* introduced thallium (III) trifluoroacetate ($Tl[tfa]_3$) as a particularly mild oxidizing agent for removing a range of *S*-protecting groups (Scheme 2A). Typically, reactions are successful with thallium reagent (1–2 Eq), in the presence of anisole (2 Eq), at 0°C for 1–2 h, using TFA as the solvent. Tryptophan and methionine residues should be protected (the latter as its sulfoxide), since under the conditions cited Trp is destroyed and Met is oxidized partially to the sulfoxide (but not sulfone). In relevant side-by-side comparisons, thallium oxidations have been found to proceed more rapidly and give cleaner results than reactions with iodine; however, it should be stressed that the storage sensitivity and toxicity of thallium precludes its use on larger than laboratory scales. Also, as with other metals, thallium can be difficult to remove entirely from sulfur-containing peptides.

Another novel oxidizing milieu (Scheme 2B), discovered independently by Kiso *(261,263)* and Fujii *(262)*, is a mixture of alkyltrichlorosilanes and sulfoxides. Typical reaction conditions involve dialkyl or diaryl sulfoxide (0.4*M*; diphenyl sulfoxide [DPSO] as shown in Scheme 2B, or DMSO or tetramethylene sulfoxide [TMSO]) and silyl chloride (1.0*M*; $MeSiCl_3$ as shown in Scheme 2B, or TmsCl, Tms triflate, or $SiCl_4$) for 4 h at 4°C, in TFA as solvent. The method is compatible with most amino acid side-chains; Trp must be used as its N^{in}-formyl derivative (later removed by rapid base treatment without affecting disulfide) because unprotected Trp becomes fully chlorinated under the reaction conditions *(263)*. An interesting variation involves treatment of peptides containing acid-labile cysteine-protecting groups such as *S*-Trt and even *S*-Mob (but not *S*-Meb) with TFA in the presence of dimethyl sulfoxide

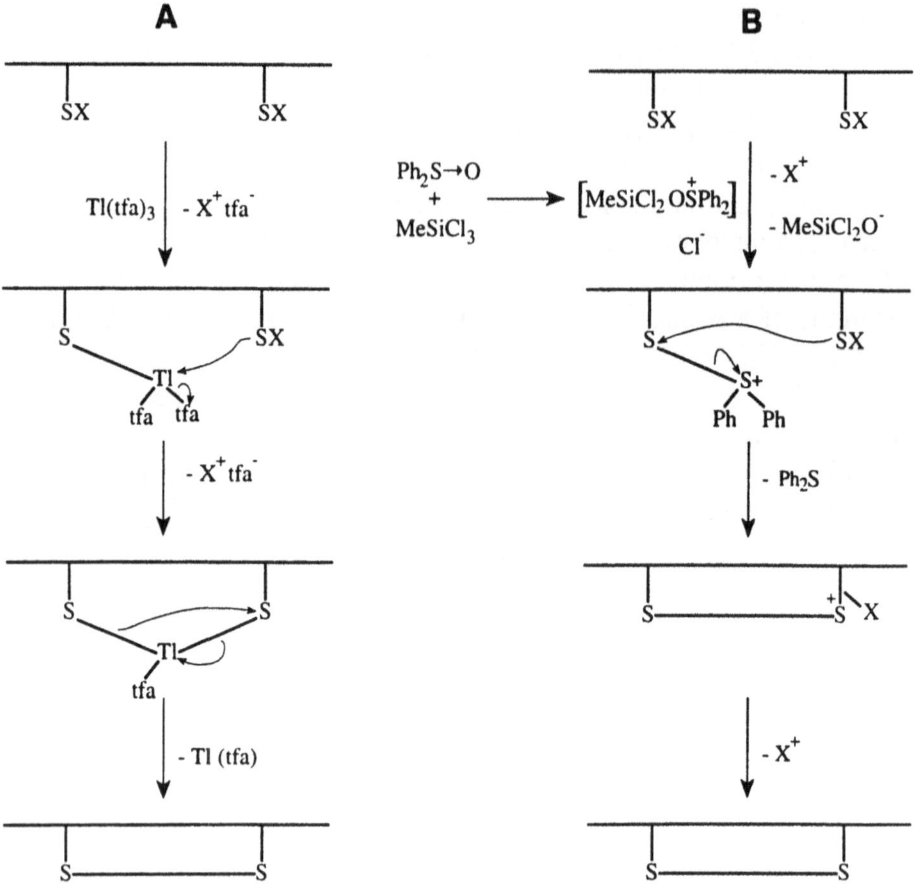

Scheme 2. Suggested mechanisms for mild oxidation methods of *S*-protected cysteine residues to form disulfide bonds. **(A)** Mediated by $Tl(tfa)_3$; adapted from ref. *183*. The initial thallium reagent is in the (III) oxidation state, and the final thallium salt coproduct is at the (I) level. Reactions are conducted typically in TFA (*183*), although DMF and acid–DMF mixtures can also be used (*124*). The chemistry has been demonstrated for X = *t*Bu, Mob, Acm, Trt, and Tmob (listed in approximative order of reactivity, from least reactive to most). **(B)** Mediated by silyl chloride/sulfoxide; adapted from ref. *261*. The chemistry has been demonstrated for X = Acm, Meb, Mob, Tacm, and *t*Bu (listed alphabetically without regard to relative reactivity). *See* refs. *261–263* and text for further discussion.

(DMSO) (10%, v/v) and anisole (1%); disulfide bonds are formed directly in good yields *(264)*.

A final example for oxidation of *S*-protected cysteine was given by Platen and Steckhan *(173)*, who showed that treatment of Cys(Mob)-

containing peptides in acetonitrile (in the presence of solid $NaHCO_3$) with the homogeneous electron-transfer agent tris(4-bromophenyl) ammoniumyl (a radical cation with antimony hexachloride counterion) (3 Eq), rapidly gives the corresponding intramolecular disulfides in high yields. Because Boc, Z, and *tert*-butyl groups are stable to these conditions, and because *S*-Mob survives manipulations of *S*-Trt (e.g., oxidation with iodine), this methodology shows considerable promise for regioselective disulfide formation.

Some of the oxidation chemistry just discussed can be transferred to the solid-phase mode (general principles already discussed in the previous section). It is preferable to use solvents that effectively swell the peptide-resin; DMF is near optimal in this regard for synthetic schemes based on either Boc or Fmoc. Typical conditions involve iodine (3–10 Eq = 0.03 – 0.08*M*) at 25°C for 1–2 h, or $Tl(tfa)_3$ (~1.2 Eq = 6 m*M*) at 0°C for 1–2 h; peptides protected with *S*-Acm, *S*-Trt, and *S*-Tmob have been oxidized this way *(48,49,124,185,265–270)*. Yields of monomeric intramolecular disulfide-bridged peptides can be as high as 60–90% by these solid-phase approaches. Everything else being equal, results are generally better using thallium rather than iodine for oxidation *(124,185,269)*, but it should be noted that thallium is not compatible with *S*-Trt *(124)*.

When acid-stable anchoring linkages, such as *p*-methylbenzhydrylamine (MBHA) or *o*-nitrobenzylamide (Nb), are used, TFA (which promotes good swelling) can be a helpful solvent for on-resin cyclization. Interestingly, the same oxidizing reagents (I_2, $Tl[tfa]_3$, DMSO) in TFA can be applied with acid-labile anchoring linkages, such as *p*-alkoxybenzyl (PAB, PAC, or "Wang" resin) or tris(alkoxy)benzylamide (PAL); in these cases, excellent yields of disulfide-cyclized peptides can be obtained *concurrently* with deprotection and release of material into solution *(185)*. In a different illustration of the general principle, iodine can be added to the dilute-acid cocktail for cleavage of 2-chlorotrityl resins; the resultant peptides include a disulfide bridge from oxidation of Cys(Trt), but other side-chain-protecting groups remain intact *(271)*.

5. Unsymmetrical Formation of Disulfides by Directed Methods

The symmetrical methods of disulfide formation discussed so far are applicable primarily to cases where the required bonds are either intramolecular cyclic or intermolecular homodimeric. When the goal is to connect two different peptide chains by a disulfide bridge, co-oxidation methods become impractical, since the desired heterodimer is often a minor product in the presence of unwanted homodimers (*see*, however, the careful work of Kamber and coworkers, ref. *182*, which shows that in appropriate solvents, co-oxidation of two linear peptides protected with Cys[Trt] and Cys[Acm] respectively, can lead predominantly to open-chain asymmetrical cystine peptides, as noted in footnote to Table 3). Control of regioselectivity is possible, in principle, by directed disulfide formation methods (Scheme 1, Approach C). Directed methods are sometimes applied to internal disulfide bridges, although their main application is to heterodimeric and multiple disulfide-containing peptides.

Selective activation of a given cysteine residue (formation of SZ* in Scheme 1) can be carried out:

1. *In situ*;
2. As a discrete step leading to an isolable, characterizable intermediate; or
3. With a choice of a cysteine-protecting group that is sufficiently stable to chain elongation conditions, yet suitably activated for disulfide formation (i.e., Z = Z*).

Initial examples for category (1) are due to the pioneering work of Hiskey *(153)*, in which thiocyanogen $(SCN)_2$ was reacted with peptides containing either a free or protected thiol. The resultant activated peptides (Z = –SCN, used without isolation) were then reacted further with second peptides containing a free thiol, to form heteromeric disulfide-bridged products. Substantial levels of homodimeric byproducts form when the activation step is either too slow or incomplete, and the high electrophilic reactivity of thiocyanogen poses significant risks for a variety of side reactions. These problems provided an impetus for efforts to devise milder chemistries according to category (2) above.

Activation to form an isolable intermediate is exemplified by *S*-sulfenylphthalimide derivatives of cysteine peptides *(272)*. This interesting advance has had limited application, perhaps because of difficulties in the preparation of the activated compounds. In contrast, *S*-alkoxycarbonylsulfenyl derivatives (e.g., Scm in Table 2 and Scheme 3A) can be prepared readily by reaction of either free or protected (e.g., *S*-Trt, *S*-Acm) thiols with the appropriate alkoxycarbonylsulfenyl chlorides *(150,205–207,224,245,273)*. These base-labile *S*-sulfenyl derivatives must be handled under neutral or slightly acidic conditions where there is, however, a risk of undesired modification of unprotected tryptophan. Forma-

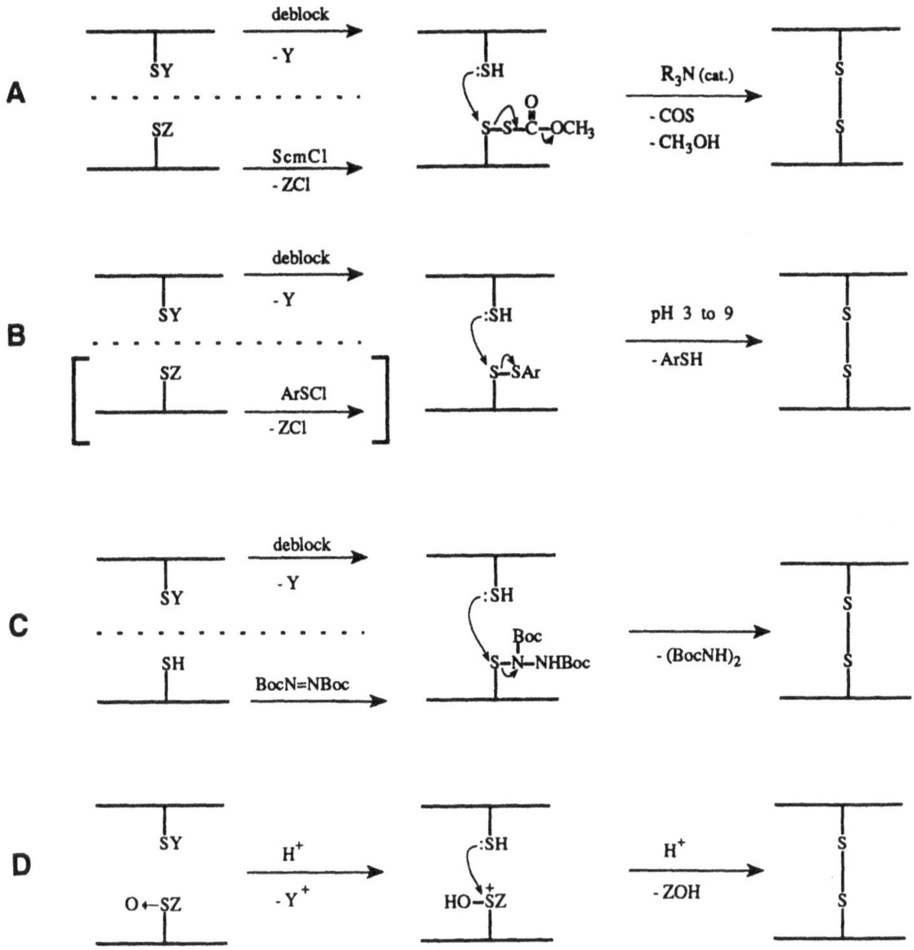

Scheme 3. Approaches for directed formation of disulfide bonds. *See* accompanying text for discussion and leading references. Reactions can be conducted when the two cysteines to be paired are on separate chains (heterodimer formation) or on the same chain (intramolecular cyclization). Transformations above and below the horizontal dotted lines are carried out separately. The square brackets for entry B indicate that the *S*-SAr intermediate is optionally prepared as shown, but can also be obtained by using Cys(SAr) as a building block in chain assembly. Numerous options exist for selective removal of protecting group Y; check Table 2 for reagents, conditions, and compatibility restrictions.

tion of the desired unsymmetrical disulfide occurs in a mild separate step catalyzed by mild base: A second free thiol displaces the *S*-alkoxycarbonylsulfenyl moieties, which are lost irreversibly as inert carbonyl sulfide

(COS) and alcohol (hence driving the reaction to completion) (Scheme 3A; ref. *273*). The references given already in this paragraph demonstrate and document the use of this chemistry for modification and disulfide-linking of cysteine-containing peptides in solution. Kemp's thiol capture method also makes use of Scm groups to facilitate connections to a 4-mercaptodibenzofuran template *(167)*. Conversions of *S*-Acm to *S*-Scm, and/or reactions of cysteine thiols with Cys(Scm) in the same or different peptides, can also be carried out in the solid-phase mode *(132,274)*. The products are intramolecular cyclic disulfides and disulfide heterodimers, respectively. Solid-phase variations have the usual advantages of simplicity, convenience of product isolation, and the possibility to drive reactions by use of excesses, but suffer from the difficulty of monitoring adequately completeness of the desired reactions and avoidance of side reactions.

A further way to activate free or protected cysteine residues involves reaction with aromatic sulfenyl halides, such as 2-nitrophenylsulfenyl chloride (Nps; Z = –S-[2-nitrophenyl])*(176,275)* or 2-pyridinesulfenyl chloride (SPyr; Z = –S-2-pyridyl) (276) (Scheme 3B). Directed disulfide formation is now driven by the low pK_a of the aromatic thiol. Unfortunately, the required sulfenyl halides can be difficult to obtain and/or suffer from extreme sensitivity to hydrolysis from atmospheric moisture. On the other hand, peptides containing cysteine in the free thiol form can be activated directly, both in solution and on the solid phase, by application of 2,2'-dithiodipyridine ($[PyrS]_2$) (*239*; *see* refs. *39,87,141,149* for examples in the peptide field). The highly specific transformation of *S*-Snm to *S*-SPyr on treatment with 2-mercaptopyridine in chloroform represents another mild avenue to the required stable *S*-arenesulfenyl intermediates *(207)*. Finally, cysteine-containing peptides can be reacted in aqueous pH 7.4 buffer with Ellman's reagent (DTNB; *238*) to provide the appropriate mixed disulfide intermediates; these react further *in situ* with a second cysteine peptide to provide significant levels of heterodimer *(277)*.

Although most of the *S*-activation procedures described in the preceding paragraphs can be performed relatively efficiently, the additional synthetic step complicates the synthetic strategy and practice both in solution and solid-phase modes. An advantageous alternative, category (3), takes advantage of the dual nature of certain groups, e.g., the 3-nitro-2-pyridinesulfenyl (Npys; Z = –S-[3-nitro-2-pyridyl; *see* Table 2) group for both protection and activation. The commercially available Boc-

Cys(Npys)-OH *(137)* derivative is incorporated readily into synthetic peptides by Boc chemistry, survives successive deprotection/coupling cycles, and is stable to final cleavage/deprotection with strong acids (e.g., HF, TFMSA) *(204)*. Asymmetric disulfide bond formation with a second peptide that contains a free thiol takes place over a wide pH range in aqueous buffers and can be monitored by spectrophotometric titration of the released 3-nitro-2-pyridinethiol *(278)*. Times for complete reaction have been reported *(137,140)* to vary from several hours at pH as low as 4, to 30 min or less at pH 8–9 (caution required at the high end of pH range, because of disproportionation, which provides symmetrical homodimers); another study gave substantially faster rates *(141,149)*. Reactions can also be carried out in the solid-phase mode *(178)*. Recent applications of the Npys group have included the synthesis of intramolecularly cyclized disulfide peptides *(178,203)*, disulfide heterodimers *(137,279)*, parallel and antiparallel bis-cystine peptides *(141,149,280)*, and conjugates of immunopeptides to carrier proteins *(140,143,145,281; see also* Chapter 10, *PAP*). Although the lability of *S*-Npys to piperidine prevents its use in Fmoc solid-phase protocols, peptides otherwise synthesized from N^{α}-Fmoc-amino acids can incorporate Boc-Cys(Npys)-OH as an *N*-terminal residue to be used subsequently in conjugation to a carrier. Additionally, internal Cys(Npys) residues can be formed after chain assembly is complete, by solution or solid-phase conversion of internal free or protected (with *S*-Acm, *S*-Trt, or *S*-*t*Bu) cysteine residues to *S*-Npys by methods covered in the previous paragraph *(148)*.

Additional approaches to directed disulfide formation exist. One attractive option, so far applied only in solution, represents in effect a controlled stepwise oxidation (Scheme 3C; refs. *138,139,282,283*). Peptides containing a free cysteine residue are dissolved in argon-saturated DMF and reacted with bis(*tert*-butyl)azodicarboxylate (2–3 Eq), for 4–12 h at 25°C, to provides isolable sulfenylhydrazides (Z = –N[Boc]NHBoc); following concentration and trituration to remove excess oxidizing agent, these are redissolved and combined under similar conditions with a *second* peptide containing deblocked cysteine to give disulfide heterodimers. Another elegant approach allows directed pairing of protected cysteine residues under strong acidic conditions (Scheme 3D; refs. *155,284,285*). One of the pair is not only protected (e.g., with Z = Mob or Acm), but oxidized further as the sulfoxide (in fact, the sulfoxide is present in the amino acid building block used during assembly of the entire peptide chain). Once the acid-labile group (e.g., Y = Mob) protecting the other Cys is removed, the liberated thiol attacks the electropositive sulfur atom of the protonated sulfoxide, and a disulfide bond results. The method has been applied for both inter- and intramolecular reactions.

6. Regioselective Formation of Disulfides

This chapter has so far presented various alternatives for selective protection and deprotection of cysteine residues, together with methods for disulfide formation. The most demanding tests of these chemistries arise when the goal is to prepare peptides containing multiple disulfide bridges. The general approach is for graduated deprotection and/or co-oxidation of pairwise half-cystine residues, as specified by the original protection scheme. The present section describes a few representative examples of viable strategies, as applied to some of the targets already presented in Table 1. The best work in the field provides unambiguous syntheses of the correct structures, although it should be noted that absolute yields are often quite low (or not reported). As with random oxidation approaches discussed earlier, the regioselective routes require considerable care in the selection of experimental conditions, particularly with regard to appropriate solvents (*see* ref. *63* for an ingenious regioselective random oxidation, based on the preference of cysteine and penicillamine residues to form mixed disulfides).

To achieve regioselective disulfide formation in single-chain peptides (Scheme 4), the desired linear protected peptide sequences are assembled by the usual solution or solid-phase methods, generally with two classes of cysteine-protecting groups. Nishiuchi and Sakakibara (Scheme 4A) used a combination of *S*-Meb and *S*-Acm groups in a synthesis of conotoxin GI that confirmed the disulfide alignment of the native peptide. The fully protected sequence prepared by solution methods was deprotected in HF and oxidized with $K_3Fe(CN)_6$ to give an intermediate retaining Acm protection at Cys^2 and Cys^7. After purification by gel filtration, the second disulfide was formed by treatment with iodine in acidic aqueous methanol. Nearly identical protection strategies have been used by other workers to prepare peptides in the conotoxin family *(42,44,46,47)*, the major differences being that initial chain assembly was by Boc solid-phase chemistry, and in some cases, *S*-Mob was used in place of *S*-Meb and oxidation at the bis(Acm), bis(thiol) stage was carried out in dilute solution with air instead of $K_3Fe(CN)_6$. The Boc/Meb/

Acm solid-phase strategy can be extended to peptides containing three disulfide bridges, as exemplified by a 13-residue *E. coli* enterotoxin fragment (Scheme 4B). In this partially regioselective approach of Shimonishi and coworkers, two of the disulfide bridges were found to pair correctly by air oxidation of a bis(Acm) precursor, and the third bridge was formed by subsequent iodine treatment. Permutations to the initial protection scheme failed to give the native folded enterotoxin, hence ruling out some alternative possible half-cystine pairings.

Other cysteine-protecting group combinations can also be used for regioselective disulfide formation together with Boc solid-phase synthesis. Ponsati and coworkers demonstrated an orthogonal protection scheme with *S*-Fm and *S*-Meb for the synthesis of a bovine pituitary peptide (Scheme 4C). The novel feature of this synthesis was formation of the first disulfide while the protected peptide remained anchored to the polymeric support (earlier sections of this chapter have discussed advantages of this approach for peptides with single disulfides). The piperidine used to promote simultaneous Fm removal and disulfide formation concomitantly deblocked a Trp(For) in the sequence. Next, the peptide-resin, as well as all remaining protecting groups, were fully cleaved by HF treatment; the strong acid did not affect the already-formed disulfide bridge. Finally, the second disulfide was formed by air oxidation at high dilution. The indicated route was more convenient and provided better results than a synthesis of the same sequence using a Boc/Acm/Meb strategy. Importantly, the unambiguous product formed by regiospecific methods coincided with the *major* thermodynamic product from random oxidation of the tetrathiol peptide, a process that also gave minor regioisomers.

The first regiospecific synthesis compatible with Fmoc chemistry was demonstrated by Atherton and coworkers (Scheme 4D). The conotoxin GI sequence was assembled on a 4-hydroxymethylbenzoyl polyamide support, using *S*-Acm and *S*-S*t*Bu protection. Ammonolysis released the protected peptide amide, and then Cys^2 and Cys^7 were selectively deblocked by phosphine-mediated reduction. The first disulfide formed by slow air oxidation, and the second was formed in the usual way with iodine in acidic aqueous acetic acid.

Akaji and coworkers used Fmoc chemistry to prepare conotoxin MI on Rink resin with *S*-Acm and *S*-*t*Bu (Scheme 4E). It should be noted that all aspects of their strategy appear to be compatible with Boc chem-

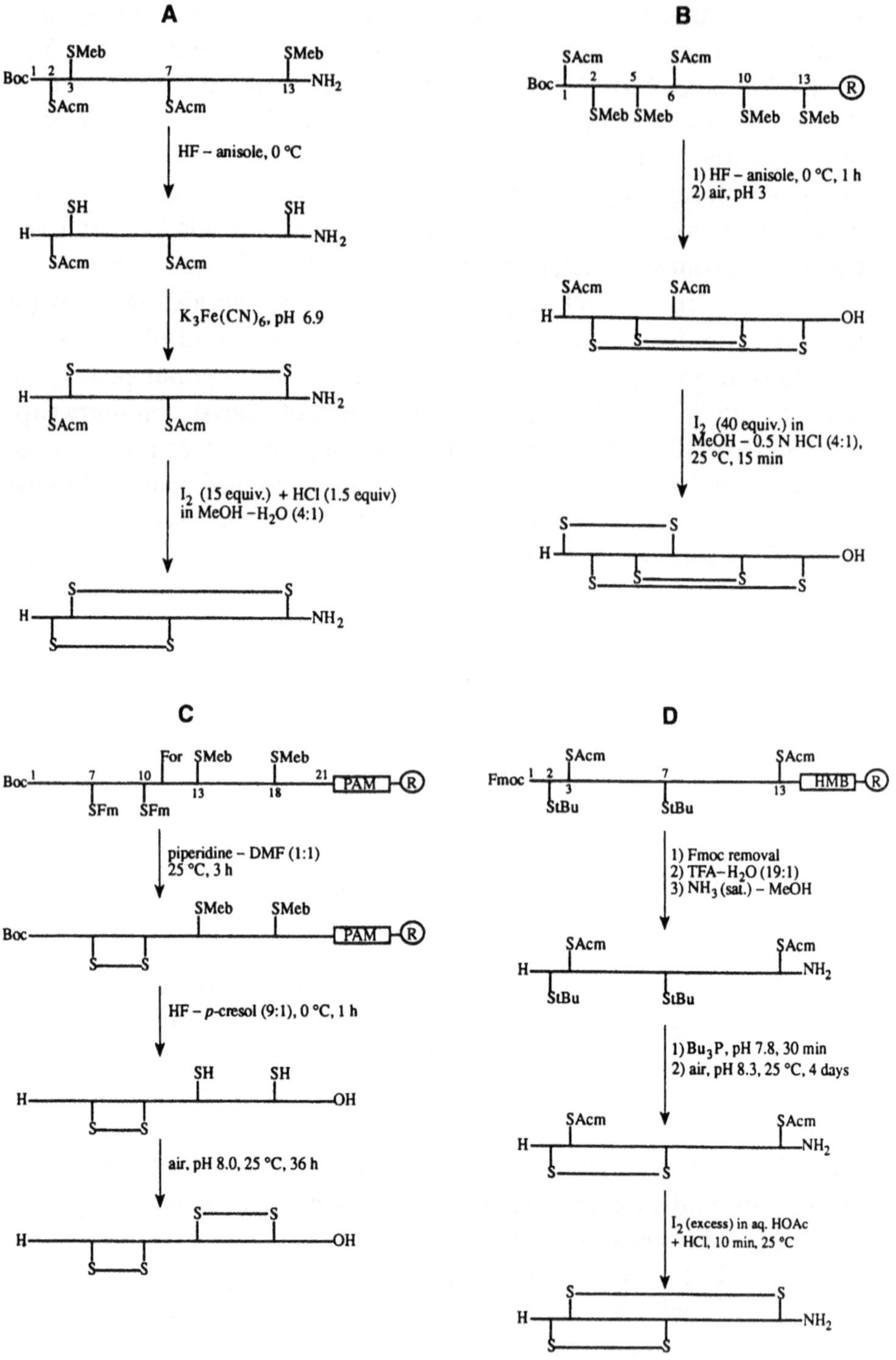
A
B
SMeb
SAcm
Boc
NH2
HF – anisole, 0 °C
SH
H
K3Fe(CN)6, pH 6.9
I2 (15 equiv.) + HCl (1.5 equiv) in MeOH –H2O (4:1)
1) HF – anisole, 0 °C, 1 h
2) air, pH 3
OH
I2 (40 equiv.) in MeOH – 0.5 N HCl (4:1), 25 °C, 15 min
C
D
For
SFm
PAM
piperidine – DMF (1:1) 25 °C, 3 h
HF – p-cresol (9:1), 0 °C, 1 h
air, pH 8.0, 25 °C, 36 h
Fmoc
StBu
HMB
1) Fmoc removal
2) TFA–H2O (19:1)
3) NH3 (sat.) – MeOH
1) Bu3P, pH 7.8, 30 min
2) air, pH 8.3, 25 °C, 4 days
I2 (excess) in aq. HOAc + HCl, 10 min, 25 °C

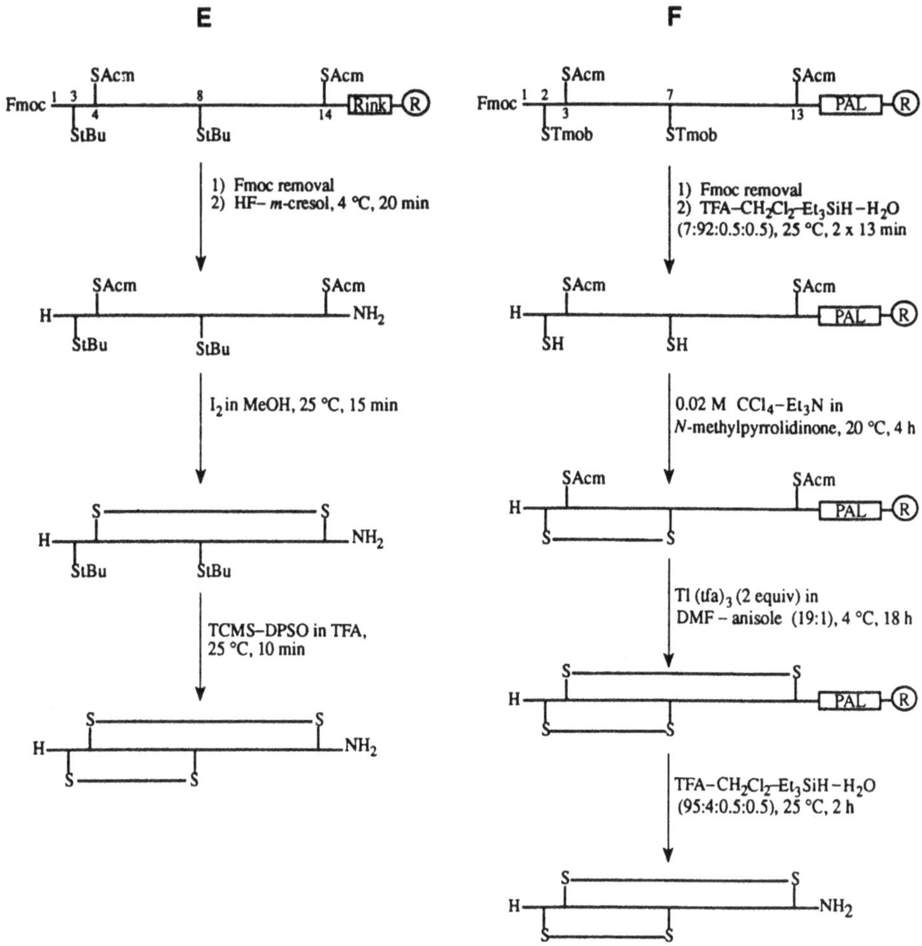

Scheme 4. Regioselective schemes for the preparation of single-chain peptides with multiple disulfide bridges. Only end-group positions and locations of half-cystine residues are shown, and lines in the schematic are not drawn to scale. *See text* for further details. **(A)** Conotoxin GI *(41)*. **(B)** Enterotoxin fragment ST_h (6-18) (*67*; note that for purposes of this scheme, residues have been renumbered). **(C)** Posterior pituitary peptide *(64)*. **(D)** Conotoxin GI *(45)*. **(E)** Conotoxin MI *(39)*. **(F)** α-Conotoxin SI *(48,49)*.

istry with appropriate HF-labile supports. In one variation (not shown in Scheme 4), cleavage with HF–anisole at 4°C for 1 h removed selectively *S-t*Bu and all side-chain protecting groups except for *S*-Acm, thus providing the bis(Acm), bis(thiol) linear chain. This intermediate is analogous to ones already described from Boc/Acm/Meb strategies, so sequential air oxidation and Acm deprotection/co-oxidation (the latter using chemistry of Scheme 2B) gave the desired bicyclic peptide. Alternatively (Scheme 4E), *both* classes of cysteine-protecting groups were stable to HF–*m*-cresol for 20 min at 4°C, under which conditions the anchoring linkage and all other side-chain-protecting groups were cleaved. In this strategy, the first disulfide bridge was formed in highly dilute solution by iodine oxidation of *S*-Acm; following purification, the silyl chloride-sulfoxide method (*see* Scheme 2B) was applied to cleave simultaneously the *t*Bu groups and form the second disulfide. Thus, a common linear precursor can be used to form regioselective disulfide bridges in either of two orders, as controlled by the deprotection conditions.

Further concepts are illustrated in several recently devised alternative routes to α-conotoxin SI (Scheme 4F). The linear sequence was assembled by Fmoc chemistry, and cysteine residues were protected by an orthogonal Tmob/Acm combination. Treatment with dilute acid removed *S*-Tmob selectively, to provide the resin-bound bis(Acm), bis(thiol) intermediate suitable for oxidation under mild conditions to close the first disulfide bridge. The resin-bound monocyclic bis(Acm) intermediate was oxidized further with $Tl(tfa)_3$ in DMF–anisole (19:1) (Scheme 2A), and finally the desired bicyclic peptide was released from the acid-labile tris(alkoxy)benzylamide (PAL) support with a TFA/scavengers cocktail. The general approach was varied to explore the order of cyclization (dictated by location of *S*-Tmob and *S*-Acm groups in initial protected sequence), and to assess the relative merits of fully solid-phase vs solution disulfide-forming strategies.

Methods for controlled formation of disulfides also come into play for the synthesis of parallel and antiparallel bis(cystine) dimers, which are worthwhile targets for a variety of applications *(149)*. For instance, some biologically active molecules, such as the hormone β-atrial natriuretic factor (β-ANP) *(36,39,286)* and the progesterone-binding protein uteroglobin *(141,287)*, consist of two identical polypeptide chains linked in antiparallel fashion by two disulfide bridges; the hinge region of immunoglobulins contains a parallel dimer *(139)*; some artificially designed dimers have interesting binding properties due to their symmetry *(280,288)*; and a parallel dimer of deamino-oxytocin has been shown recently to act as a long-lasting prohormogen, presumably by slow disproportionation under physiological conditions *(256)*. Given the regularity with which dimers arise as unwanted byproducts of syntheses directed at monomeric cyclic disulfide peptides, it is interesting how challenging the intentional synthesis of dimers can be (for example, disulfide exchange during formation of the second bridge can give the corresponding cyclic monomer; *see* refs. *256* and *289*).

In a prototype example of the kinds of approaches that are possible for the preparation of dimers, Ruiz-Gayo and coworkers carried out Boc solid-phase syntheses of a linear heptapeptide with acid-stable *S*-Fm and *S*-Acm protection (Scheme 5A,B). This was straightforward when the goal was the parallel dimer. For the antiparallel synthesis, the two chains with inverted protection were required; these were obtained after simultaneous elongation carried out on two polymeric supports with different flotation properties. To obtain the parallel dimer, the single linear chain was released from the support by HF, then *S*-Fm removal and simultaneous oxidation was carried in solution by treatment with piperidine–DMF (1:1), and lastly, codeprotection/oxidation of *S*-Acm with iodine in aqueous acetic acid gave the desired product. The orthogonal alternative of oxidizing *S*-Acm first was considered less desirable because of the poor solubility of the resultant bis(Fm) dimer intermediate. To prepare the antiparallel dimer, both of the monomeric chains obtained after HF cleavage were treated with piperidine in the presence of β-mercaptoethanol to furnish the corresponding mono(Acm) peptides with free thiols. One of these chains was activated to furnish its *S*-pyridyl derivative, which was purified and reacted with the second chain in its free thiol form to give the first disulfide (chemistry of Scheme 3B). The second disulfide was closed as before with iodine.

Another set of (chronologically earlier) examples came from Sakakibara's laboratory, where the linear protected sequences of α-human atrial natriuretic peptide (α-hANP) were prepared by solution Boc chemistry (Scheme 5C,D). With the goal of preparing the parallel dimer, *S*-Acm- and *S*-Meb-protecting groups were used, whereas for the antiparallel dimer, the *S*-Npys group (stable to conditions of chain assembly) was needed as well. As in other syntheses described earlier, HF deprotections of these peptides gave either the mono(Acm), mono(thiol) chain, or the mono(Acm), mono(Npys) chain. For the parallel molecule, the first of these chains was oxidized sequentially with $K_3Fe(CN)_6$ and treated with iodine. For the antiparallel molecule, the first of these chains (with free thiol) attacked the second one (with Npys activation) to form a disulfide bridge, and again iodine closed the second bridge. Unfortunately,

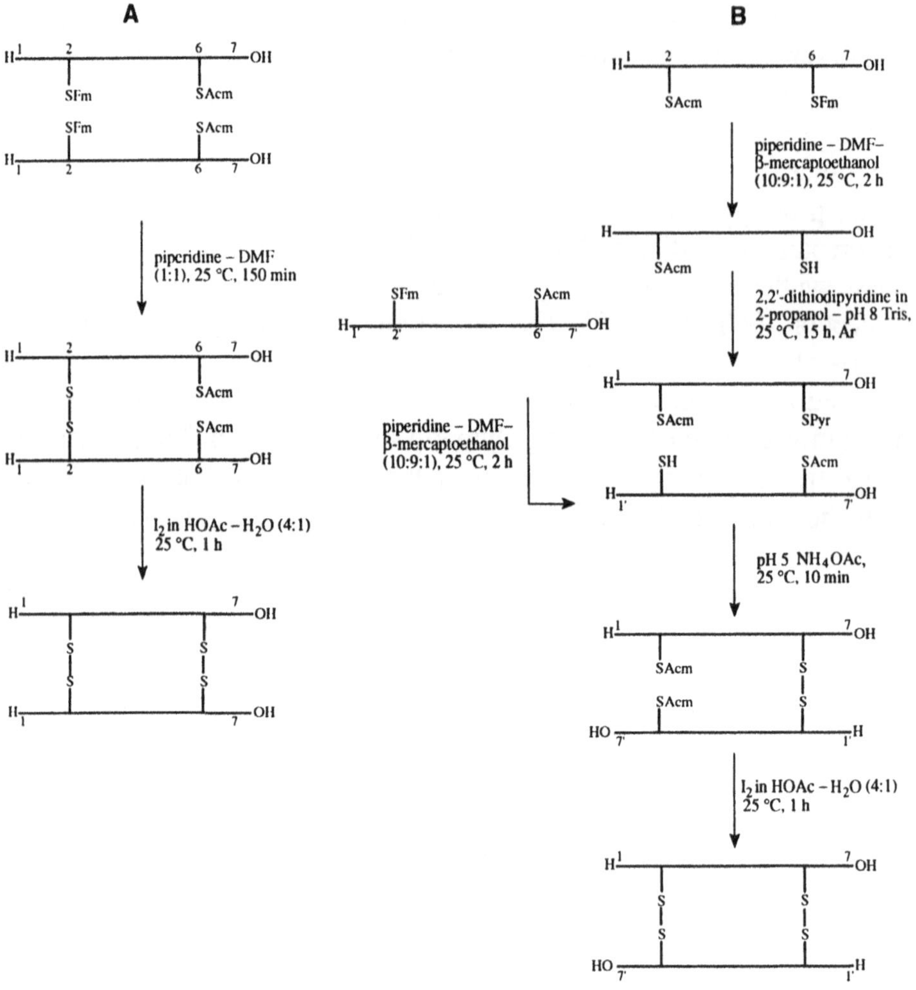

Scheme 5. Regioselective schemes for the preparation of representative dimers (single peptide sequence, connected at two points by disulfide bridges). Only end-group positions and locations of half-cystine residues are shown, and lines in the schematic are not drawn to scale. Chains are numbered from *N*- to *C*-terminal, and the second chain has a prime (') numbering. *See text* for further details. **(A)** Parallel uteroglobin-like cavities *(141,149)*. **(B)** Antiparallel uteroglobin-like cavities *(141,149)*.

the oxidation was accompanied by several side reactions, including Met(O) formation. A potentially milder route to disulfide cyclic dimers was developed by Wünsch and coworkers *(139)* (chemistry not drawn as part of Scheme 5, but protection scheme chosen similar to Scheme 4D).

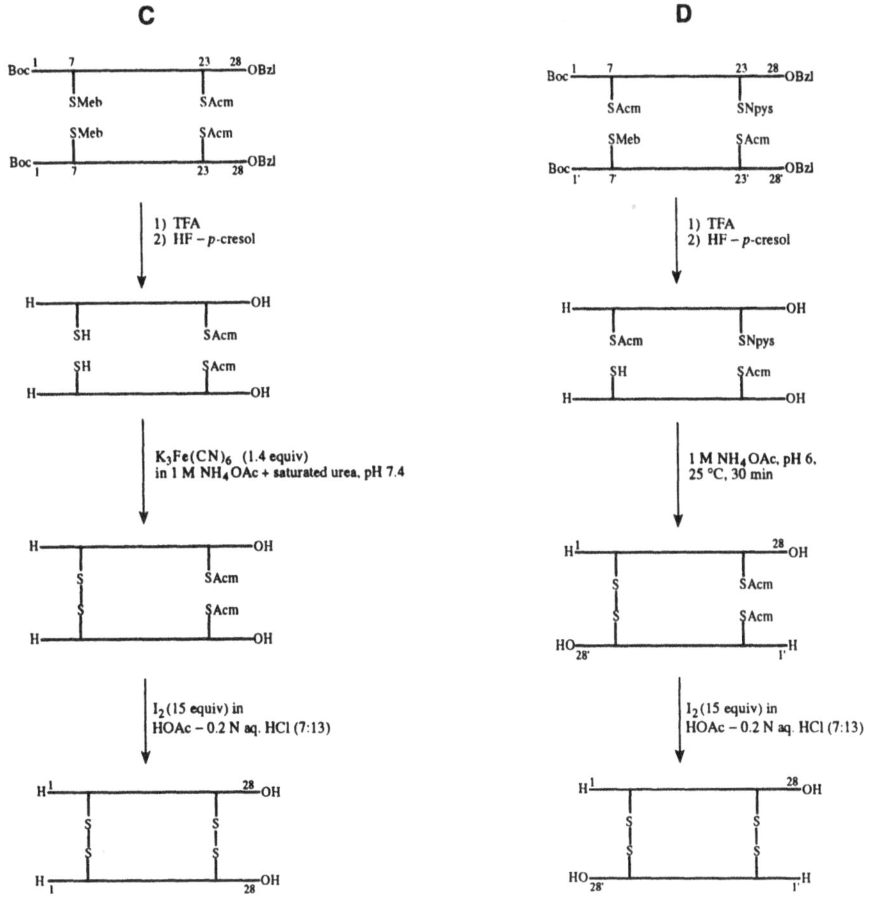

Scheme 5. *(continued)* **(C)** Parallel α-human atrial natriuretic peptide dimer *(36)*. **(D)** Antiparallel dimer of α-human natriuretic peptide dimer (synthesis in ref. *36*), which is in fact β-human atrial natriuretic peptide *(286)(continued)*.

The required linear chains prepared by solution methods with *S*-Acm and *S*-*St*Bu protection were deblocked selectively with tri-*n*-butyl-phosphine, and the azodicarboxylate oxidation procedure (Scheme 3C) was used to form the first disulfide bridge. The second bridge was closed in the usual way with iodine, and the proper design of precursors allowed preparation of the targets in both the parallel and antiparallel alignments (note, the latter compounds rearranged readily to the former under conditions conducive to thiol–disulfide interchange).

Dimers can also be prepared in conjunction with Fmoc chemistry for linear chain synthesis on *p*-alkoxybenzyl ester and related supports. The

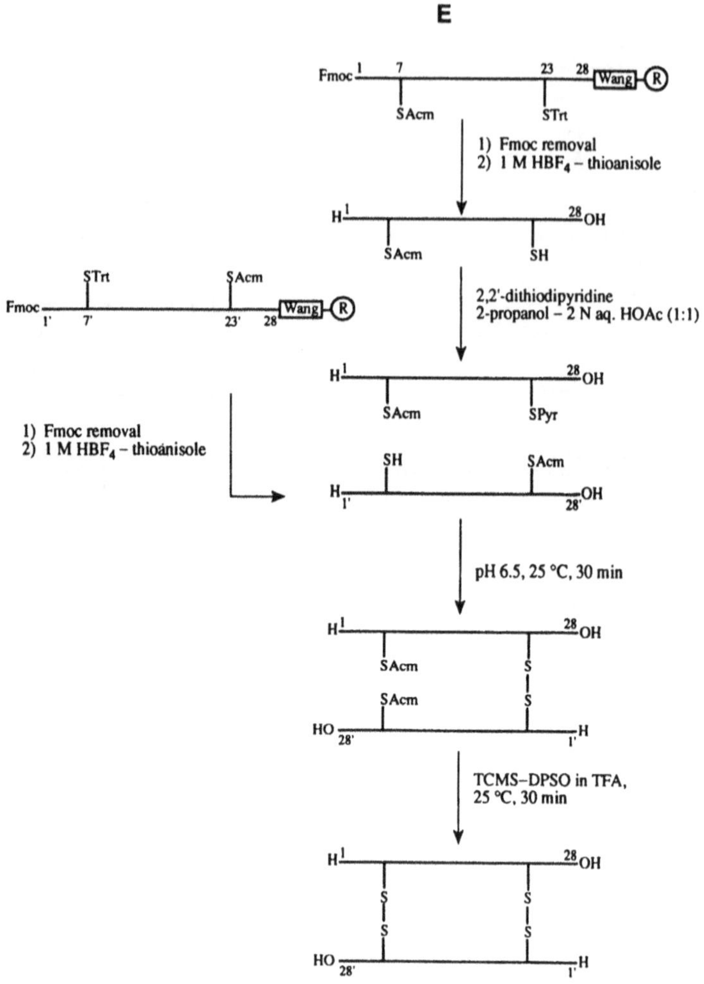

Scheme 5. *(continued)* **(E)** Antiparallel dimer: β-human atrial natriuretic peptide dimer (synthesis in ref. *39*); sequence identical to molecule prepared in (D).

β-human atrial natriuretic peptide already mentioned (Scheme 5D) was prepared in an alternative way by Kiso and coworkers (Scheme 5E). The chains were synthesized twice, with inverted *S*-Acm/*S*-Trt protection on cysteine. Cleavage and partial deprotection with 1*M* HBF_4–thioanisole gave monomeric precursors each containing one Cys(Acm) and one Cys as the free thiol. One of the chains was activated by conversion to the *S*-pyridyl derivative, which was attacked by the free thiol of the other chain. Next, the silyl chloride/sulfoxide method (Scheme 2B) served to remove *S*-Acm and form the second bridge simultaneously.

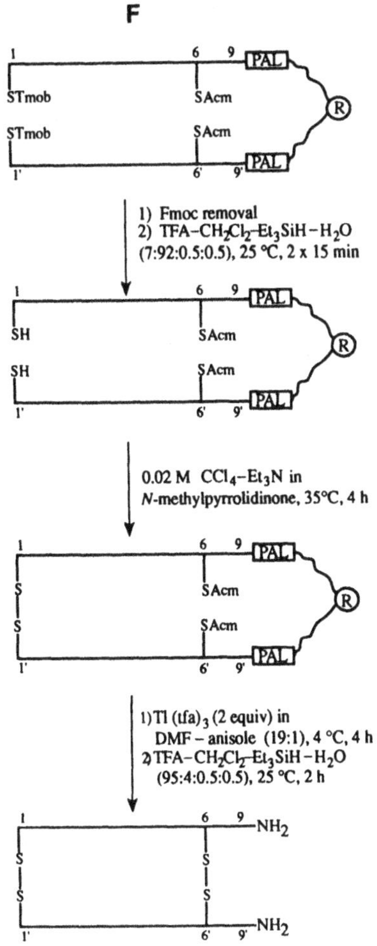

Scheme 5. *(continued)* **(F)** Parallel deamino-oxytocin dimer *(256)*.

A recent orthogonal synthesis of a parallel dimer of deamino-oxytocin relied on two intermolecular polymer-supported reactions to form disulfide bridges (Scheme 5F). The protection scheme involved acid-labile *S*-Tmob for the *N*-terminal Mpa[1] residue, and *S*-Acm for the internal Cys[6] residue. Selective removal of *S*-Tmob and mild oxidation of the free thiols using CCl_4–Et_3N precluded formation of any intramolecular byproducts, and thus gave the corresponding pseudo-cyclic intermediate in high yield and purity. This resin-bound intermediate was treated with thallium to remove vicinal *S*-Acm groups oxidatively and form the desired target.

The peptide targets described so far have consisted of either a single or else two identical polypeptide chain(s), with regioselective chemistry controlling two disulfide bridges. By contrast, the impressive solution synthesis of crystalline human insulin by Sieber and coworkers from Ciba-Geigy provides a remarkable juxtaposition of regioselective disulfide-forming reactions to connect two chains with two intermolecular and a third intramolecular bridge (Scheme 6). The initial A(20-21)-B(17-20) heterodimer was formed *(205)* by the sulfenyl thiocarbonate fragmentation method (Scheme 3A), i.e., activating the *S*-Trt (or *S*-Acm)-protected B tetrapeptide with ScmCl, and carrying out reaction with the A chain dipeptide counterpart in its free thiol form. Establishment of the two remaining disulfide bonds of insulin took advantage of earlier discoveries by the same research team on the selective iodine oxidation of the *S*-Trt and *S*-Acm groups (Table 3). Thus, the key A(6-13) fragment was formed by intramolecular cyclization of two Cys(Trt) residues in the presence of Cys(Acm) (ref. *182* describes direct synthesis of corresponding A[1-13] fragment). A further noteworthy feature of the synthesis relates to the N^{α}-amino protection scheme and novel deprotection conditions, which facilitated couplings of protected insulin segments containing pre-established disulfide bridges. Similar methods provided two unnatural disulfide bond isomers, that had chromatographic, physico-chemical, and biological properties distinct from the natural material *(94)*.

The bombyxins, peptides of the insulin superfamily isolated from the silkworm, have been targets of recent synthetic efforts using regioselective disulfide formation strategies by Suzuki and coworkers. In one case (bombyxin IV, *see* Scheme 7), the A and B chains were assembled separately by Fmoc solid-phase synthesis on Wang resins and released into solution on treatment with TFA while cleaving simultaneously most of the side-chain-protecting groups. For the A chain, the internal disulfide bridge between positions 6 and 11 (both originally protected as *S*-Trt) formed spontaneously on lengthy stirring of the peptide at high dilution; meanwhile, the Cys(Acm) at position 7 and Cys(*t*Bu) at position 20 were unaffected. Acidolytic cleavage of a Cys(Acm)^{B10}, Cys(Trt)^{B22}-protected peptide-resin gave a B chain intermediate retaining *S*-Acm protection, for which the free thiol was temporarily *S*-pyridylated to prevent unwanted dimerization. Selective deprotection of *S*-*t*Bu on the aforementioned A chain intermediate already containing a single disulfide proved challenging, even in the presence of a variety of

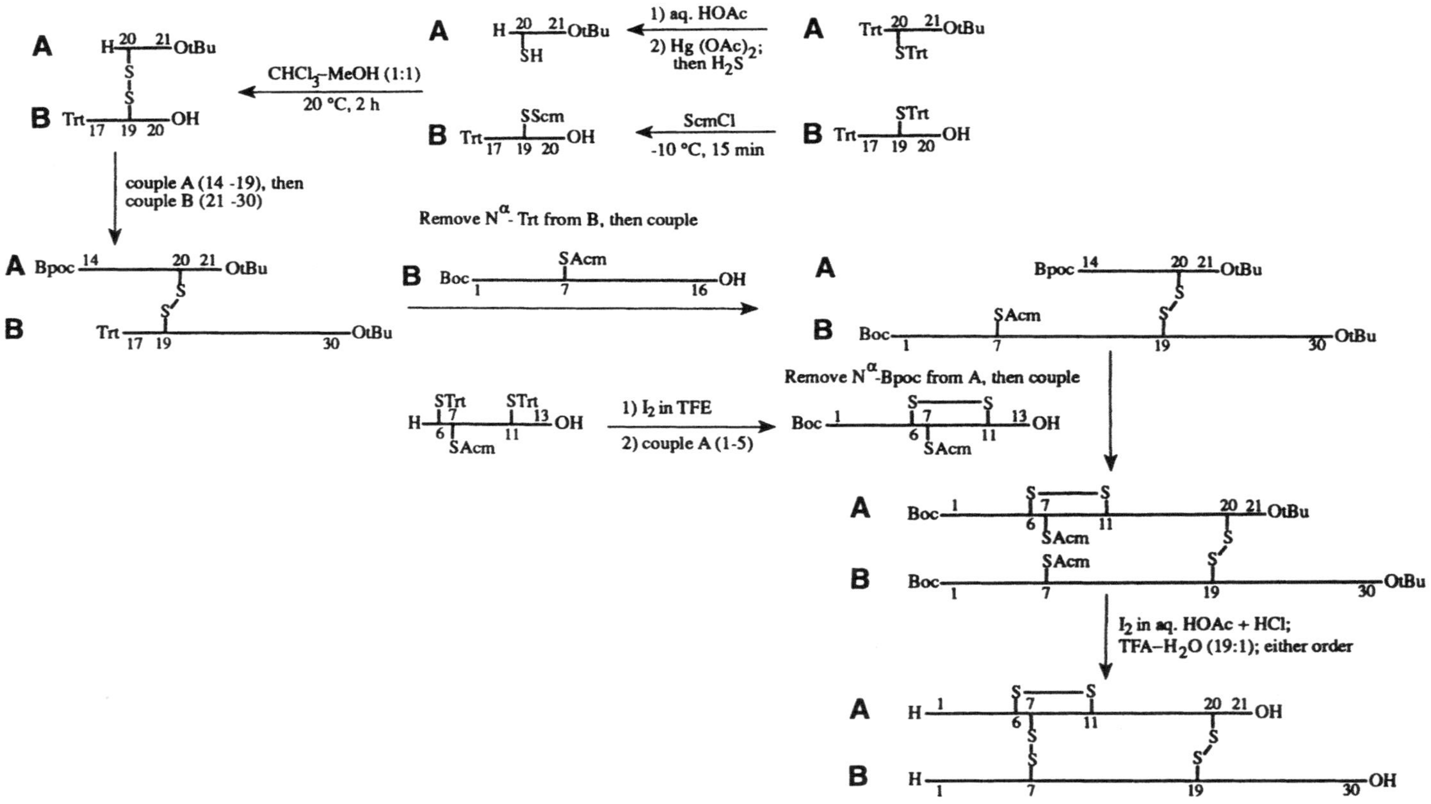

Scheme 6. Regioselective scheme for the synthesis of human insulin. Adapted from refs. *92* and *93*. Only end-group positions and locations of half-cystine residues are shown, and lines in the schematic are not drawn to scale. A and B chains are numbered from *N*- to *C*-terminal. *See text* for further details.

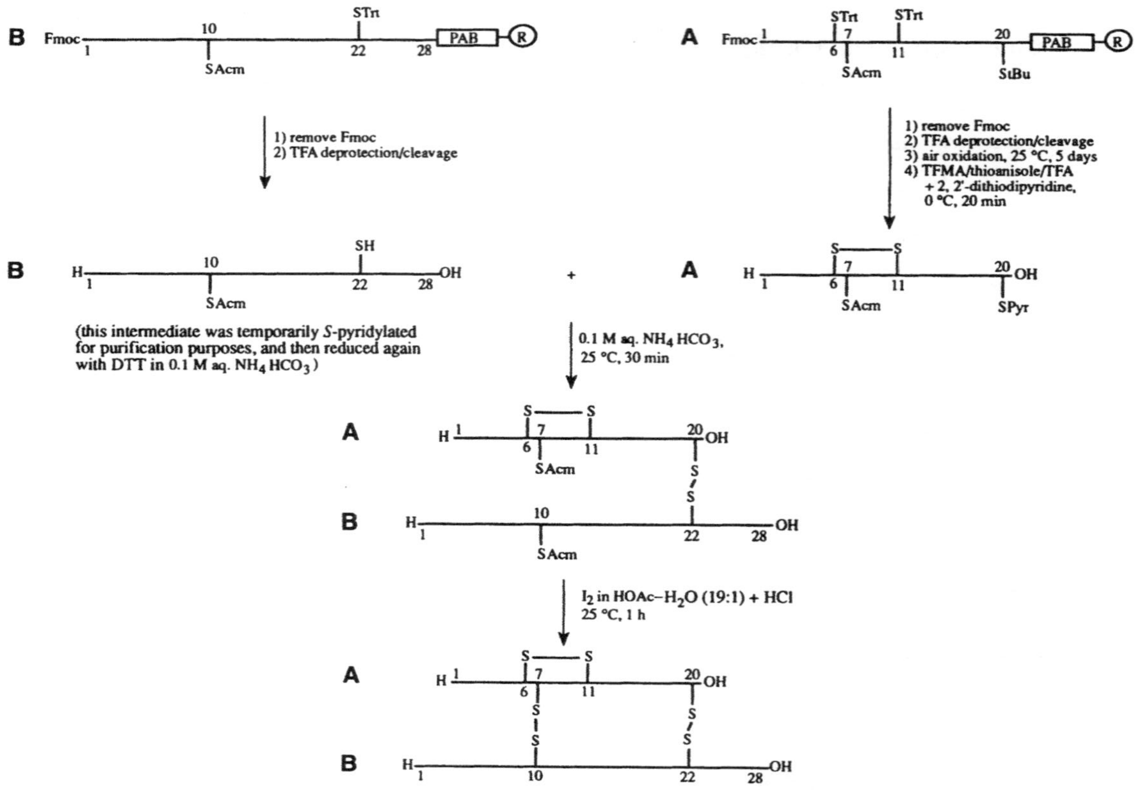

Scheme 7. Regioselective scheme for the synthesis of bombyxin IV. Adapted from ref. *87*; *see* refs. *88* and *89* for related work by the same authors. Only end-group positions and locations of half-cystine residues are shown, and lines in the schematic are not drawn to scale. A and B chains are numbered from *N*- to *C*-terminal. *See text* for further details.

scavengers. Consequently, a one-pot acid deprotection/simultaneous *S*-pyridylation procedure was devised for the A chain. The *S*-pyridyl-A chain intermediate was then mixed with the free thiol form of the B chain intermediate, thus forming the first interchain disulfide connecting positions A20 and B22 (chemistry of Scheme 3B). The resultant [Cys(Acm)^{A7,B10}] bombyxin IV was purified, and the second interchain disulfide was formed by iodine oxidation under acidic conditions similar to the Ciba-Geigy insulin synthesis.

Human relaxin, another insulin-like peptide, has been prepared by regioselective methods as well, in an elegant study by Büllesbach and Schwabe *(96)*. The A and B chains were assembled by Fmoc and Boc solid-phase chemistry, respectively, and in all, four thiol-protecting groups were used: *S*-Trt, *S*-Acm, *S*-Meb, and *S*-Npys. The A chain was cleaved first with TFA–thiophenol, exposing free thiols at positions 10 and 15 (originally blocked by *S*-Trt), whereas the Cys(Acm) at position 11 and Cys(Meb) at position 24 remained intact. Intramolecular oxidation with iodine formed a cyclic disulfide in the A chain intermediate, which was treated further with HF. The stronger acid step now cleaved selectively *S*-Meb, exposing a free thiol to form an intermolecular bridge by reaction with the *S*-Npys group at position 23 of the HF-cleaved B chain intermediate. Finally, as with the two strategies already described (Schemes 6 and 7), the relaxin synthesis involved iodine oxidation to link Cys(Acm) residues (positions A11 and B11) for the third disulfide bridge. There remained deprotection steps for the Trp(HCO) and Met(O) residues present in the B chain; these were accomplished by applications of aqueous NaOH and ammonium iodide in TFA–water (9:1) respectively, without destroying the assembled disulfide array.

7. Analytical Methods to Establish Disulfide Pairing

The accurate and efficient assignment of cystine connectivities in natural and synthetic peptides is an essential final step for determination and/or proof of structure, yet poses a variety of analytical challenges (for analyses of synthetic peptides, availability of material is generally not a limiting factor, and the order of amino acids does not usually require extensive verification). In the classic insulin work of Sanger and coworkers *(290)*, and in numerous studies carried out since then *(291)*, the plan has been as follows: (1) work out the linear sequence(s) by using materials in which the disulfide bridges have been cleaved and irreversibly modified, typically by reduction/*S*-alkylation (e.g., dithiothreitol [DTT] or tri-*n*-butylphosphine, followed by iodoacetic acid or 4-vinylpyridine; *see* ref. *292* for protocols and leading references), or by performic acid oxidation; and (2) separately, perform one or more rounds of proteolysis (or less commonly, chemical cleavage) on the natural unmodified material with the goal of obtaining relatively small fragments that retain intact disulfide bridges (Table 4). Once such fragments are isolated in pure form, a process often expedited by diagonal electrophoretic or chromatographic methods *(277,304,319–323)* and by disulfide-specific spectrophotometric, fluorescence, or electrochemical assays *(324–327)*, determinations of their compositions (via amino acid analysis), masses (via mass spectrometry, as described in greater detail below), and/or partial sequences (via Edman degradation) invariably provide more than adequate information to relate each fragment unambiguously to its source(s) in the primary sequence. The positions of the disulfide bridges are then readily deduced (Scheme 8).

Interpretations of fragmentation/peptide mapping data become more intricate when the linear sequence includes consecutive (or nearly so) half-cystines that are paired to different partners; it may be difficult or impossible to cut between such residues, and specialized approaches are necessary (for examples, *see* refs. *67,77,147,300,328*). Another potentially challenging task is to distinguish intermolecularly cyclized parallel and antiparallel dimers: The problem is solved by cutting specifically *between* the relevant half-cystines on *each* chain. The parallel arrangement gives two *different* homodimers, whereas antiparallel structures give a *single* heterodimer (for examples, *see* refs. *141,149,286,329*).

Edman sequential degradation is usually carried out on peptides or proteins in which the disulfide bridges have been modified. However, degradations on *unmodified* materials, either complete or after fragmentation, can be instructive with regard to the locations of multiple disulfides *(47,72,294,314,315)*. When there is a single free *N*-terminus, the cycle corresponding to the first of two paired half-cystines should give a null result, i.e., no phenylthiohydantoin (PTH) derivative is observed. Further in the sequencing process, bis(PTH)-cystine should be noted at the cycle corresponding to the *second* of the paired half-cystines. When there are two chains connected by a disulfide bridge and two *N*-termini, each Edman cycle gives primarily two PTH-derivatives in approximately equimolar amounts. Formation of a significant level of bis(PTH)-cystine

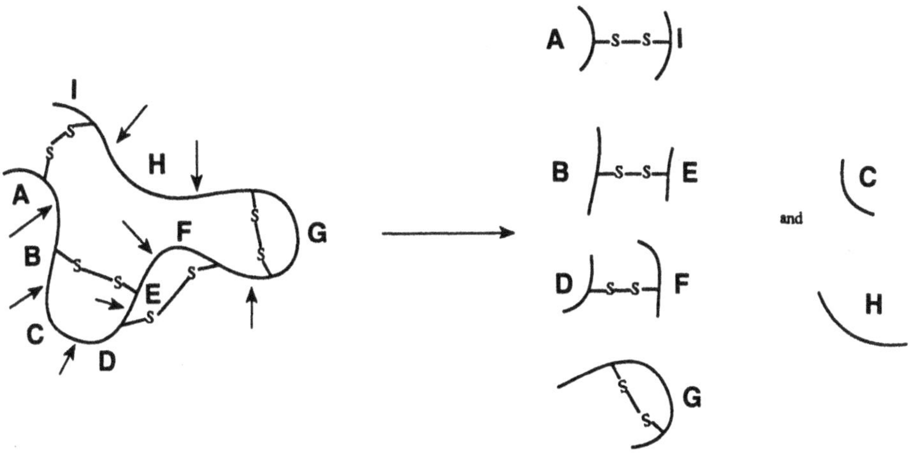

Scheme 8. Classical approach for determination of disulfide alignments in peptides and proteins. A hypothetical single-chain polypeptide with four intramolecular bridges is subjected to fragmentation under conditions that do not promote disulfide scrambling. For clear results, there should be at least one cleavage site between successive half-cystines in the linear sequence. A number of disulfide-bridged peptides (including one with an intact loop) are isolated, as are peptides that lack cysteine. Diagonal methods are based on the premise that after reductive or oxidative cleavage, disulfide-containing peptides give two new peptides (unless the disulfide is in an intramolecular loop), whereas the electrophoretic or chromatographic mobilities of peptides that lack disulfide bridging will not change following such treatments.

at one such cycle allows the conclusion that the appropriate half-cystines located the same number of residues into the sequence on both chains are paired (a situation that occurs quite frequently in natural structures, or that can be arranged by judicious choices of enzymes or reagents for fragmentation). Otherwise, bis(PTH)-cystine will appear at the cycle corresponding to the *later* of the paired half-cystines. While it should be stressed that the experimental design and interpretation requires considerable caution (particularly with regard to stepwise yields; preview and carry-over effects; avoidance of the reducing agents typically added to sequencing reactions; and identification of alternative cystine-derived PTHs, particularly adducts to PTH-dehydroalanine), the approaches outlined have been used successfully to differentiate various isomeric disulfide arrangements connecting as many as three linear chains (references at beginning of this paragraph).

Table 4
Methods for Fragmentation of Peptides and Proteins While Retaining Intact Disulfide Bridges[a]

Name	Specificity[b]	Conditions[c,d]	References[a,e]
Proteolytic enzymes			
Chymotrypsin[f]	Tyr-Y, Phe-Y, Trp-Y (Leu-Y, Met-Y, Ala-Y)	pH 6.5–8.5, 37°C, 2–24 h	*73,77,112, 293–297*
Elastase	Ala-Y, Gly-Y broad specificity	pH 8.0–8.5, 37°C, 1 m*M* $CaCl_2$, 24 h	*295,298,299*
Endoproteinase Glu-C (V8 protease)	Glu-Y (Asp-Y)[b]	pH 4–4.5 (*or* 7.8), 37°C, 2–24 h	*135,293,297, 300*
Endoproteinase Lys-C[f]	Lys-Y	pH 6.5–7.0, 37°C, 2–24 h	*65,72,112, 301*
Endoproteinase Asp-N	X-Asp	pH 7.0–8.0, 37°C, 6–24 h	*299,302*
Pepsin	Phe-Y, Leu-Y pairs of nonpolar residues	pH 1.0–2.5, 37°C, 2–16 h	*301,303–306*
Thermolysin	X-Leu, X-Phe other nonpolar residues	pH 6.5–7.5, 37–65°C, 1 m*M* $CaCl_2$, 2–12 h	*88,107, 113,136, 305–309*
Trypsin[f]	Arg-Y, Lys-Y	pH 6.5–8.5, 37°C, 2–24 h	*73,77,293,294, 297,298,301, 303,310*
Chemical cleavage			
Cyanogen bromide (CNBr)	Met-Y	70% aq. HCOOH *or* 0.1*N* HC1; 25°C, 24 h	*293,296,298, 311*
Edman degradation	N-terminal AA (sequential)	(1) PhNCS in semi-aq. media, "pH" 9–10, 25–50°C, 5 min to 1 h; (2) anhydrous TFA, 40°C, 15 min	*118,296,300, 312–315*
Mild acid cleavage	Asp-Pro; other sites depending on peptide structure and acid conditions	70% aq. HCOOH; 37–40°C, 24 h *or* 2% aq. HCOOH, 110°C, 4 h	*118,290,291, 293,316*

[a]Modified and expanded from tables compiled by Carrey *(317)*. The references provided are examples from the protein analytical literature or from analyses of synthetic peptides with multiple disulfide bridges; they are intended to be representative rather than exhaustive. Conditions are listed that minimize scrambling or disproportionation of disulfides. Thiol proteases, such as papain, are not recommended under any circumstances. *See* ref. *318* for warning about transpeptidation during proteolysis as a possible source of disulfide misassignments.

Another general kind of approach to working out disulfide alignments involves partial reduction of the peptide or protein, followed by peralkylation of the freed sulfhydryls with reagent RX, followed by complete reduction, followed by peralkylation with a different reagent R'X (Scheme 9). The groups R and R' serve to "tag" paired half-cystines, and are located after one or more specific steps to fragment the linearized reduced/alkylated chain. Such approaches are closely allied, in a reciprocal sense, to methodologies being used to identify disulfide-bridged folding intermediates *(4,9,44,330–333)*, and hence subject to the same potential ambiguities with regard to partial scrambling under conditions of alkylation and/or fragmentation.

Recent advances in mass spectrometry have been a major boon to structural work on peptides and small proteins (*291,334–337*, *see also* Chapter 7, *PAP*). When the appropriate instrumentation is on hand and operating well, results are obtained quickly and with relatively little material. An obvious benefit of mass spectrometric techniques is the accurate elucidation of molecular masses of charged biomolecules in the range of up to 10,000 amu for fast atom bombardment mass spectrometry (FABMS), and in excess of 100,000 amu for electrospray mass spectrometry (ESMS) and matrix-assisted laser desorption mass spectrometry (MALD). Furthermore, modern instruments make it possible to analyze *mixtures* (direct probe insertion or on HPLC/MS), and to deduce sequences based on the fragmentation of molecular ions (directly or by tandem MS/MS techniques), hence shortcutting tedious fractionation and selective cleavage or degradation steps that characterize wet protein analytical chemistry (compare to Schemes 8 and 9; *see* refs. *338–343* for specific examples).

The methods for ionization and determination of molecular mass, as listed in the previous paragraph, are readily applicable for the study of

---

[b]Primary cleavage sites are indicated, followed where appropriate by secondary cleavage sites in parentheses. Digestion with V8 protease at acidic pH is specific adjacent to Glu, whereas at alkaline pH, the specificity is extended to both Glu and Asp peptide bonds.

[c]For proteolysis, the enzyme:substrate ratio is typically 1:5 to 1:100 (w/w); this table specifies reaction pH, salt if critical, temperature, and time. Cyanogen bromide cleavage is carried out with a 30- to 100-fold molar excess of reagent over Met cleavage sites.

[d]For safe digestion on the higher pH end, an alkylating agent, such as iodoacetate or iodoacetamide, is often added to inhibit disulfide scrambling (e.g., refs. *301,305*).

[e]In many of the references cited, more than one enzyme or chemical reagent is used to achieve secondary cleavage(s) of isolated peptide fragments.

[f] Often, trypsin and chymotrypsin are used in combination.

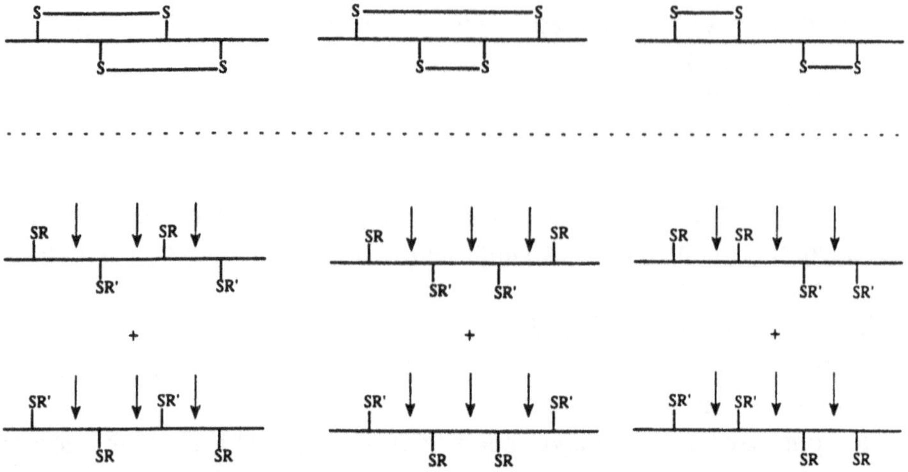

Scheme 9. A general approach for distinguishing alternative possible disulfide arrays. In the example schematized above, a hypothetical single-chain polypeptide with two intramolecular bridges is first subjected to *partial* reduction and then alkylated to introduce group R'. There follows complete reduction and further alkylation to introduce group R. To distinguish R and R', they should have different charge and/or mass and/or radiolabels, e.g., R = [^{14}C]-CH_2CONH_2 and R' = [^{3}H]-CH_2CO_2H, as was chosen by Gray et al. *(44)*. Below the horizontal dotted line are given two possible linear products (not shown are the linear sequence with four R, corresponding to complete reduction, or the linear sequence with four R', corresponding to no reduction, since these will provide no information concerning the original disulfide array). The arrows show desirable sites of cleavage after single or serial cleavages by enzymes or chemical means. Application of one or more of these cleavage procedures will clearly distinguish the alternative disulfide pairings.

disulfide-containing synthetic polypeptides. Clearly, the material to be analyzed needs to be dissolved in a *nonreducing* matrix, e.g., glycerol, *acidified* thioglycerol, *m*-nitrobenzyl alcohol, and mixtures thereof; beyond that, the standard technology can be used *(296,344,345)*. Oxidized and reduced forms of *intramolecularly* linked peptides and proteins are easily distinguished, since the mass differences (2 amu × number of disulfides) are covered by the resolution of routine spectrometers set up for FABMS (provides MH^+) and for ESMS (provides families of ions with $m/z = [M + nH]^+/n$). For asymmetrical intermolecular disulfides, it is difficult to avoid reduction under FABMS conditions, which leads to weaker quasi-molecular ions (and sometimes pairing information *en passant*; *see* ref. *291* and discussion below), but reduction is *not* a problem with ESMS (e.g., refs. *106,346*). In fact, ESMS can be used to "count" the numbers of free cysteine residues and disulfide-bridged half-cystines, based on mass measurements on unmodified, alkylated, and reduced/alkylated materials *(347)*.

Mass spectrometry can be applied to work out disulfide pairing. In the mass spectrometric analog of diagonal methodology (*see* legend to Scheme 8), disulfide-containing peptides are recognized easily in FABMS or ESMS by the weight change after performic acid oxidation *(348)* or reduction *(349,350)*. Reduction can be carried out as a discrete chemical procedure, or it can occur *in situ* with matrices such as alkaline dithiothreitol *(349,351,352)*. Scission of an intramolecular bridge leaves a single polypeptide, whereas cleavage of intermolecular disulfides gives two daughter peptides. Moreover, in the tandem mass spectrometry mode, intact intermolecular disulfides give rise to strong disulfide cleavage peaks (generally a triplet, for RS^+, RSH^+, and RSH_2^+; *see* ref. *352*). In all of these cases, it then becomes possible to obtain at least partial sequence information on the linearized forms of the previously bridged peptide fragment(s), and this is often enough to make unambiguous assignments.

An elegant approach to disulfide alignment developed by Hidaka and Shimonishi combines chemical synthesis and mass spectrometry (Scheme 10). Deuterio-cysteine is incorporated at specific positions in the linear chain, followed by acid hydrolysis and determination of the isotope distribution in cystine.

Finally, it should be noted that when experimentally feasible, it is desirable to apply NMR or X-ray structural analysis to confirm disulfide connectivities deduced by the chemical and mass spectrometric methods covered in this section.

8. Representative Experimental Procedures

This chapter has been organized in a way that the range of deprotection and oxidation conditions used for the preparation of disulfide-containing peptides and proteins are indicated and discussed extensively within the appropriate sections of the text. For the convenience of the reader who would like a starting point for her or his own research applications, we offer in this section a *few* selected procedures that have been used successfully in our laboratories and/or are reported in the literature (refer-

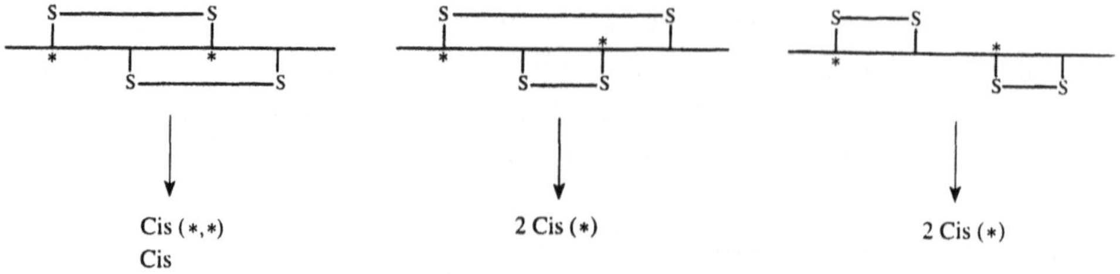

Scheme 10. A synthetic/mass spectrometric strategy for determining disulfide pairings. Adapted from Hidaka and Shimonishi *(353,354)*. Regular (unlabeled) and heavy (*) cysteine is incorporated into the linear polypeptide by chemical synthesis, and the chain is then allowed to oxidize and fold spontaneously. There follows total acid hydrolysis, under conditions shown in control experiments to minimize (<20%) scrambling of disulfides. Amino acids are then converted to their Fmoc derivatives, and bis(Fmoc)-cystine is isolated by HPLC and analyzed by FABMS. Bis(Fmoc)-cystine gives a monoisotopic MH^+ at 685. Observation of an ion at 689 corresponding to bis(Fmoc)-cystine (*,*) is taken as evidence for the disulfide array at the left of the scheme. Observation of ions at 687 is consistent with the middle and right disulfide arrays. The experiment can be set up in other ways to differentiate further among various possibilities and to solve the problem for molecules with more than four half-cystines.

ences given in the following are not necessarily the primary ones; also for transformations not included, the reader is referred to the publications cited throughout this chapter). We must reiterate and stress that structural and conformational factors are often highly critical, as are solubility considerations and solvent effects. Consequently, experimental conditions for disulfide bond formation usually need to be optimized on a case-by-case basis. Also, workup and purification procedures will vary widely, depending on the properties of the target peptide or protein. Finally, the reader is reminded of the need for careful analytical work to characterize intermediates and completed products *(131,166,292)*.

8.1. Removal of S-Acm with Hg(II), Followed by Air Oxidation

This text was adapted from refs. *55* and *187*.

1. Prepare a pH 4.0 solvent by adding a few drops of HOAc to water, and then carry out degassing.
2. Use 2 mL of this degassed solvent to dissolve 20 µmol of an *S*-Acm-protected peptide, in a screw-cap test tube.

3. Add solid $Hg(OAc)_2$ (mol wt 318.7) corresponding to 2.5 Eq/*S*-Acm function. A gray precipitate is noted.
4. The reaction mixture is purged by bubbling through N_2, the tube is sealed and covered with aluminum foil, and stirring is carried out for 70 min at 25°C.
5. 0.45 mL of β-mercaptoethanol is added (at this point the precipitate becomes solubilized), and the mixture is stirred overnight at 25°C.
6. Gel filtration chromatography on a Bio-Gel P2 column (1.2 × 100 cm) equilibrated with 0.1*M* aqueous HOAc removes excess mercury salt and thiol.
7. The column fractions corresponding to the unprotected, reduced peptide are poured directly into 2 L of 0.1*M* Tris-HCl buffer, pH 8.0, and air oxidation is allowed to proceed with slow stirring. Aliquots are removed at different times to monitor the oxidation by HPLC (*see* Chapter 3, *PAP*), Ellman's assay *(238)* and/or biological assay.
8. When oxidation is judged to be complete (e.g., after 2 d), the peptide is purified further by ion-exchange chromatography (*see* Chapters 2 and 5, *PAP*).

8.2. Simultaneous Deprotection/Oxidation of S-Acm with Iodine

This text was adapted from refs. *37* and *149*; *see* also ref. *182* for alternative procedures.

1. Prepare a mixture of HOAc and water in a 4:1 ratio (optionally add 1.5 Eq of HCl/*S*-Acm function).
2. Use 100 mL of the aforementioned mixture to dissolve 100 μmol of an *S*-Acm-protected peptide.
3. Add solid iodine (mol wt 253.8) corresponding to 5 Eq/*S*-Acm function. The reaction proceeds with vigorous mixing at 25°C.
4. Quench the reaction after 10 min to 1 h (based on HPLC monitoring) by dilution with 100 mL of water.
5. Extract with CCl_4 (4 × 50 mL) to remove the iodine.
6. Lyophilize the aqueous phase, which contains the oxidized (disulfide cyclized) peptide.

8.3. Simultaneous Deprotection/Oxidation of S-Acm with Tl(III)

This text was adapted from refs. *49* and *288*.

1. Prepare a mixture of anisole and TFA in a 1:19 ratio.
2. Use 5 mL of this mixture to dissolve 5 μmol of an *S*-Acm-protected peptide, in a screw-cap test tube. Chill to 4°C in an ice bath.
3. Add solid $Tl(tfa)_3$ (mol wt 543.4) **(Caution: handle this reagent with care)** corresponding to 0.6 Eq/S-Acm function.

4. Stir the reaction for 1 h at 4°C.
5. Add 15 mL of ethyl ether to precipitate the peptide, and triturate for 2 min.
6. The peptide is collected by centrifugation, the ether is decanted, and the trituration/centrifugation cycle is repeated two more times to ensure complete removal of the toxic thallium salt.

8.4. Intramolecular Disulfide Cyclization by Potassium Ferricyanide Oxidation

This text was adapted from refs. *24*, *26*, and *33*.

1. The peptide dithiol obtained after a cleavage/deprotection procedure that leaves Cys residues deblocked is dissolved to a concentration of 0.1–1 mg/mL in a suitable buffer, pH 7–8.
2. Over a 30-min period, the peptide solution is titrated at 25°C with 0.01*M* aqueous $K_3Fe(CN)_6$ solution, until a slight yellow color persists (the oxidation can be followed by Ellman analysis, ref. *238*). The amount of oxidant used is typically in 20% or so excess over theory. (An inverse addition procedure at pH 7 is advocated to prevent intermolecular disulfide bond formation by keeping the concentration of sulfhydryl as low as possible during the oxidation.)
3. The pH is adjusted to 5 with 50% aqueous HOAc.
4. The oxidant is removed with AG-3 anion-exchange resin.

8.5. Disulfide Formation Mediated by Glutathione Redox Buffers

This text was adapted from refs. *75, 79, 90, 91, 101*, and *134*.

1. A buffer of 0.1*M* Tris-HCl, pH 8.0–8.5, is prepared, and used to dissolve both reduced (1–10 m*M*) and oxidized (0.1–1.0 m*M*) glutathione (the molar ratio of reduced to oxidized glutathione is *typically* 10:1, but other ratios can be tried as well).
2. The aforementioned redox buffer is used to dissolve a poly(thiol) peptide that has been previously reduced (and preferably purified, including removal of excess salts and reducing agents). The final concentration should be approx 50 µg/mL.
3. The oxidation reaction is allowed to proceed at 25°C (higher temperature reaction, at 30–35°C, has also been described), and monitored by HPLC.
4. Once an end point is determined, typically in 16 h to 2 d, the oxidized peptide is concentrated by lyophilization, and purified by gel filtration chromatography on Sephadex G-10 or G-25, developed with aqueous buffers at acidic pH. (It is sometimes noted that during this procedure, perma-

nently intractable precipitates form. In such cases, an alternative folding procedure is recommended, in which the peptide is oxidized against a series of redox buffers with a slow pH gradient from 2.2 to 7.0 or 8.0.)

8.6. DMSO-Mediated Disulfide Formation

This text was adapted from ref. *81*.

1. Use HOAc and water (as required) to dissolve 50–100 µmol of a crude peptide that is obtained directly from a cleavage/deprotection procedure that leaves Cys residues deblocked (e.g., HF reaction, followed by extractions to remove organic scavengers).
2. Bring the overall volume to 250 mL, so that the final ratio of HOAc to water is 1:19.
3. Use $(NH_4)_2CO_3$ to bring the pH to 6, and add 50 mL of DMSO.
4. The oxidation reaction is run for 1–4 h at 25°C, and its progress is monitored by HPLC.
5. The reaction mixture is diluted with 2 vol of CH_3CN–0.05% aqueous TFA (1:19) (Buffer A), and loaded directly onto a preparative reversed-phase HPLC column (Vydac, 10 × 25 cm, 5µ particle size).
6. The HPLC purification is developed with a linear gradient of Buffer A and CH_3CN-0.04% aqueous TFA (3:2) (Buffer B).

8.7. Solid-Phase Simultaneous Deprotection/ Oxidation of S-Fm

This text was adapted from refs. *64* and *124*.

1. Following completion of solid-phase chain assembly, 0.5 g of a peptide-resin with a substitution level of 0.4–0.6 mmol/g is placed into a reaction vessel for manual synthesis and swollen by repeated washes with CH_2Cl_2 and DMF (10 mL/wash).
2. Five milliliters of a freshly prepared 1:1 mixture of piperidine and DMF are added.
3. Reaction is allowed to proceed for 3 h at 25°C (monitored by a convenient qualitative solid-phase adaptation of Ellman analysis: A 3–5 mg resin aliquot is treated with 1 mL of a 1:1 mixture of DTNB in pH 8.0 phosphate buffer with DMF; a soluble yellow color indicates that free thiol groups remain on the resin, i.e., oxidation is still incomplete; both positive and negative [i.e., with protected Cys] peptide-resin controls should be run concurrently).
4. The anchoring linkage connecting the peptide to the support is cleaved by appropriate methods, and the disulfide-containing peptide is released into solution and isolated further.

9. Conclusions and Future Challenges

The goal of this chapter has been to illustrate different chemical approaches to form one or more disulfide bonds in synthetic peptides and small proteins. Current success in this area (Table 1) can be attributed to a repertoire of cysteine-protecting groups (Table 2) that are applied to selective synthetic strategies. Peptides with a single disulfide bond, e.g., cyclic monomers, homodimers, and heterodimers, can be obtained via a range of the approaches discussed in these pages. For some cyclic structures, the ring size is such (either too small or too large) that the desired intramolecular process is slow with respect to unwanted intermolecular dimerization and oligomerization. Nevertheless, modifications in synthetic strategies and reaction conditions usually suffice to provide at least modest yields of the target molecules. The synthesis of peptides with two disulfides, in any of the three possible intramolecular arrangements as well as parallel or antiparallel dimers, requires considerable experimental skill for optimal results. Regioselective methods for disulfide bond formation have been especially useful in this field. Although it would be premature to conclude that successful results within the endothelin, apamin, conotoxin, and other families can be readily generalized to any given two-disulfide structure, the substantial collective experience acquired to date justifies a certain measure of optimism. By far the greatest challenges to peptide chemists are posed by molecules with three or more disulfide bridges. Increasingly the literature provides accounts of syntheses of this complexity, mostly relying on simultaneous air oxidation for the crucial folding steps. Purifications are often tedious and result in relatively low overall yields, and some published work has been difficult to reproduce in other systems and/or laboratories. This state of affairs punctuates the lack of generality of the modern art, and supports a compelling urgency for the development of improved chemistries, for example multiorthogonal cysteine-protection schemes. With the continued discovery and characterization of biologically active peptides containing multiple disulfide bridges *(135,136,294,295,306,355–358)*, there can be little doubt that the synthetic challenge for peptide chemists will remain undiminished in the coming years.

Acknowledgments

We thank our coworkers and colleagues for sharing their experiences on peptide disulfide chemistry, and single out Miriam Royo and Robert

P. Hammer for especially valuable discussions. Preparation of this chapter and the underlying experimental work from our Barcelona and Minneapolis laboratories were supported by CICYT (PB89-257, PB91-266, PB91-283, and SAL90-828), NIH (GM 28934 and 43552), Commission of the European Communities (SC1-CT91-0748), NATO (Collaborative Research Grants 0841/88 and 92095), and CIRIT (travel grant EE92/1-142).

Abbreviations

Abbreviations used for amino acids and the designations of peptides follow the rules of the IUPAC-IUB Commission of Biochemical Nomenclature in *J. Biol. Chem.* (1972) **247,** 977–983. The following additional abbreviations are used: Acm, acetamidomethyl; ANF or ANP, atrial natriuretic factor (peptide); Ar, aryl; Boc, *tert*-butyloxycarbonyl; Bpoc, 2-(4-biphenylyl)propyl(2)oxycarbonyl; Bzl, benzyl; DBU, 1,8-diazabicyclo[5.4.0]undec-7-ene; DEAD, diethyl azodicarboxylate; DMF, *N,N*-dimethylformamide; DMSO, dimethyl sulfoxide; Dnpe, 2-(2,4-dinitrophenyl)ethyl; DPSO, diphenyl sulfoxide; DTNB, 5,5'-dithiobis(2-nitrobenzoic acid); DTT, dithiothreitol; EDTA, ethylenediaminetetracetic acid; ESMS, electrospray mass spectrometry; FABMS, fast atom bombardment mass spectrometry; Fm, 9-fluorenylmethyl; Fmoc, 9-fluorenylmethyloxycarbonyl; For, formyl; Gu, guanidine; HF, hydrogen flouride; HFIP, hexafluoroisopropanol; HMB, hydroxymethylbenzoyl linker; HOAc, acetic acid; HOBt, 1-hydroxybenzotriazole; HPLC, high-performance liquid chromatography; MALD, matrix-assisted laser desorption mass spectrometry; MBHA, 4-methylbenzhydrylamine (resin); Meb, 4-methylbenzyl; MeOH, methanol; Mob, 4-methoxybenzyl; Moz, 4-methoxybenzyloxycarbonyl; Mpa, β-mercaptopropionic acid; Nb, *o*-nitrobenzylamide; NMM, *N*-methylmorpholine; NMP, *N*-methylpyrrolidinone; Nps, 2-nitrophenylsulfenyl; Npys, 3-nitro-2-pyridinesulfenyl; PAB or PAC, *p*-alkoxybenzyl (ester) linker (Wang resin); PAL, tris(alkoxy)benzylamide linker [5-(4-FMOC-)aminomethyl-3,5-dimethoxyphenoxy)valeric acid]; PAM, phenylacetamidomethyl (resin); Ph, phenyl; Phacm, phenylacetamidomethyl; Pixyl, 9-phenylxanthen-9-yl; PTH, phenylthiohydantoin; Pyr, 2-pyridyl; Rink, 4-(2',4'-dimethoxyphenylaminomethyl)phenoxy acid linker; Scb, *tert*-butyloxycarbonylsulfenyl; Sce, ethyloxycarbonylsulfenyl; Scm, *S*-methyloxycarbonylsulfenyl; Snm, (*N'*-methyl-*N'*-phenylcarbamoyl)sulfenyl; S*t*Bu, *tert*-butylmercapto; Sz, benzyloxycarbonylsulfenyl; Tacm, trimethylacetamidomethyl; *t*Bu, *tert*-butyl; TCEP, tris(2-carboxyethyl)phosphine; TCMS, trichloromethylsilane; TFA, trifluoroacetic acid; TFE, trifluoroethanol; TFMSA, trifluoromethanesulfonic acid; $Tl(tfa)_3$, thallium (III) trifluoroacetate; Tmob, 2,4,6-trimethoxybenzyl; TMTr, 4,4',4''-trimethoxytriphenylmethyl; Trt, triphenylmethyl; Z, benzyloxycarbonyl. Unless stated otherwise, amino acid symbols denote the L-configuration, and all solvent ratios and percentages are vol/vol.

References

1. Schulz, G. E. and Schirmer, R. H. (1979) *Principles of Protein Structure.* Springer-Verlag, New York, pp. 53–55.
2. Richardson, J. S. (1981) The anatomy and taxonomy of protein structure. *Adv. Prot. Chem.* **34,** 167–339.
3. Thornton, J. M. (1981) Disulfide bridges in globular proteins. *J. Mol. Biol.* **151,** 261–287.
4. Creighton, T. E. (1986) Disulfide bonds as probes of protein folding pathways. *Methods Enzymol.* **131,** 83–106.
5. Creighton, T. E. (1988) Disulphide bonds and protein stability. *BioEssays* **8,** 57–63.
6. Srinivasan, N., Sowdhamini, R., Ramakrishnan, C., and Balaram, P. (1990) Conformations of disulfide bridges in proteins. *Int. J. Peptide Protein Res.* **36,** 147–155, and references cited therein.
7. Branden, C. and Tooze, J. (1991) *Introduction to Protein Structure.* Garland, New York.
8. Rizo, J. and Gierasch, L. M. (1992) Constrained peptides: models of bioactive peptides and protein substructures. *Annu. Rev. Biochem.* **61,** 387–418.
9. Creighton, T. E. (ed.) (1992) *Protein Folding.* W. H. Freeman, New York, especially Chapter 7, Folding pathways determined using disulfide bonds, by Creighton, T. E., pp. 301–351.
10. Creighton, T. E. (1993) *Proteins–Structure and Molecular Properties,* 2nd ed., W. H. Freeman, New York.
11. Schiller, P. W., Eggimann, B., DiMaio, J., Lemieux, C., and Nguyen, T. M.-D. (1981) Cyclic enkephalin analogs containing a cystine bridge. *Biochem. Biophys. Res. Commun.* **101,** 337–343.
12. Hruby, V. J. (1982) Conformational restrictions of biologically active peptides via amino acid side chain groups. *Life Sciences* **31,** 189–199.
13. Hruby, V. J., Al-Obeidi, F., and Kazmierski, W. (1990) Emerging approaches in the molecular design of receptor-selective peptide ligands: conformational, topographical and dynamic considerations. *Biochem J.* **268,** 249–262.
14. Bolin, D. R., Cottrell, J., Garippa, R., O'Neill, N., Simko, B., and O'Donnell, M. (1993) Structure-activity studies of vasoactive intestinal peptide (VIP): cyclic disulfide analogs. *Int. J. Peptide Protein Res.* **41,** 124–132, and references cited therein.
15. Pohl, M., Ambrosius, D., Grötzinger, J., Kretzschmar, T., Saunders, D., Wollmer, A., Brandenburg, D., Bitter-Suermann, D., and Höcker, H. (1993) Cyclic disul-

fide analogues of the complement component C3a. Synthesis and conformational investigations. *Int. J. Peptide Protein Res.* **41,** 362–375.

16. Wetzel, R. (1987) Harnessing disulfide bonds using protein engineering. *Trends Biochem. Sci.* **12,** 478–482, and references cited therein.
17. Hodges, R. S., Zhou, N. E., Kay, C. M., and Semchuk, P. D. (1990) Synthetic model proteins: contribution of hydrophobic residues and disulfide bonds to protein stability. *Peptide Res.* **3,** 123–137, and other recent contributions from this research team.
18. Kanaya, S., Katsuda, C., Kimura, S., Nakai, T., Kitakuni, E., Nakamura, H., Katayanagi, K., Morikawa, K., and Ikehara, M. (1991) Stabilization of *Escherichia coli* ribonuclease H by introduction of an artificial disulfide bond. *J. Biol. Chem.* **266,** 6038–6044, and references cited therein.
19. Shortle, D. (1992) Mutational studies of protein structures and their stabilities. *Quart. Rev. Biophys.* **25,** 205–250, and references cited therein.
20. du Vigneaud, V., Ressler, C., Swan, J. M., Roberts, C. W., Katsoyannis, P. G., and Gordon, S. (1953) The synthesis of a peptide with the hormonal activity of oxytocin. *J. Am. Chem. Soc.* **75,** 4879–4880.
21. Kikuchi, T., Némethy, G., and Scheraga, H. A. (1986) Spatial geometric arrangements of disulfide-crosslinked loops in proteins. *J. Comp. Chem.* **7,** 67–88.
22. Mao, B. (1989) Molecular topology of multiple-disulfide polypeptide chains. *J. Am. Chem. Soc.* **111,** 6132–6136.
23. Warne, N. W. and Laskowski, Jr., M. (1990) All fifteen possible arrangements of three disulfide bridges in proteins are known. *Biochem. Biophys. Res. Commun.* **172,** 1364–1370.
24. Hope, D. B., Murti, V. V. S., and du Vigneaud, V. (1962) A highly potent analogue of oxytocin, desamino-oxytocin. *J. Biol. Chem.* **237,** 1563–1566.
25. Hruby, V. J., Upson, D. A., and Agarwal, N. S. (1977) Comparative use of benzhydrylamine and chloromethylated resins in solid-phase synthesis of carboxamide terminal peptides. Synthesis of oxytocin derivatives. *J. Org. Chem.* **42,** 3552–3556.
26. Live, D. H., Agosta, W. C., and Cowburn, D. (1977) A rapid, efficient synthesis of oxytocin and 8-arginine-vasopressin. Comparison of benzyl, *p*-methoxybenzyl, and *p*-methylbenzyl as protecting groups for cysteine. *J. Org. Chem.* **42,** 3556–3561.
27. Hruby, V. J. and Smith, C. W. (1987) Structure-activity relationships of neurohypophyseal peptides, in *The Peptides–Analysis, Synthesis, Biology,* vol. 8 (Udenfriend, S. and Meienhofer, J., eds.; Smith, C. W., vol. ed.), Academic, New York, pp. 77–207.
28. Jost, K., Lebl, M., and Brtník, F. (eds.) (1987) *Handbook of Neurohypophyseal Hormone Analogs,* vols. I and II, CRC, Boca Raton, FL.
29. Manning, M. and Sawyer, W. H. (1993) Design, synthesis and some uses of receptor-specific agonists and antagonists of vasopressin and oxytocin. *J. Receptor Res.* **13,** 195–214.
30. Yamashiro, D. and Li, C. H. (1973) Synthesis of a peptide with full somatostatin activity. *Biochem. Biophys. Res. Commun.* **54,** 882–888.

31. Coy, D. H., Coy, E. J., Arimura, A., and Schally, A. V. (1973) Solid phase synthesis of a growth hormone-release inhibiting factor. *Biochem. Biophys. Res. Commun.* **54,** 1267–1273.
32. Rivier, J. E. F. (1974) Somatostatin. Total solid phase synthesis. *J. Am. Chem. Soc.* **96,** 2986–2992.
33. Rivier, J., Kaiser, R., and Galyean, R. (1978) Solid-phase synthesis of somatostatin and glucagon-selective analogs in gram quantities. *Biopolymers* **17,** 1927–1938, and references cited therein.
34. Chang, C.-D., Felix, A. M., Jimenez, M. H., and Meienhofer, J. (1980) Solid-phase synthesis of somatostatin using mild base cleavage of N^{α}-9-fluorenylmethyloxycarbonylamino acids. *Int. J. Peptide Protein Res.* **15,** 485–494.
35. Moroder, L., Gemeiner, M., Goehring, W., Jaeger, E., Thamm. P., and Wünsch, E. (1981) New synthesis of somatostatin according to the *S-tert*-butylthiocysteine procedure. *Biopolymers* **20,** 17–37.
36. Chino, N., Yoshizawa-Kumagaye, K., Noda, Y., Watanabe, T. X., Kimura, T., and Sakakibara, S. (1986) Synthesis and biological properties of antiparallel and parallel dimers of α-human atrial natriuretic peptide. *Biochem. Biophys. Res. Commun.* **141,** 665–672.
37. Wade, J. D., Fitzgerald, S. P., McDonald, M. R., McDougall, J. G., and Tregear, G. W. (1986) Solid-phase synthesis of α-human atrial natriuretic factor: comparison of the Boc-polystyrene and Fmoc-polyamide methods. *Biopolymers* **25,** S21–S37.
38. Lyle, T. A., Brady, S. F., Ciccarone, T. M., Colton, C. D., Paleveda, W. J., Veber, D. F., and Nutt, R. F. (1987) Chemical synthesis of rat atrial natriuretic factor by fragment assembly on a solid support. *J. Org. Chem.* **52,** 3752–3759, and references cited therein.
39. Akaji, K., Fujino, K., Tatsumi, T., and Kiso, Y. (1992) Regioselective double disulfide formation using silylchloride-sulfoxide system. *Tetrahedron Lett.* **33,** 1073–1076.
40. Sieber, P., Brugger, M., Kamber, B., Riniker, B., and Rittel, W. (1968) Menschliches calcitonin. IV. Die synthese von calcitonin M. *Helv. Chim. Acta* **51,** 2057–2061.
41. Nishiuchi, Y. and Sakakibara, S. (1982) Primary and secondary structures of conotoxin GI, a neurotoxic tridecapeptide from a marine snail. *FEBS Lett.* **148,** 260–262.
42. Gray, W. R., Rivier, J. E., Galyean, R., Cruz, L. J., and Olivera, B. M. (1983) Conotoxin MI: disulfide bonding and conformational states. *J. Biol. Chem.* **258,** 12,247–12,251.
43. Nishiuchi, Y. and Sakakibara, S. (1984) Synthesis of conotoxin MI and GII: structure-activity relationship of conotoxins, in *Peptide Chemistry 1983* (Munekata, E., ed.), Protein Research Foundation, Osaka, Japan, pp. 191–196.
44. Gray, W. R., Luque, F. A., Galyean, R., Atherton, E., Sheppard, R. C., Stone, B. L., Reyes, A., Alford, J., McIntosh, M., Olivera, B. M., Cruz, L. J., and Rivier, J. (1984) Conotoxin GI: disulfide bridges, synthesis, and preparation of iodinated derivatives. *Biochemistry* **23,** 2796–2802.
45. Atherton, E., Sheppard, R. C., and Ward, P. (1985) Peptide synthesis. Part 7. Solid-phase synthesis of conotoxin G1. *J. Chem. Soc., Perkin Trans.* **I,** 2065–2073.

46. Myers, R. A., Zafaralla, G. C., Gray, W. R., Abbott, J., Cruz, L. J., and Olivera, B. M. (1991) α-Conotoxins, small peptide probes of nicotinic acetylcholine receptors. *Biochemistry* **30,** 9370–9377.
47. Zhang, R. and Snyder, G. H. (1991) Factors governing selective formation of specific disulfides in synthetic variants of α-conotoxin. *Biochemistry* **30,** 11,343–11,348, and references cited therein.
48. Albericio, F., Royo, M., Munson, M. C., Solé, N. A., Van Abel, R. J., Alsina, J., García-Echeverría, C., Slomczyńska, U., Eritja, R., Pons, M., Giralt, E., and Barany, G. (1993) Strategies and tactics for the solid-phase synthesis of cystine-containing peptides, in *Peptide Chemistry 1992* (Yanaihara, N., ed.), Escom, Leiden, The Netherlands, pp. 19–23.
49. Munson, M. C. and Barany, G. (1993) Synthesis of α-conotoxin SI, a bicyclic tridecapeptide amide with two disulfide bridges: illustration of novel protection schemes and oxidation strategies. *J. Am. Chem. Soc.* **115,** 10,203–10,216.
50. Van Rietschoten, J., Granier, C., Rochat, H., Lissitzky, S., and Miranda, F. (1975) Synthesis of apamin, a neurotoxic peptide from bee venom. *Eur. J. Biochem.* **56,** 35–40.
51. Van Rietschoten, J., Pedroso Muller, E., and Granier, C. (1977) Cysteine protection in solid phase synthesis of apamin, in *Peptides: Proceedings of the Fifth American Peptide Symposium* (Goodman, M. and Meienhofer, J., eds.), Wiley, New York, pp. 522–524.
52. Cosand, W. L. and Merrifield, R. B. (1977) Concept of internal structural controls for evaluation of inactive synthetic peptide analogs: synthesis of $[Orn^{13,14}]$apamin and its guanidination to an apamin derivative with full neurotoxic activity. *Proc. Natl. Acad. Sci. USA* **74,** 2771–2775.
53. Granier, C., Pedroso Muller, E., and Van Rietschoten, J. (1978) Use of synthetic analogs for a study on the structure-activity relationship of apamin. *Eur. J. Biochem.* **82,** 293–299, and references cited therein.
54. Sandberg, B. E. B. and Ragnarsson, U. (1978) Solid phase synthesis of apamin, the principal neurotoxin in bee venom. Isolation and characterization of acetamidomethyl apamin. *Int. J. Peptide Protein Res.* **11,** 238–245.
55. Albericio, F., Granier, C., Labbé-Juillié, C., Seagar, M., Couraud, F., and Van Rietschoten, J. (1984). Solid phase synthesis and HPLC purification of the protected 1-12 sequence of apamin for rapid synthesis of apamin analogues differing in the C-terminal region. *Tetrahedron* **40,** 4313–4326.
56. Koide, T., Otaka, A., Arai, H., Funakoshi, S., Fujii, N., and Yajima, H. (1990) Synthetic studies on cystine-containing peptides using regioselective disulfide bond forming reactions, in *Peptide Chemistry 1989* (Yanaihara, N., ed.), Protein Research Foundation, Osaka, Japan, pp. 171–174.
57. Shih, H. (1993) New approaches to the synthesis of cystine peptides using *N*-iodosuccinimide in the construction of disulfide bridges. *J. Org. Chem.* **58,** 3003–3008.
58. Kumagaye, S.-I., Kuroda, H., Nakajima, K., Watanabe, T. X., Kimura, T., Masaki, T., and Sakakibara, S. (1988) Synthesis and disulfide structure determination of porcine endothelin: an endothelium-derived vasoconstricting peptide. *Int. J. Peptide Protein Res.* **32,** 519–526.

59. Immer, H., Eberle, I., Fischer, W., and Moser, E. (1989) Solution-synthesis of endothelin, in *Peptides 1988: Proceedings of the Twentieth European Peptide Symposium* (Jung, G. and Bayer, E., eds.), de Gruyter, Berlin, pp. 94–96.
60. Morimoto, H., Fujii, N., Otaka, A., Kuramochi, K., and Yajima, H. (1989) Silver trifluoromethanesulphonate, as an Acm-deprotecting reagent and its application to the synthesis of endothelin and endothelin-like peptide, in *Peptide Chemistry 1988* (Ueki, M., ed.), Protein Research Foundation, Osaka, Japan, pp. 211–214.
61. Tam, J. P., Liu, W., Zhang, J.-W., Galantino, M., and de Castiglione, R. (1991) D-Amino acid and alanine scans of endothelin: an approach to study refolding intermediates, in *Peptides 1990: Proceedings of the Twenty-First European Peptide Symposium* (Giralt, E. and Andreu, D., eds.), Escom, Leiden, The Netherlands, pp. 160–163.
62. Casaretto, R. and Nyfeler, R. (1991) Isolation, structure and activity of a side product from the synthesis of human endothelin, in *Peptides 1990: Proceedings of the Twenty-First European Peptide Symposium* (Giralt, E. and Andreu, D., eds.), Escom, Leiden, The Netherlands, pp. 181–182.
63. Hunt, J. T., Lee, V. G., Liu, E. C. K., Moreland, S., McMullen, D., Webb, M. L., and Bolgar, M. (1994) Control of peptide disulfide regioisomer formation by mixed cysteine-penicillamine bridges: application to endothelin-1. *Int. J. Pept. Prot. Res.* **42,** 249–258.
64. Ponsati, B., Giralt, E., and Andreu, D. (1990) Solid-phase approaches to regiospecific double disulfide formation. Application to a fragment of bovine pituitary peptide. *Tetrahedron* **46,** 8255–8266, and references cited therein.
65. Aimoto, S., Hojoh, H., and Takasaki, C. (1990) Studies on the disulfide bridges of sarafotoxins. Chemical synthesis of sarafotoxin S6B and its homologue with different disulfide bridges. *Biochemistry Int.* **21,** 1051–1057.
66. Gariépy, J., Judd, A. K., and Schoolnik, G. K. (1987) Importance of disulfide bridges in the structure and activity of *Escherichia coli* enterotoxin ST1b. *Proc. Natl. Acad. Sci. USA* **84,** 8907–8911.
67. Shimonishi, Y., Hidaka, Y., Koizumi, M., Hane, M., Aimoto, S., Takeda, T., Miwatani, T., and Takeda, Y. (1987) Mode of disulfide bond formation of a heat-stable enterotoxin (ST_h) produced by a human strain of enterotoxigenic *Escherichia coli. FEBS Lett.* **215,** 165–170.
68. Hidaka, Y., Kubota, H., Yoshimura, S., Ito, H., Takeda, Y., and Shimonishi, Y. (1988) Disulfide linkages in a heat-stable enterotoxin (ST_p) produced by a porcine strain of enterotoxigenic *Escherichia coli. Bull. Chem. Soc. Jpn.* **61,** 1265–1271, and other contributions from this research team.
69. Cruz, L. J., Kupryszewski, G., LeCheminant, G. W., Gray, W. R., Olivera, B. M., and Rivier, J. (1989) μ-Conotoxin GIIIA, a peptide ligand for muscle sodium channels: chemical synthesis, radiolabeling, and receptor characterization. *Biochemistry* **28,** 3437–3442.
70. Becker, S., Atherton, E., and Gordon, R. D. (1989) Synthesis and characterization of μ-conotoxin IIIa. *Eur. J. Biochem.* **185,** 79–84.

71. Hatanaka, Y., Yoshida, E., Nakayama, H., and Kanaoka, Y. (1990) Synthesis of μ-conotoxin GIIIA: a chemical probe for sodium channels. *Chem. Pharm. Bull.* **38,** 236–238.
72. Kubo, S., Chino, N., Watanabe, T. X., Kimura, T., and Sakakibara, S. (1993) Solution synthesis of μ-conotoxin GIIIB: optimization of the oxidative folding reaction. *Peptide Res.* **6,** 66–72.
73. Nishiuchi, Y., Kumagaye, K., Noda, Y., Watanabe, T. X., and Sakakibara, S. (1986) Synthesis and secondary-structure determination of ω-conotoxin GVIA: a 27-peptide with three intramolecular disulfide bonds. *Biopolymers* **25,** S61–S68, and references cited therein.
74. Rivier, J., Galyean, R., Gray, W. R., Azimi-Zonooz, A., McIntosh, J. M., Cruz, L. J., and Olivera, B. M. (1987) Neuronal calcium channel inhibitors. Synthesis of ω-conotoxin GVIA and effects on ^{45}Ca uptake by synaptosomes. *J. Biol. Chem.* **262,** 1194–1198.
75. Pennington, M. W., Festin, S. M., and Maccecchini, M. L. (1991) Comparison of folding procedures on synthetic ω-conotoxin, in *Peptides 1990: Proceedings of the Twenty-First European Peptide Symposium* (Giralt, E. and Andreu, D., eds.), Escom, Leiden, The Netherlands, pp. 164–166.
76. Sabo, T., Gilon, C., Shafferman, A., and Elhanaty, E. (1992) Structure-activity studies of ω-conotoxin: the importance of disulfide bridges for biological activity, in *Peptides–Chemistry and Biology: Proceedings of the Twelfth American Peptide Symposium* (Smith, J. A. and Rivier, J. E., eds.), Escom, Leiden, The Netherlands, pp. 159–160.
77. Pennington, M. W., Festin, S. M., Maccecchini, M. L., and Kem, W. R. (1992) Synthesis and characterization of a disulfide bond isomer of omega-conotoxin GVIA. *Toxicon.* **30,** 755–764.
78. Le-Nguyen, D., Nalis, D., and Castro, B. (1989) Solid phase synthesis of a trypsin inhibitor isolated from the Cucurbitaceae *Ecballium elaterium. Int. J. Peptide Protein Res.* **34,** 492–497.
79. Kupryszewski, G., Ragnarsson, U., Rolka, K., and Wilusz, T. (1986) Solid-phase synthesis of trypsin inhibitor CMTI III from squash seeds *(Cucurbita maxima). Int. J. Peptide Protein Res.* **27,** 245–250.
80. Fujii, N., Okamachi, A., Funakoshi, S., Kuroda, M., Hayashi, Y., Yajima, H., Fukada, J., Imura, H., Bessalle, R., and Fridkin, M. (1990) Synthesis and biological activities of rabbit corticostatin (rCS), in *Peptides–Chemistry, Structure and Biology: Proceedings of the Eleventh American Peptide Symposium* (Rivier, J. E. and Marshall, G. R., eds.), Escom, Leiden, The Netherlands, pp. 241–243.
81. Tam, J. P., Wu, C.-R., Liu, W., and Zhang, J.-W. (1991) Disulfide bond formation in peptides by dimethyl sulfoxide. Scope and applications. *J. Am. Chem. Soc.* **113,** 6657–6662, and references cited therein.
82. Chow, M.-S., Tran, D., Dobbs, C. H., Rao, V. S. V., Hsieh, M. M., Harwig, S. S. L., and Selsted, M. E. (1992) Fmoc solid phase peptide synthesis of defensins, in *Abstracts of the Sixth Symposium of the Protein Society*, p. 135.
83. Rao, A. G., Rood, T., Maddox, J., and Duvick, J. (1992) Synthesis and characterization of defensin NP-1. *Int. J. Peptide Protein Res.* **40,** 507–514.

84. Lambert, P., Kuroda, H., Chino, N., Watanabe, T. X., Kimura, T., and Sakakibara, S. (1990) Solution synthesis of charybdotoxin (ChTX), a K^+ channel blocker. *Biochem. Biophys. Res. Commun.* **170,** 684–690.
85. Cotton, R., Dutta, A. S., Giles, M. B., and Hayward, C. F. (1992) Solid phase synthesis of a number of venom toxins containing two to six cysteine residues, in *Peptides–Chemistry and Biology: Proceedings of the Twelfth American Peptide Symposium* (Smith, J. A. and Rivier, J. E., eds.), Escom, Leiden, The Netherlands, pp. 639–640.
86. Vita, C., Bontems, F., Roumestand, C., Ménez, A., and Toma, F. (1993) Solid-phase synthesis: structural and functional characterisation of charybdotoxin and of two truncated analogues, in *Peptides 1992: Proceedings of the Twenty-Second European Peptide Symposium* (Schneider, C. H. and Eberle, A. N., eds.), Escom, Leiden, The Netherlands, pp. 641–642.
87. Maruyama, K., Nagata, K., Tanaka, M., Nagasawa, H., Isogai, A., Ishizaki, H., and Suzuki, A. (1992) Synthesis of bombyxin-IV, an insulin superfamily peptide from the silkworm, *Bombyx mori*, by stepwise and selective formation of three disulfide bridges. *J. Protein Chem.* **11,** 1–12.
88. Maruyama, K., Nagasawa, H., Isogai, A., Ishizaki, H., and Suzuki, A. (1992) Determination of disulfide bond arrangement in bombyxin-IV, an insulin superfamily peptide from the silkworm, *Bombyx mori*, by combination of thermolysin digestion of natural peptide and selective synthesis of disulfide bond isomers. *J. Protein Chem.* **11,** 13–20.
89. Nagata, K., Marayuma, K., Nagasawa, H., Urushibata, I., Isogai, A., Ishizaki, H., and Suzuki, A. (1992) Bombyxin-II and its disulfide bond isomers: synthesis and activity. *Peptides* **13,** 653–662.
90. Pennington, M. W., Kem, W. R., Norton, R. S., and Dunn, B. M. (1990) Chemical synthesis of a neurotoxic polypeptide from the sea anemone *Stichodactyla helianthus. Int. J. Peptide Protein Res.* **36,** 335–343.
91. Pennington, M. W., Kem, W. R., and Dunn, B. M. (1990). Synthesis and biological activity of six monosubstituted analogs of a sea anemone polypeptide neurotoxin. *Peptide Res.* **3,** 228–232.
92. Sieber, P., Kamber, B., Hartmann, A., Jöhl, A., Riniker, B., and Rittel, W. (1974) Totalsynthese von humaninsulin unter gezielter bildung der disulfidbindungen. *Helv. Chem. Acta* **57,** 2617–2621.
93. Sieber, P., Kamber, B., Hartmann, A., Jöhl, A., Riniker, B., and Rittel, W. (1977) Totalsynthese von humaninsulin. IV. Beschreibung der endstufen. *Helv. Chem. Acta* **60,** 27–37, and preceding papers in this series.
94. Sieber, P., Eisler, K., Kamber, B., Riniker, B., Rittel, W., Märki, F., and De Gasparo, M. (1978) Synthesis and biological activity of two disulfide bond isomers of human insulin: [A7-A11, A6-B7-cystine]- and [A6-A7, A11-B7-cystine]insulin (human). *Hoppe-Seyler's Z. Physiol. Chem.* **359,** 113–123.
95. Birr, C. and Pipkorn, R. (1979) Fully active insulin by selective formation of the disulfide bridges between a synthetic A-chain and natural B-chain. *Angew. Chem. Int. Ed. Engl.* **18,** 536–538.

96. Büllesbach, E. E. and Schwabe, C. (1991) Total synthesis of human relaxin and human relaxin derivatives by solid-phase peptide synthesis and site-directed chain combination. *J. Biol. Chem.* **266,** 10,754–10,761.
97. Wade, J. D., Vandlen, R. L., and Tregear, G. W. (1992) Continuous flow Fmoc-polyamide solid phase synthesis of human gene 1 relaxin, in *Innovation and Perspectives in Solid Phase Synthesis: Peptides, Polypeptides and Oligonucleotides. 1992* (Epton, R., ed.), Intercept Ltd., Andover, England, pp. 485–487.
98. Tam, J. P., Marquardt, H., Rosberger, D. F., Wong, T. W., and Todaro, G. J. (1984) Synthesis of biologically active rat transforming growth factor I. *Nature* **309,** 376–378.
99. Heath, W. F. and Merrifield, R. B. (1986) A synthetic approach to structure-function relationships in the murine epidermal growth-factor molecule. *Proc. Natl. Acad. Sci. USA* **83,** 6367–6371.
100. Tam, J. P., Sheikh, M. A., Solomon, D. S., and Ossowski, L. (1986) Efficient synthesis of human type α transforming growth factor: its physical and biological characterization. *Proc. Natl. Acad. Sci. USA* **83,** 8082–8086.
101. Tam, J. P. (1987) Synthesis of biologically active transforming growth factor alpha. *Int. J. Peptide Protein Res.* **29,** 421–431.
102. Lin, Y.-Z., Caporaso, G., Chang, P.-Y., Ke, X.-H., and Tam, J. P. (1988) Synthesis of biological active tumor growth factor from the predicted DNA sequence of Shope fibroma virus. *Biochemistry* **27,** 5640–5645.
103. Tam, J. P., Lin, Y.-Z., Liu, W., Wang, D.-X., Ke, X.-H., and Zhang, J.-W. (1991) Mapping the receptor-recognition site of human transforming growth factor-α. *Int. J. Peptide Protein Res.* **38,** 204–211.
104. Tan, N. H. and Kaiser, E. T. (1976) Studies on the solid-phase synthesis of bovine pancreatic trypsin inhibitor (Kunitz) and the characterization of the synthetic material. *J. Org. Chem.* **41,** 2787–2793.
105. Aimoto, S., Mizoguchi, N., Hojo, H., and Yoshimura, S. (1989) Development of a facile method for polypeptide synthesis. Synthesis of bovine pancreatic trypsin inhibitor (BPTI). *Bull. Chem. Soc. Jpn.* **62,** 524–531.
106. Ferrer, M., Woodward, C., and Barany, G. (1992) Solid-phase synthesis of bovine pancreatic trypsin inhibitor (BPTI) and two analogues. *Int. J. Peptide Protein Res.* **40,** 194–207.
107. Lin, Y.-Z., Isaac, D. D., and Tam, J. P. (1990) Synthesis and properties of cholecystokinin-releasing peptide (monitor peptide), a 61-residue trypsin inhibitor. *Int. J. Peptide Protein Res.* **36,** 433–439.
108. Chino, N., Kubo, S., Nishiuchi, Y., Kumagaye, S.-I., Yoshizawa-Kumagaye, K., Takai, M., Kimura, T., and Sakakibara, S. (1988) Synthesis of porcine C5a anaphylatoxin by the solution procedure and confirmation of the reported structure. *Biochem. Biophys. Res. Commun.* **161,** 1285–1292.
109. Kimura, T., Chino, N., Kumagaye, S.-I., Kuroda, H., Emura, J., and Sakakibara, S. (1990) Total synthesis of human angiogenin by the solution procedure, in *Peptide Chemistry 1989* (Yanaihara, N., ed.), Protein Research Foundation, Osaka, Japan, pp. 303–308.

110. Nishio, H., Kumagaye, K. Y., Chen, Y.-N., Kimura, T., and Sakakibara, S. (1994) Synthesis and disulfide structure determination of ω-agatoxin IV, in *Peptides–Chemistry, Structure and Biology: Proceedings of the Thirteenth American Peptide Symposium* (Hodges, R. S. and Smith, J. A., eds.), Escom, Leiden, The Netherlands, pp. 31–33.
111. Garsky, V. M., Lumma, P. K., Freidinger, R. M., Pitzenberger, S. M., Randall, W. C., Veber, D. F., Gould, R. J., and Friedman, P. A. (1989) Chemical synthesis of echistatin, a potent inhibitor of platelet aggregation from *Echis carinatus*: synthesis and biological activity of selected analogs. *Proc. Natl. Acad. Sci. USA*, **86,** 4022–4026.
112. Nishio, H., Kumagaye, S., Kuroda, H., Chino, N., Emura, J., Kimura, T., and Sakakibara, S. (1992) Solution synthesis of Na^+, K^+-ATPase inhibitor-1 (SPAI-1). *Peptide Res.* **5,** 227–244.
113. Tsunemi, M., Kato, H., Nishiuchi, Y., Kumagaye, S.-I., and Sakakibara, S. (1992) Synthesis and structure-activity relationships of elafin, an elastase-specific inhibitor. *Biochem. Biophys. Res. Commun.* **185,** 967–973.
114. Gutte, B. and Merrifield, R. B. (1971) The synthesis of ribonuclease A. *J. Biol. Chem.* **246,** 1922–1941.
115. Yajima, H. and Fujii, N. (1981) Studies on peptides. 103. Chemical synthesis of a crystalline protein with the full enzymatic activity of ribonuclease A. *J. Am. Chem. Soc.* **103,** 5867–5871.
116. Bodanszky, M. and Stahl, G. L. (1974) The structure and synthesis of malformin A. *Proc. Natl. Acad. Sci. USA* **71,** 2791–2794.
117. Kao, P. N. and Karlin, A. (1986) Acetylcholine receptor binding site contains a disulfide cross-link between adjacent half-cystinyl residues. *J. Biol. Chem.* **261,** 8085–8088.
118. Marti, T., Rösselet, S. J., Titani, K., and Walsh, K. A. (1987) Identification of disulfide-bridged substructures within human von Willebrand factor. *Biochemistry* **26,** 8099–8109, and references cited therein.
119. Mosckovitz, R. and Gershon, J. M. (1988) Three possible disulfides in the acetylcholine receptor α-subunit. *J. Biol. Chem.* **263,** 1017–1022.
120. Prorok, M. and Lawrence, D. S. (1990) An affinity label of absolute peptidic origin. *J. Am. Chem. Soc.* **112,** 8626–8627.
121. Sukumaran, D. K., Prorok, M., and Lawrence, D. S. (1991) A molecular constraint that generates a cis peptide bond. *J. Am. Chem. Soc.* **113,** 706–707, and references cited therein.
122. Brady, S. F., Paleveda, Jr., W. J., Arison, B. H., Saperstein, R., Brady, E. J., Raynor, K., Reisine, T., Veber, D. F., and Freidinger, R. M. (1993) Aproaches to peptidomimetics which serve as surrogates for the *cis* amide bond: novel disulfide-constrained bicyclic hexapeptide analogs of somatostatin. *Tetrahedron* **49,** 3449–3466.
123. Zhang, R. and Snyder, G. H. (1989) Dependence of formation of small disulfide loops in two-cysteine peptides on the number and types of intervening amino acids. *J. Biol. Chem.* **264,** 18,472–18,479, and references cited therein.
124. Albericio, F., Hammer, R. P., García-Echevarría, C., Molins, M. A., Chang, J. L., Munson, M. C., Pons, M., Giralt, E., and Barany, G. (1991) Cyclization of

disulfide-containing peptides in solid-phase synthesis. *Int. J. Peptide Protein Res.* **37,** 402–413, and references cited therein.
125. Falcomer, C. M., Meinwald, Y. C., Choudhary, I., Talluri, S., Milburn, P. J., Clardy, J., and Scheraga, H. A. (1992) Chain reversals in model peptides: studies of cystine-containing cyclic peptides. 3. Conformational free energies of cyclization of tetrapeptides of sequence Ac-Cys-Pro-X-Cys-NHMe. *J. Am. Chem. Soc.* **114,** 4036–4062, and references cited therein.
126. Creighton, T. E. and Goldenberg, D. P. (1984) Kinetic role of a meta-stable native-like two-disulphide species in the folding transition of bovine pancreatic trypsin inhibitor. *J. Mol. Biol.* **179,** 497–526.
127. Altmann, K.-H. and Scheraga, H. A. (1990) Local structure in ribonuclease A. Effect of amino acid substitutions on the preferential formation of the native disulfide loop in synthetic peptides corresponding to residues Cys^{58}-Cys^{72} of bovine pancreatic ribonuclease A. *J. Am. Chem. Soc.* **112,** 4926–4931, and references cited therein.
128. Noszál, B., Guo, W., and Rabenstein, D. L. (1992) Characterization of the macroscopic and microscopic acid-base chemistry of the native disulfide and reduced dithiol forms of oxytocin, arginine-vasopressin, and related peptides. *J. Org. Chem.* **57,** 2327–2334, and references cited therein.
129. Fiser, A., Cserzö, M., Tüdös, E., and Simon, I. (1992) Different sequence environments of cysteines and half cystines in proteins. Application to predict disulfide forming residues. *FEBS Lett.* **302,** 117–120, and references cited therein.
130. Mazur, S. and Jayalekshmy, P. (1979) Chemistry of polymer-bound *o*-benzyne. Frequency of encounter between substituents on cross-linked polystyrenes. *J. Am. Chem. Soc.* **101,** 677–683.
131. Barany, G. and Merrifield, R. B. (1979) Solid-phase peptide synthesis, in *The Peptides–Analysis, Synthesis, Biology,* vol. 2 (Gross, E. and Meienhofer, J., eds.), Academic, New York, pp. 1–284.
132. Mott, A. W., Slomczyńska, U., and Barany, G. (1986) Formation of sulfur-sulfur bonds during solid-phase peptide synthesis: application to the synthesis of oxytocin, in *Forum Peptides Le Cap d'Agde 1984* (Castro, B. and Martinez, J., eds.), Les Impressions Dohr, Nancy, France, pp. 321–324.
133. Tsuji, T., Nakagawa, R., Sugimoto, N., and Fukuhara, K.-I. (1987) Characterization of disulfide bonds in recombinant proteins: reduced human interleukin 2 in inclusion bodies and its oxidative refolding. *Biochemistry* **26,** 3129–3134, and references cited therein.
134. Sabatier, J.-M., Darbon, H., Fourquet, P., Rochat, H., and Van Rietschoten, J. (1987) Reduction and reoxidation of the neurotoxin II from the scorpion *Androctonus australis Hector. Int. J. Peptide Protein Res.* **30,** 125–134.
135. Iwai, M., Kobayashi, M., Tamura, K., Ishii, Y., Yamada, H., and Niwa, M. (1989) Direct identification of disulfide bond linkages in human insulin-like growth factor I (IGF) by chemical synthesis. *J. Biochem.* **106,** 949–951.
136. Kataoka, H., Li, J. P., Lui, A. S. T., Kramer, S. J., and Schooley, D. A. (1992) Complete structure of eclosion hormone of *manduca sexta.* Assignment of disulfide bond location. *Int. J. Peptide Protein Res.* **39,** 29–35.

137. Bernatowicz, M. S., Matsueda, R., and Matsueda, G. R. (1986) Preparation of Boc-[*S*-(3-nitro-2-pyridinesulfenyl)]-cysteine and its use for unsymmetrical disulfide bond formation. *Int. J. Peptide Protein Res.* **28,** 107–112.
138. Romani, S., Moroder, L., Göhring, W., Scharf, R., Wünsch, E., Barde, Y. A., and Thoenen, H. (1987). Synthesis of the trypsin fragment 10-25/75-88 of mouse nerve growth factor. *Int. J. Peptide Protein Res.* **29,** 107–117.
139. Wünsch, E., Moroder, L., Göhring-Romani, S., Musiol, H.-J., Göhring, W., and Bovermann, G. (1988) Synthesis of the bis-cystinyl-fragment 225-232/225'-232' of the human IgG1 hinge region. *Int. J. Peptide Protein Res.* **32,** 368–383.
140. Drijfhout, J. W., Perdijk, E. W., Weijer, W. J., and Bloemhoff, W. (1988) Controlled peptide-protein conjugation by means of 3-nitro-2-pyridinesulfenyl protection-activation. *Int. J. Peptide Protein Res.* **32,** 161–166.
141. Ruiz-Gayo, M., Albericio, F., Pons, M., Royo, M., Pedroso, E., and Giralt, E. (1988) Uteroglobin-like peptide cavities I. Synthesis of antiparallel and parallel dimers of bis-cysteine peptides. *Tetrahedron Lett.* **29,** 3845–3848.
142. Ponsati, B., Giralt, E., and Andreu, D. (1989) A synthetic strategy for simultaneous purification-conjugation of antigenic peptides. *Anal. Biochem.* **181,** 389–395.
143. Albericio, F., Andreu, D., Giralt, E., Navalpotro, C., Pedroso, E., Ponsati, B., and Ruiz-Gayo, M. (1989) Use of the Npys thiol protection in solid phase peptide synthesis. Application to direct peptide-protein conjugation through cysteine residues. *Int. J. Peptide Protein Res.* **34,** 124–128.
144. Chaturvedi, S. and Bahl, O. P. (1990) Synthesis of cystine peptides 21-25/70-73 and 35-39/56-59 of the β-subunit of human choriogonadotropin. *Int. J. Peptide Protein Res.* **35,** 133–140.
145. Drijfhout, J. W. and Bloemhoff, W. (1991) A new synthetic functionalized antigen carrier. *Int. J. Peptide Protein Res.* **37,** 27–32.
146. Tam, J. P. and Shen, Z.-Y. (1992) Efficient approach to synthesis of two-chain asymmetric cysteine analogs of receptor-binding region of transforming growth factor-α. *Int. J. Peptide Protein Res.* **39,** 464–471.
147. Fischer, P. M. and Howden, M. E. H. (1993) Application of anti-peptide antibodies to the assignment of the inter-chain disulphide bond in tetanus toxin. *Int. J. Peptide Protein Res.* **41,** 415–419, and references cited therein.
148. Simmonds, R. G., Tupper, D. E., and Harris, J. R. (1994) Synthesis of disulfide-bridged fragments of ω-conotoxins GVIA and MVIIA. Use of Npys as a protecting/activating group for cysteine in Fmoc syntheses. *Int. J. Peptide Protein Res.* **43,** 363–366.
149. Ruiz-Gayo, M., Royo, M., Fernández, I., Albericio, F., Giralt, E., and Pons, M. (1993) Unequivocal synthesis and characterization of a parallel and an antiparallel bis-cystine peptide. *J. Org. Chem.* **58,** 6319–6328.
150. Kullmann, W. and Gutte, B. (1978) Synthesis of an open-chain asymmetrical cystine peptide corresponding to the sequence A^{18-21}–B^{19-26} of bovine insulin by solid phase fragment condensation. *Int. J. Peptide Protein Res.* **12,** 17–26, and references cited therein.

151. Büllesbach, E. E. and Schwabe, C. (1992) Sequential synthesis of an unsymmetrical two-chain disulfide peptide on solid-phase. *Tetrahedron Lett.* **33,** 5881–5884, and references cited therein.
152. Photaki, I. (1976) The role of sulfur in amino acid protective group chemistry, in *Topics in Sulfur Chemistry*, vol. 1 (Senning, A., ed.), Georg Thieme, Stuttgart, pp. 111–183.
153. Hiskey, R. G. (1981) Sulfhydryl group protection in peptide synthesis, in *The Peptides–Analysis, Synthesis, Biology*, vol. 3 (Gross, E. and Meienhofer, J., eds.), Academic, New York, pp. 137–167.
154. König, W. and Geiger, R. (1981) Selective formation of the disulfide bond in peptide synthesis, in *Perspectives in Peptide Chemistry* (Eberle, A., Geiger, R., and Wieland, T., eds.), S. Karger, Basel, pp. 31–44.
155. Yajima, H., Fujii, N., Funakoshi, S., Watanabe, T., Murayama, E., and Otaka, A. (1988) New strategy for the chemical synthesis of proteins. *Tetrahedron* **44,** 805–819.
156. Cavelier, F., Daunis, J., and Jacquier, R. (1989) Les réactions régiosélectives de formation de ponts disulfure en synthèse de peptides contenant des cystines. *Bull. Soc. Chim. Fr.*, 788–798.
157. Cavelier, F., Daunis, J., and Jacquier, R. (1990) Etude bibliographique et critique de la protection de la fonction thiol en synthèse peptidique (mise au point). *Bull. Soc. Chim. Fr.*, 210–225.
158. Büllesbach, E. E. (1992) Site-directed disulfide formation in peptide synthesis. *Kontakte (Darmstadt)* **1,** 21–29.
159. Oae, S. (ed.) (1977) *Organic Chemistry of Sulfur*, Plenum, New York, especially Chapter 4, Thiols, by Ohno, A. and Oae, S., pp. 119–187, and Chapter 7, Disulfides and polysulfides, by Field, L., pp. 303–382.
160. Oae, S. (1991) *Organic Sulfur Chemistry: Structure and Mechanism.* CRC, Boca Raton, FL.
161. Wünsch, E. (ed.) (1974) Synthese von Peptiden, in *Houben-Weyl's Methoden der Organischen Chemie*, vol. 15, 4th ed., parts 1 and 2 (Müller, E., ed.), Georg Thieme, Stuttgart.
162. Bodanszky, M. (1984) *Principles of Peptide Synthesis.* Springer-Verlag, Berlin.
163. Bodanszky, M. and Bodanszky, A. (1984) *The Practice of Peptide Synthesis.* Springer-Verlag, Berlin.
164. Stewart, J. M. and Young, J. D. (1984) *Solid Phase Peptide Synthesis*, 2nd ed., Pierce Chemical Co., Rockford, IL.
165. Atherton, E. and Sheppard, R. C (1989) *Solid Phase Peptide Synthesis: A Practical Approach.* IRL, Oxford.
166. Fields, G. B., Tian, Z., and Barany, G. (1992) Principles and practice of solid-phase peptide synthesis, in *Synthetic Peptides: A User's Guide* (Grant, G. A., ed.), W. H. Freeman, New York, pp. 77–183.
167. Fotouhi, N., Galakatos, N. G., and Kemp, D. S. (1989) Peptide synthesis by prior thiol capture. 6. Rates of the disulfide bond forming capture reaction and demonstration of the overall strategy by synthesis of the C-terminal 29-peptide sequence

of BPTI. *J. Org. Chem.* **54,** 2803–2817, and previous contributions from this research team.
168. du Vigneaud, V., Audrieth, L. F., and Loring, H. S. (1930) The reduction of cystine in liquid ammonia by metallic sodium. *J. Am. Chem. Soc.* **52,** 4500–4504.
169. Erickson, B. W. and Merrifield, R. B. (1973) Acid stability of several benzylic protecting groups used in solid-phase peptide synthesis. Rearrangement of *O*-benzyltyrosine to 3-benzyltyrosine. *J. Am. Chem. Soc.* **95,** 3750–3756.
170. Heath, W. F., Tam, J. P., and Merrifield, R. B. (1986) Improved deprotection of cysteine-containing peptides in HF. *Int. J. Peptide Protein Res.* **28,** 498–507, and references cited therein.
171. Akabori, S., Sakakibara, S., Shimonishi, Y., and Nobuhara, Y. (1964) A new method for the protection of the sulfhydryl group during peptide synthesis. *Bull. Chem. Soc. Jpn.* **37,** 433–434.
172. Nishimura, O., Kitada, C., and Fujino, M. (1978) A new method for removing the *S-p*-methoxybenzyl and *S-t*-butyl groups of cysteine residues with mercuric trifluoroacetate. *Chem. Pharm. Bull.* **26,** 1576–1585.
173. Platen, M. and Steckhan, E. (1984) Oxidative deblocking of the 4-methoxybenzyl thioether protecting group: application to the directed synthesis of poly-cystinyl peptides. *Liebigs Ann. Chem.,* 1563–1576.
174. Fujii, N., Otaka, A., Watanabe, T., Okamachi, A., Tamamura, H., Yajima, H., Inagaki, Y., Nomizu, M., and Asano, K. (1989) Silver trifluoromethanesulphonate as an S-deprotecting reagent for the synthesis of cystine peptides. *J. Chem. Soc., Chem. Comm.,* 283–284.
175. Sakakibara, S., Shimonishi, Y., Kishida, Y., Okada, M., and Sugihara, H. (1967) Use of anhydrous hydrogen fluoride in peptide synthesis. I. Behavior of various protective groups in anhydrous hydrogen fluoride. *Bull. Chem. Soc. Jpn.* **40,** 2164–2167.
176. Pastuszak, J. J. and Chimiak, A. (1981) *tert*-Butyl group as thiol protection in peptide synthesis. *J. Org. Chem.* **46,** 1868–1873, and references cited therein.
177. Atherton, E., Pinori, M., and Sheppard, R. C. (1985) Peptide synthesis. Part 6. Protection of the sulphydryl group of cysteine in solid-phase synthesis using N_α-fluorenylmethoxycarbonylamino acids. Linear oxytocin derivatives. *J. Chem. Soc., Perkin Trans. I,* 2057–2064.
178. Ploux, O., Chassaing, G., and Marquet, A. (1987) Cyclization of peptides on a solid support. Application to cyclic analogs of substance P. *Int. J. Peptide Protein Res.* **29,** 162–169.
179. McCurdy, S. N. (1989) The investigation of Fmoc-cysteine derivatives in solid phase peptide synthesis. *Peptide Res.* **2,** 147–152.
180. Kamber, B. and Rittel, W. (1968) Eine neue, einfache methode zur synthese von cystinpeptiden. *Helv. Chim. Acta* **51,** 2061–2064.
181. Photaki, I., Taylor-Papadimitriou, J., Sakarellos, C., Mazarakis, P., and Zervas, L. (1970) On cysteine and cystine peptides. Part V. *S*-trityl- and *S*-diphenylmethyl-cysteine and -cysteine peptides. *J. Chem. Soc. (C),* 2683–2689, and other papers from this research team.
182. Kamber, B., Hartmann, A., Eisler, K., Riniker, B., Rink, H., Sieber, P., and Rittel, W. (1980) The synthesis of cystine peptides by iodine oxidation of *S*-trityl-cysteine

and *S*-acetamidomethyl-cysteine peptides. *Helv. Chim. Acta* **63,** 899–915, and references cited therein.
183. Fujii, N., Otaka, A., Funakoshi, S., Bessho, K., Watanabe, T., Akaji, K., and Yajima, H. (1987) Studies on peptides. CLI. Syntheses of cystine-peptides by oxidation of *S*-protected cysteine-peptides with thallium (III) trifluoroacetate. *Chem. Pharm. Bull.* **35,** 2339–2347, and references cited therein.
184. Frankel, A. D., Biancalana, S., and Hudson, D. (1989) Activity of synthetic peptides from the Tat protein of human immunodeficiency virus type 1. *Proc. Natl. Acad. Sci. USA* **86,** 7397–7401.
185. Munson, M. C., García-Echeverría, C., Albericio, F., and Barany, G. (1992) *S*-2,4,6-Trimethoxybenzyl (Tmob): a novel cysteine protecting group for the N^{α}-9-fluorenylmethyloxycarbonyl (Fmoc) strategy of peptide synthesis. *J. Org. Chem.* **57,** 3013–3018, and references cited therein.
186. Munson, M. C., García-Echeverría, C., Albericio, F., and Barany, G. (1992) Novel cysteine protecting groups for the N^{α}-9-fluorenylmethyloxycarbonyl (Fmoc) strategy of peptide synthesis, in *Peptides–Chemistry and Biology: Proceedings of the Twelfth American Peptide Symposium* (Smith, J. A. and Rivier, J. E., eds.), Escom, Leiden, The Netherlands, pp. 605–606.
187. Veber, D., Milkowski, J. D., Varga, S. L., Denkewalter, R. G., and Hirschmann, R. (1972) Acetamidomethyl. A novel thiol protecting group for cysteine. *J. Am. Chem. Soc.* **94,** 5456–5461.
188. Kamber, B. (1971) Cystinpeptide aus (*S*-acetamidomethyl-cystein)-peptiden durch oxydation mit jod: die synthese von *cyclo*-L-cystin. *Helv. Chim. Acta* **54,** 927–930.
189. Liu, W., Shiue, G. H., and Tam, J. P. (1989) A novel strategy for the deprotection of *S*-acetamidomethyl containing peptides: an approach to the efficient synthesis of endothelin, in *Peptides–Chemistry, Structure and Biology: Proceedings of the Eleventh American Peptide Symposium* (Rivier, J. E. and Marshall, G. R., eds.), Escom, Leiden, The Netherlands, pp. 271–272.
190. Kiso, Y., Yoshida, M., Kimura, T., Fujiwara, Y, and Shikomura, M. (1989) A new thiol protecting trimethylacetamidomethyl group. Synthesis of a new porcine brain natriuretic peptide using the *S*-trimethylacetamidomethyl-cysteine. *Tetrahedron Lett.* **30,** 1979–1982.
191. Greiner, G. and Hermann, P. (1991) Application of penicillin amidohydrolase (PAH) for sulphur and amino deprotection in peptide synthesis, in *Peptides 1990: Proceedings of the Twenty-First European Peptide Symposium* (Giralt, E. and Andreu, D., eds.), Escom, Leiden, The Netherlands, pp. 277–278.
192. Hermann, P. and Schillings, T. (1993) Enzymatic manipulation of protecting groups, in *Peptides 1992: Proceedings of the Twenty-Second European Peptide Symposium* (Schneider, C. H. and Eberle, A. N., eds.), Escom, Leiden, The Netherlands, pp. 411–412, and references cited therein.
193. Royo, M., Alsina, J., Giralt, E., Slomczyńska, U., and Albericio, F. (1994). *S*-Phenylacetamidomethyl (Phacm). A versatile cysteine protecting group for Boc and Fmoc solid-phase synthesis strategies, in *Peptides–Chemistry, Structure and Biology: Proceedings of the Thirteenth American Peptide Symposium* (Hodges, R. S. and Smith, J. A., eds.), Escom, Leiden, The Netherlands, pp. 116–118.

194. Bodanszky, M. and Bednarek, M. A. (1982) Derivatives of S-9-fluorenylmethyl-L-cysteine. *Int. J. Peptide Protein Res.* **20,** 434–437.
195. Ruiz-Gayo, M., Albericio, F., Pedroso, E., and Giralt, E. (1986) (*S*)-9-Fluorenylmethyl-L-cysteine, a useful HF-stable derivative for peptide synthesis. *J. Chem. Soc., Chem. Commun.,* 1501–1502.
196. Albericio, F., Nicolás, E., Rizo, J., Ruiz-Gayo, M., Pedroso, E., and Giralt, E. (1990) Convenient syntheses of fluorenylmethyl-based side chain derivatives of glutamic and aspartic acids, lysine, and cysteine. *Synthesis,* 119–122.
197. Royo, M., García-Echeverría, C., Giralt, E., Eritja, R., and Albericio, F. (1992) *S*-2-(2,4-dinitrophenyl)ethyl-L-cysteine: a new derivative for solid-phase peptide synthesis. *Tetrahedron Lett.* **33,** 2391–2394.
198. Weber, U. and Hartter, P. (1970) *S*-Alkylmercapto-gruppen zum schutz der SH-funktion des cysteins, I: synthese und stabilität einiger *S*-(alkylmerkapto)cysteine. *Hoppe-Seyler's Z. Physiol. Chem.* **351,** 1384–1388.
199. Wünsch, E. and Spangenberg, R. (1971) Eine neue *S*-schutzgruppe für cystein, in *Peptides 1969* (Scoffone, E., ed.), North-Holland Publ., Amsterdam, pp. 30–34.
200. Eritja, R., Ziehler-Martin, J. P., Walker, P. A., Lee, T. D., Legesse, K., Albericio, F., and Kaplan, B. E. (1987) On the use of *S-t*-butylsulphenyl group for protection of cysteine in solid-phase peptide synthesis using Fmoc-amino acids. *Tetrahedron* **43,** 2675–2680.
201. Hemmasi, B., Zeng, W., Waidelich, D., and Bayer E. (1993) Reaction of HF with peptides containing Cys(StBu) in the presence of different scavengers, in *Peptides 1992: Proceedings of the Twenty-Second European Peptide Symposium* (Schneider, C. H. and Eberle, A. N., eds.), Escom, Leiden, The Netherlands, pp. 202–204, and references cited therein.
202. Matsueda, R., Kimura, T., Kaiser, E. T., and Matsueda, G. R. (1981) 3-Nitro-2-pyridinesulfenyl group for protection and activation of the thiol function of cysteine. *Chem. Lett.* **6,** 737–740.
203. Ridge, R. J., Matsueda, G. R., Haber, E., and Matsueda, R. (1982) Sulfur protection with the 3-nitro-2-pyridinesulfenyl group in solid-phase peptide synthesis. Synthesis of lysine^8-vasopressin. *Int. J. Peptide Protein Res.* **19,** 490–498.
204. Matsueda, R., Higashida, S., Albericio, F., and Andreu, D. (1992) Compatibility of the *S*-(3-nitro-2-pyridinesulfenyl) protecting group with DCC/HOBt coupling chemistry. *Peptide Res.* **5,** 262–263, and references cited therein.
205. Kamber, B. (1973) Die gezielte synthese offenkettiger asymmetrischer cystinpeptide mittels thiol-induzierter fragmentierung von sulfenylthiocarbonaten. Insulinfragmente mit intakter disulfidbrücke A^{20}-B^{19}. *Helv. Chim. Acta* **56,** 1370–1381.
206. Hiskey, R. G., Muthukumaraswamy, N., and Vunnam, R. R. (1975) Sulfur-containing polypeptides XVII. The *S*-carbomethoxysulfenyl derivative as a protective group for cysteine. *J. Org. Chem.* **40,** 950–953.
207. Schroll, A. L. and Barany, G. (1989) A new protecting group for the sulfhydryl function of cysteine. *J. Org. Chem.* **54,** 244–247, and references cited therein.
208. Atherton, E., Hardy, P. M., Harris, D. E., and Matthews, B. H. (1991) Racemization of C-terminal cysteine during peptide assembly, in *Peptides 1990: Proceed-*

ings of the Twenty-First European Peptide Symposium (Giralt, E. and Andreu, D., eds.), Escom, Leiden, The Netherlands, pp. 243–244.
209. Ueki, M. and Shinozaki, K. (1983). Phosphinyl and phosphinothioyl amino acids and peptides. VII. The use of the dimethylphosphinothioyl group as a thio protecting group of cysteine. *Bull Chem. Soc. Jpn.* **56,** 1187–1191.
210. Galakatos, N. G. and Kemp, D. S. (1985) New S-protection from known N-protection: thio esters of *N*-urethanyl-*N*-methyl-γ-aminobutyric acid as a class of protective groups for thiols in peptide synthesis. *J. Org. Chem.* **50,** 1302–1304, and references cited therein.
211. Otaka, A., Morimoto, H., Fujii, N., Koide, T., Funakoshi, S., and Yajima, H. (1989) S-Benzyloxymethylcysteine, its properties and application in the synthesis of porcine brain natriuretic peptide (pBNP). *Chem. Pharm. Bull.* **37,** 526–528, and references cited therein.
212. Blake, J., Woodworth, B. A., Litzi-Davis, L., and Cosand, W. L. (1992) Ethylcarbamoyl protection for cysteine in the preparation of peptide-conjugate immunogens. *Int. J. Peptide Protein Res.* **40,** 62–65, and references cited therein.
213. Hallinan, E. A. (1991) Formation of a dehydroalanyl residue from S-benzylcysteine upon HF cleavage of a [Sar^1, Cys^8]-angiotensin II peptide resin. *Int. J. Peptide Protein Res.* **38,** 601–602.
214. Echner, H. and Voelter, W. (1992) 9-Phenylxanthen-9-yl- (Pixyl): a new thiol protecting group and its use in solid phase peptide chemistry, in *Innovation and Perspectives in Solid Phase Synthesis: Peptides, Polypeptides and Oligonucleotides. 1992* (Epton, R., ed.), Intercept Ltd., Andover, England, pp. 371–375.
215. King, D. S., Fields, C. G., and Fields, G. B. (1990) A cleavage method which minimizes side reactions following Fmoc solid phase peptide synthesis. *Int. J. Peptide Protein Res.* **36,** 255–266.
216. Albericio, F., Kneib-Cordonier, N., Biancalana, S., Gera, L., Masada, R. I., Hudson, D., and Barany, G. (1990) Preparation and application of the 5-(4-(9-fluorenylmethyloxycarbonyl)aminomethyl-3,5-dimethoxyphenoxy)valeric acid (PAL) handle for the solid-phase synthesis of *C*-terminal peptide amides under mild conditions. *J. Org. Chem.* **55,** 3730–3743.
217. Pearson, D. A., Blanchette, M., Baker, M. L., and Guindon, C. A. (1989) Trialkylsilanes as scavengers for the trifluoroacetic acid deblocking of protecting groups in peptide synthesis. *Tetrahedron Lett.* **30,** 2739–2742.
218. Solé, N. A. and Barany, G. (1992) Optimization of solid-phase synthesis of [Ala^8]-dynorphin A. *J. Org. Chem.* **57,** 5399–5403.
219. Brady, S., Paleveda, W. J., and Nutt, R. (1989) Studies of acetamidomethyl as cysteine protection: application in synthesis of ANF analogs, in *Peptides–Chemistry and Biology: Proceedings of the Tenth American Peptide Symposium* (Marshall, G. R., ed.), Escom, Leiden, The Netherlands, pp. 192–194.
220. Mendelson, W. L., Tickner, A. M., Holmes, M. M., and Lantos, I. (1990) Efficient solution phase synthesis of [1-(β-mercapto-β,β-cyclopentamethylenepropionic acid)-2-(*O*-ethyl-D-tyrosine)-4-valine-9-desglycine]arginine vasopressin. *Int. J. Peptide Protein Res.* **35,** 249–257.

221. Lamthanh, H., Roumestand, C., Deprun, C., and Ménez, A. (1993) Side reaction during the deprotection of (*S*-acetamidomethyl)cysteine in a peptide with a high serine and threonine content. *Int. J. Peptide Protein Res.* **41,** 85–95, and references cited therein.
222. Inukai, N., Nakano, K., and Murakami, M. (1967) The peptide synthesis. I. Use of the *S*-ethylmercapto group for the protection of the thiol function of cysteine. *Bull. Chem. Soc. Jpn.* **40,** 2913–2918.
223. Inukai, N., Nakano, K., and Murakami, M. (1968) The peptide synthesis. II. Use of the phenol resin for the peptide synthesis. *Bull. Chem. Soc. Jpn.* **41,** 182–186.
224. Nokihara, K. and Berndt, H. (1978) Studies on sulfur-containing peptides: *tert*-butyloxycarbonylsulfenyl and benzyloxycarbonylsulfenyl derivatives as protecting groups for cysteine. *J. Org. Chem.* **43,** 4893–4895.
225. Glass, J. D. (1987) Enzymatic manipulation of protecting groups in peptide synthesis, in *The Peptides–Analysis, Synthesis, Biology*, vol. 9 (Udenfriend, S. and Meienhofer, J., eds.), Academic, San Diego, pp. 167–184.
226. Jaenicke, R. and Rudolph, R. (1989) Folding proteins, in *Protein Structure: A Practical Approach* (Creighton, T. E., ed.), IRL, Oxford, pp. 191–223, and references cited therein.
227. Narhi, L. O., Arakawa, T., McGinley, M. D., Rohde, M. F., and Westcott, K. R. (1992) Circular dichroism of reduced and oxidized recombinant human epidermal growth factor. *Int. J. Peptide Protein Res.* **39,** 182–187.
228. Singh, R. and Whitesides, G. M. (1991) A reagent for reduction of disulfide bonds in proteins that reduces disulfide bonds faster than does dithiothreitol. *J. Org. Chem.* **56,** 2332–2337.
229. Ranganathan, S. and Jayaraman, N. (1991) Highly efficient propane-1,3-dithiol mediated thiol-disulphide interchange: a facile and clean methodology for S–S reduction in peptides. *J. Chem. Soc., Chem. Commun.*, 934–936.
230. Lamoureux, G. V. and Whitesides, G. M. (1993) Synthesis of dithiols as reducing agents for disulfides in neutral aqueous solution and comparison of reduction potentials. *J. Org. Chem.* **58,** 633–641.
231. Rüegg, U. T. and Rudinger, J. (1977) Reductive cleavage of cystine disulfides with tributylphosphine. *Methods Enzymol.* **47,** 111–116.
232. Burns, J. A., Butler, J. C., Moran, J., and Whitesides, G. M. (1991) Selective reduction of disulfides by tris(2-carboxyethyl)phosphine. *J. Org. Chem.* **56,** 2648–2650.
233. Weldon, M. A. and Packman, L. C. (1992) Disulphide bridge formation in synthetic peptides—rapid monitoring of the process by capillary electrophoresis, in *Innovation and Perspectives in Solid Phase Synthesis: Peptides, Polypeptides and Oligonucleotides. 1992* (Epton, R., ed.), Intercept Ltd., Andover, England, pp. 489–493.
234. Chance, R. E., Hoffmann, J. A., Kroeff, E. P., Johnson, M. G., Schirmer, E. W., and Bromer, W. W. (1981) The production of human insulin using recombinant DNA technology and a new chain combination procedure, in *Peptides–Synthesis, Structure, Function: Proceedings of the Seventh American Peptide Symposium* (Rich, D. H. and Gross, E., eds.), Pierce Chemical Co., Rockford, IL, pp. 721–728.

235. Tarbell, D. S. (1961) The mechanism of oxidation of thiols to disulfides, in *Organic Sulfur Compounds,* vol. 1 (Kharasch, N., ed.), Pergamon, Oxford, pp. 97–102, and references cited therein.
236. Jocelyn, P. C. (1972) *Biochemistry of the SH Group.* Academic, London, especially Chapter 4, Oxidation of thiols, pp. 94–115, and references cited therein.
237. Adamson, J. G. and Lajoie, G. A. (1994) Guanidine hydrochloride assists intramolecular disulfide bond formation in cysteine peptides, in *Peptides–Chemistry, Structure and Biology: Proceedings of the Thirteenth American Peptide Symposium* (Hodges, R. S. and Smith, J. A., eds.), Escom, Leiden, The Netherlands, pp. 44,45.
238. Ellman, G. L. (1959) Tissue sulfhydryl groups. *Arch. Biochem. Biophys.* **82,** 70–77.
239. Grassetti, D. R. and Murray, J. F., Jr. (1967) Determination of sulfhydryl groups with 2,2'- or 4,4'-dithiodipyridine. *Arch. Biochem. Biophys.* **119,** 41–49.
240. Riddles, P. W., Blakeley, R. L., and Zerner, B. (1979) Ellman's reagent: 5,5'-Dithiobis(2-nitrobenzoic acid)—a reexamination. *Anal. Biochem.* **94,** 75–81.
241. Ashworth, M. R. F. (1976) *The Determination of Sulphur-Containing Groups, vol. 2, Analytical Methods for Thiol Groups.* Academic, London.
242. Zonta, A. and Adermann, K. (1993) Disulfide bridge formation in synthetic urodilatin (CDD/ANP-95-126), in *Peptides 1992: Proceedings of the Twenty-Second European Peptide Symposium* (Schneider, C. H. and Eberle, A. N., eds.), Escom, Leiden, The Netherlands, pp. 397–398.
243. Wallace, T. J. and Mahon, J. J. (1965) Reactions of thiols with sulfoxides. III. Catalysis by acids and bases. *J. Org. Chem.* **30,** 1502–1506, and references cited therein.
244. Fujii, N., Otaka, A., Okamachi, A., Watanabe, T., Arai, H., Tamamura, H., Funakoshi, S., and Yajima, H. (1989) Synthetic studies on cystine-containing peptides, in *Peptides 1988: Proceedings of the Twentieth European Peptide Symposium* (Jung, G. and Bayer, E., eds.), Walter de Gruyter & Co., Berlin, pp. 58–60.
245. Le-Nguyen, D. and Rivier, J. (1986) Use of carboethoxysulfenyl chloride for disulfide bond formation. *Int. J. Peptide Protein Res.* **27,** 285–292.
246. Hantgan, R. R., Hammes, G. G., and Scheraga, H. A. (1974) Pathways of folding of reduced bovine pancreatic ribonuclease. *Biochemistry* **13,** 3421–3431.
247. Karim Ahmed, A., Schaffer, S. W., and Wetlaufer, D. B. (1975) Nonenzymic reactivation of reduced bovine pancreatic ribonuclease by air oxidation and by glutathione oxidoreduction buffers. *J. Biol. Chem.* **250,** 8477–8482.
248. Rothwarf, D. M. and Scheraga, H. A. (1993) Regeneration of bovine pancreatic ribonuclease A. 1. Steady-state distribution. *Biochemistry* **32,** 2671–2679, and three successive articles in same journal.
249. Pigiet, V. P. and Schuster, B. J. (1986) Thioredoxin-catalyzed refolding of disulfide-containing proteins. *Proc. Natl. Acad. Sci. USA* **83,** 7643–7647.
250. Hillson, D. A., Lambert, N., and Freedman, R. B. (1984) Formation and isomerization of disulfide bonds in proteins: protein disulfide-isomerase. *Methods Enzymol.* **107,** 281–294.
251. Edman, J. C., Ellis, L., Blacher, R. W., Roth, R. A., and Rutter, W. J. (1985) Sequence of protein disulphide isomerase and implications of its relationship to thioredoxin. *Nature* **317,** 267–270, and references cited therein.

252. Lyles, M. M. and Gilbert, H. F. (1991) Catalysis of the oxidative folding of ribonuclease A by protein disulfide isomerase: dependence of the rate on the composition of the redox buffer. *Biochemistry* **30,** 613–619, and references cited therein.
253. Bondi, E., Fridkin, M., and Patchornik, A. (1968) The use of polymeric carriers in the synthesis of cyclic peptides containing S–S bonds. *Isr. J. Chem.* **6,** 22p.
254. Buchta, R., Bondi, E., and Fridkin, M. (1986) Peptides related to the calcium binding domains II and III of calmodulin. Synthesis and calmodulin-like features. *Int. J. Peptide Protein Res.* **28,** 289–297.
255. Wenschuh, E., Heydenreich, M., Runge, R., and Fischer, S. (1989) CCl_4 als mildes oxidans in der schwefelchemie: oxidation von thiolen zu disulfiden. *Sulfur Lett.* **8,** 251–260.
256. Munson, M. C., Lebl, M., Slaninová, J., and Barany, G. (1993) Solid-phase synthesis and biological activity of the parallel dimer of deamino-oxytocin. *Peptide Res.* **6,** 155–159.
257. Ruiz-Gayo, M., Albericio, F., Royo, M., García-Echevarría, C., Pedroso, E., Pons, M., and Giralt, E. (1989) A convenient procedure for synthesis of cystine peptides via iodine oxidation of S-acetamidomethyl-cysteine peptides. *An. Quim.* **85C,** 116–118.
258. Van Vliet, A., Smulders, R. H. P. H., Rietman, B. H., and Tesser, G. I. (1992) Protected peptide intermediates using a trityl linker on a solid support, in *Innovation and Perspectives in Solid Phase Synthesis: Peptides, Polypeptides and Oligonucleotides. 1992* (Epton, R., ed.), Intercept, Andover, England, pp. 475–477.
259. Sieber, P., Kamber, B., Riniker, B., and Rittel, W. (1980) Iodine oxidation of *S*-trityl- and *S*-acetamidomethyl-cysteine-peptides containing tryptophan: conditions leading to the formation of tryptophan-2-thioethers. *Helv. Chem. Acta* **63,** 2358–2363.
260. Bishop, P. and Chmielewski, J. (1992) Cyanogen iodide: a new reagent for disulfide bond formation in peptides. *Tetrahedron Lett.* **42,** 6263–6266.
261. Akaji, K., Tatsumi, T., Yoshida, M., Kimura, T., Fujiwara, Y., and Kiso, Y. (1991) Synthesis of cystine-peptide by a new disulfide bond-forming reaction using the silyl chloride–sulphoxide system. *J. Chem. Soc., Chem. Commun.,* 167–169.
262. Koide, T., Otaka, A., Suzuki, H., and Fujii, N. (1991) Selective conversion of *S*-protected cysteine derivatives to cystine by various sulphoxide–silyl compound / trifluoroacetic acid systems. *Syn. Lett.,* 345–346.
263. Akaji, K., Tatsumi, T., Yoshida, M., Kimura, T., Fujiwara, Y., and Kiso, Y. (1992) Disulfide bond formation using the silyl chloride–sulfoxide system for the synthesis of a cystine peptide. *J. Am. Chem. Soc.* **114,** 4137–4143, and references cited therein.
264. Otaka, A., Koide, T., Shide, A., and Fujii, N. (1991) Application of dimethylsulphoxide (DMSO) / trifluoroacetic acid (TFA) oxidation to the synthesis of cystine-containing peptide. *Tetrahedron Lett.* **32,** 1223–1226.
265. Funakoshi, S., Murayama, E., Guo, L., Fujii, N., and Yajima, H. (1988) A modified benzhydrylamine as a handle reagent for the solid phase synthesis of peptide amides based on the fluorenylmethoxycarbonyl method. *J. Chem. Soc., Chem Commun.,* 382–384.

266. Seidel, C., Klein, C., Empl, B., Bayer, H., Lin, M., and Batz, H.-G. (1990) Synthesis of cyclic peptide for diagnostic use by oxidation of its resin-bound intermediate, in *Peptides 1990: Proceedings of the Twenty-First European Peptide Symposium* (Giralt, E. and Andreu, D., eds.), Escom, Leiden, The Netherlands, pp. 236–237.
267. Albrecht, E., Harada, Y., Cooper, G. J. S., Jones, H., and Lehman de Gaeta, L. S. (1992) Synthesis and activity of human amylin and analogues, in *Peptides–Chemistry and Biology: Proceedings of the Twelfth American Peptide Symposium* (Smith, J. A. and Rivier, J. E., eds.), Escom, Leiden, The Netherlands, pp. 441–442.
268. Canas, M., Jodas, G., Albericio, F., Andreu, D., García-Antón, J. M., Parente, A., and Ponsati, B. (1993) Optimization of iodine oxidation methods for S-Acm, S-Trt dicysteinyl-peptides on the resin, in *Peptides 1992: Proceedings of the Twenty-Second European Peptide Symposium* (Schneider, C. H. and Eberle, A. N., eds.), Escom, Leiden, The Netherlands, pp. 401–402.
269. Edwards, W. B., Anderson, C. J., Welch, M. J., Fields, C. G., and Fields, G. B. (1994) The synthesis of DTPA-D-Phe[1]-octreotide by solid-phase synthesis. *J. Labelled Compds. Radiopharm.* **35,** 359–361.
270. Deadman, U., Lu, X., Moreno, A., Rahman, S., Chino, N., Claeson, G., Kakkar, V. V., and Williams, J. A. (1993) Novel procedure for on-resin cyclisation via disulphide bond formation in large rings. Abstract P314, Thirteenth American Peptide Symposium, Edmonton, Canada.
271. Barlos, K., Gatos, D., Kutsogianni, S., Papaphotiou, G., Poulos, C., and Tsegenidis, T. (1991) Solid phase synthesis of partially protected and free peptides containing disulphide bonds by simultaneous cysteine oxidation-release from 2-chlorotrityl resin. *Int. J. Peptide Protein Res.* **38,** 562–568.
272. Harpp, D. N. and Back, T. G. (1971) The synthesis of some new cysteine-containing unsymmetrical disulfides. *J. Org. Chem.* **36,** 3828–3829.
273. Brois, S. J., Pilot, J. F., and Barnum, H. W. (1970) A new pathway to unsymmetrical disulfides. The thiol-induced fragmentation of sulfenyl thiocarbonates. *J. Am. Chem. Soc.* **92,** 7629–7631.
274. Ten Kortenaar, P. B. W. and van Nispen, J. W. (1988) Formation of open-chain asymmetrical cystine peptides on a solid support. Synthesis of pGlu-Asn-Cyt-Pro-Arg-Gly-OH. *Coll. Czech. Chem. Commun.* **53,** 2537–2541.
275. Moroder, L., Marchiori, F., Borin, G., and Scoffone, E. (1973) Studies on cytochrome c. Part II. Synthesis of the protected heptapeptide (sequence 17–23) of baker's yeast iso-1-cytochrome c. *Biopolymers* **12,** 493–505.
276. Castell, J. V. and Tun-Kyi, A. (1979) The removal of *S*-cysteine protection by means of 2-pyridine sulfenyl chloride and the subsequent formation of disulfide bonds. *Helv. Chim. Acta* **62,** 2507–2510
277. Crimmins, D. L. (1989) Analysis of disulfide-linked homo- and hetero-peptide dimers with strong cation-exchange sulfoethyl aspartamide column. *Peptide Res.* **2,** 395–401, and references cited therein.
278. Kimura, T., Matsueda, R., Nakagawa, Y., and Kaiser, E. T. (1982) New reagents for the introduction of the thiomethyl group at sulfhydryl residues of proteins with

concomitant spectrophotometric titration of the sulfhydryls: methyl 3-nitro-2-pyridyl disulfide and methyl 2-pyridyl disulfide. *Anal. Biochem.* **122,** 274–282.

279. Ponsati, B., Ruiz-Gayo, M., Giralt, E., Albericio, F., and Andreu, D. (1990) Solid-phase-mediated peptide heterodisulfide formation. *J. Am. Chem. Soc.* **112,** 5345–5347.
280. Royo, M., Albericio, F., Giralt, E., and Pons, M. (1993) Use of cyclic dimers of palindromic peptides for the study of coiled-coils, in *Peptides 1992: Proceedings of the Twenty-Second European Peptide Symposium* (Schneider, C. H. and Eberle, A. N., eds.), Escom, Leiden, The Netherlands, pp. 487–488.
281. Baleux, F. and Dubois, P. (1992) Novel version of multiple antigenic peptide allowing incorporation on a cysteine functionalized lysine tree. *Int. J. Peptide Protein Res.* **40,** 7–12
282. Mukaiyama, T. and Takahashi, K. (1968) A convenient method for the preparation of unsymmetrical disulfides by the use of diethyl azodicarboxylate. *Tetrahedron Lett.* **56,** 5907–5908.
283. Wünsch, E. and Romani, S. (1982) A new method for the selective synthesis of unsymmetrical cystine peptides. *Hoppe Seyler's Z. Physiol. Chem.* **363,** 449–453.
284. Fujii, N., Otaka, A., Watanabe, T., Arai, H., Funakoshi, S., Bessho, K., and Yajima, H. (1987) Sulphoxide-directed disulphide bond-forming reaction for the synthesis of cystine peptides. *J. Chem. Soc., Chem. Commun.,* 1676–1678.
285. Fujii, N., Watanabe, T., Aotake, T., Otaka, A., Yamamoto, I., Konishi, J., and Yajima, H. (1988) Studies on peptides. CLXII. Synthesis of chicken calcitonin-gene-related peptide (cCGRP) by application of sulphoxide-directed disulfide-bond-forming reaction. *Chem. Pharm. Bull.* **36,** 3304–3311, and other contributions from this research team.
286. Kangawa, K., Fukuda, A., and Matsuo, H. (1985) Structural identification of β- and γ-human atrial natriuretic polypeptides. *Nature* **313,** 397–400.
287. Nieto, A., Postingl, H., and Beato, M. (1977) Purification and quaternary structure of the hormonally induced protein uteroglobin. *Arch. Biochem. Biophys.* **180,** 82–92.
288. García-Echeverría, C., Albericio, F., Giralt, E., and Pons, M. (1993) Design, synthesis, and complexing properties of (^{1}Cys-$^{1'}$Cys, ^{4}Cys-$^{4'}$Cys)-dithiobis(Ac-L-^{1}Cys-L-Pro-D-Val-L-^{4}Cys-NH_2). The first example of a new family of ion-binding peptides. *J. Am. Chem. Soc.* **115,** 11,663–11,670.
289. Hiskey, R. G., Davis, G. W., Safdy, M. E., Inui, T., Upham, R. A., and Jones, W. C., Jr. (1970) Sulfur-containing polypeptides. XIII. Bis cystine peptide derivatives. *J. Org. Chem.* **35,** 4148–4156.
290. Ryle, A. P., Sanger, F., Smith, L. F., and Kitai, R. (1955) The disulphide bonds of insulin. *Biochem. J.* **60,** 541–556.
291. Smith, D. L. and Zhou, Z. (1990) Strategies for locating disulfide bridges in proteins. *Methods Enzymol.* **193,** 374–389, and references cited therein.
292. Inglis, A. S. (1983) Single hydrolysis method for all amino acids, including cysteine and tryptophan. *Methods Enzymol.* **91,** 26–36.
293. McMullen, B. A., Fujikawa, K., and Davie, E. W. (1991) Location of the disulfide bonds in human plasma prekallikrein: the presence of four novel apple domains in the amino-terminal portion of the molecule. *Biochemistry* **30,** 2050–2056.

294. Sardana, M., Sardana, V., Rodkey, J., Wood, T., Ng, A., Vlasuk, G. P., and Waxman, L. (1991) Determination of disulfide bond pairs and stability in recombinant tick anticoagulant peptide. *J. Biol. Chem.* **266,** 13,560–13,563.
295. Calvete, J. J., Schäfer, W., Soszka, T., Lu, W., Cook, J. J., Jameson, B. A., and Niewiarowski, S. (1991) Identification of the disulfide bond pattern in albolabrin, an RGD-containing peptide from the venom of *trimeresurus albolabris*: significance for the expression of platelet aggregation inhibitory activity. *Biochemistry* **30,** 5225–5229.
296. Ishibashi, Y., Kikuchi, T., Wakimasu, M., Mizuta, E., and Fujino, M. (1991) Assignment of disulfide bonds in synthetic endothelin-1 isomers by fast atom bombardment mass spectrometry. *Biol. Mass Spectrometry* **20,** 703–708.
297. Bendixen, E., Halkier, T., Magnusson, S., Sottrup-Jensen, L., and Kristensen, T. (1992) Complete primary structure of bovine β_2-glycoprotein I: localization of the disulfide bridges. *Biochemistry* **31,** 3611–3617.
298. Töpfer-Petersen, E., Calvete, J., Schäfer, W., and Henschen, A. (1990) Complete localization of the disulfide bridges and glycosylation sites in boar sperm acrosin. *FEBS Lett.* **275,** 139–142.
299. Calvete, J. J., Wang, Y., Mann, K., Schäfer, W., Niewiarowski, S., and Stewart, G. J. (1992) The disulfide bridge pattern of snake venom disintegrins, flavoridin and echistatin. *FEBS Lett.* **309,** 316–320, and references cited therein.
300. Axelsson, K., Johansson, S., Eketorp, G., Zazzi, H., Hemmendorf, B., and Gellefors, P. (1992) Disulfide arrangement of human insulin-like growth factor I derived from yeast and plasma. *Eur. J. Biochem.* **206,** 987–994, and references cited therein.
301. Taniyama, Y., Yamamoto, Y., Kuroki, R., and Kikuchi, M. (1990) Evidence for difference in the roles of two cysteine residues involved in disulfide bond formation in the folding of human lysozyme. *J. Biol. Chem.* **265,** 7570–7575.
302. Fabri, L., Nice, E. C., Ward, L. D., Maruta, H., Burgess, A. W., and Simpson, R. J. (1992) Characterization of bovine heparin-binding neurotrophic factor (HBNF): assignment of disulfide bonds. *Biochem. Internat.* **228,** 1–9.
303. Smith, M. C., Cook, J. A., Furman, T. C., and Occolowitz, J. L. (1989) Structure and activity dependence of recombinant human insulin-like growth factor II on disulfide bond pairing. *J. Biol. Chem.* **264,** 9314–9321.
304. Carr, C., Aykent, S., Kimack, N. M., and Levine, A. D. (1991) Disulfide assignments in recombinant mouse and human interleukin 4. *Biochemistry* **30,** 1515–1523.
305. Hess, D., Schaller, J., and Rickli, E. E. (1991) Identification of the disulfide bonds of human complement C1s. *Biochemistry* **30,** 2827–2833.
306. Ng, N. F. L. and Hew, C. L. (1992) Structure of an antifreeze polypeptide from the sea raven. Disulfide bonds and similarity to lectin-binding proteins. *J. Biol. Chem.* **267,** 16,069–16,075.
307. Kumazaki, T. and Ishii, S.-I. (1990) Disulfide bridge structure of ascidian trypsin inhibitor I: similarity to Kazal-type inhibitors. *J. Biochem.* **107,** 414–419.
308. Lepage, P., Bitsch, F., Roecklin, D., Keppi, E., Dimarcq, J.-L., Reichhart, J.-M., Hoffman, J. A., Roitsch, C., and Van Dorsselaer, A. (1991) Determination of

disulfide bridges in natural and recombinant insect defensin A. *Eur. J. Biochem.* **196,** 735–742.
309. Violand, B. N., Tou, J. S., Vineyard, B. D., Siegel, N. R., Smith, C. E., Pyla, P. D., Zobel, J. F., Toren, P. C., and Kolodziej, E. W. (1991) Determination of the disulfide bond pairings in bovine transforming growth factor-α. *Int. J. Peptide Protein Res.* **37,** 463–467.
310. Huth, J. R., Mountjoy, K., Perini, F., and Ruddon, R. W. (1992) Intracellular folding pathway of human chorionic gonadotropin β subunit. *J. Biol. Chem.* **267,** 8870–8879.
311. Gross, E. and Witkop, B. (1962) Nonenzymatic cleavage of peptide bonds: the methionine residues in bovine pancreatic ribonuclease. *J. Biol. Chem.* **237,** 1856–1860.
312. Edman, P. and Henschen, A. (1975) Sequence determination, in *Protein Sequence Determination: A Sourcebook of Methods and Techniques* (Needleman, S. B., ed.), Springer-Verlag, New York, pp. 232–279.
313. Hunkapiller, M. W. (1988) Gas phase sequence analysis of proteins/peptides, in *Protein/Peptide Sequence Analysis: Current Methodologies* (Brown, A. S., ed.), CRC, Boca Raton, FL, pp. 87–117.
314. Burman, S., Wellner, D., Chait, B., Chaudhary, T., and Breslow, E. (1989) Complete assignment of neurophysin disulfides indicates pairing in two separate domains. *Proc. Natl. Acad. Sci. USA* **86,** 429–433.
315. Haniu, M., Acklin, C., Stoney, K., Kenney, W. C., and Rohde, M. F. (1994) Direct assignment of disulfide bonds by Edman degradation of selected peptide fragments. *Int. J. Peptide Protein Res.* **43,** 81–86 and references cited therein.
316. Landon, M. (1977) Cleavage at aspartyl-prolyl bonds. *Methods Enzymol.* **47,** 145–149.
317. Carrey, E. A. (1989) Peptide mapping, in *Protein Structure: A Practical Approach* (Creighton, T. E., ed.), IRL, Oxford, pp. 117–144, and references cited therein.
318. Canova-Davis, E., Kessler, T. J., and Ling, V. T. (1991) Transpeptidation during the analytical proteolysis of proteins. *Anal. Biochem.* **196,** 39–45.
319. Brown, J. R. and Hartley, B. S. (1966) Location of disulphide bridges by diagonal paper electrophoresis. The disulphide bridges of bovine chymotrypsinogen A. *Biochem. J.* **101,** 214–228.
320. Walsh, K. A., McDonald, R. M., and Bradshaw, R. A. (1970) Automatic systems for detecting cystine and cystinyl peptides during column chromatography. *Anal. Biochem.* **35,** 193–202.
321. Thannhauser, T. W., McWherter, C. A., and Scheraga, H. A. (1985) Peptide mapping of bovine pancreatic ribonuclease A by reverse-phase high-performance liquid chromatography. II. A two-dimensional technique for determination of disulfide pairings using a continuous-flow disulfide-detection system. *Anal. Biochem.* **149,** 322–330.
322. Creighton, T. E. (1989) Disulphide bonds between cysteine residues, in *Protein Structure: A Practical Approach* (Creighton, T. E., ed.), IRL, Oxford, pp. 155–167, and references cited therein.
323. Stone, K. L., Elliott, J. I., Peterson, G., McMurray, W., and Williams, K. R. (1990) Reversed-phase high-performance liquid chromatography for fractionation of

enzymatic digests and chemical cleavage products of proteins. *Methods Enzymol.* **193,** 389–412.
324. Thannhauser, T. W., Konishi, Y., and Scheraga, H. A. (1984) Sensitive quantitative analysis of disulfide bonds in polypeptides and proteins. *Anal. Biochem.* **138,** 181–188.
325. Sueyoshi, T., Miyata, T., Iwanaga, S., Toyo'oka, T., and Imai, K. (1985) Application of a fluorogenic reagent, ammonium 7-fluorobenzo-2-oxa-1,3-diazole-4-sulfonate for detection of cystine-containing peptides. *J. Biochem.* **97,** 1811–1813.
326. Lazure, C., Rochemont, J., Seidah, N. G., and Chrétien, M. (1985) Novel approach to rapid and sensitive localization of protein disulfide bridges by high-performance liquid chromatography and electrochemical detection. *J. Chromatography* **326,** 339–348.
327. Sun, Y., Smith, D. L., and Shoup, R. E. (1991) Simultaneous detection of thiol- and disulfide-containing peptides by electrochemical high-performance liquid chromatography with identification by mass spectrometry. *Anal. Biochem.* **197,** 69–76, and references cited therein.
328. Selsted, M. E. and Harwig, S. S. L. (1989) Determination of the disulfide array in the human defensin HNP-2. A covalently cyclized peptide. *J. Biol. Chem.* **264,** 4003–4007, and references cited therein.
329. Hiskey, R. G., Li, C.-D., and Vunnam, R. R. (1975) Sulfur-containing polypeptides. XVIII. Unambiguous synthesis of the parallel and antiparallel isomers of some bis-cystine peptides. *J. Org. Chem.* **40,** 3697–3703.
330. Weissman, J. S. and Kim, P. S. (1991) Reexamination of the folding of BPTI: predominance of native intermediates. *Science* **253,** 1386–1393, and references cited therein.
331. Chatrenet, B. and Chang, J.-Y. (1992) The folding of hirudin adopts a mechanism of trial and error. *J. Biol. Chem.* **267,** 3038–3043.
332. Andersson, M., Östman, A., Bäckström, G., Hellman, U., George-Nascimento, C., Westermark, B., and Heldin, C.-H. (1992) Assignment of interchain disulfide bonds in platelet-derived growth factor (PDGF) and evidence for agonist activity of monomeric PDGF. *J. Biol. Chem.* **267,** 11,260–11,266.
333. Huth, J. R., Mountjoy, K., Perini, F., Bedows, E., and Ruddon, R. W. (1992) Domain-dependent protein folding is indicated by the intracellular kinetics of disulfide bond formation of human chorionic gonadotropin β subunit. *J. Biol. Chem.* **267,** 21,396–21,403.
334. Fenn, J. B., Mann, M., Meng, C. K., Wong, S. F., and Whitehouse, C. M. (1989) Electrospray ionization for mass spectrometry of large biomolecules. *Science* **246,** 64–71, and references cited therein.
335. Desiderio, D. M. (ed.) (1991) *Mass Spectrometry of Peptides.* CRC, Boca Raton, FL.
336. Burlingame, A. L., Baillie, T. A., and Russell, D. H. (1992) Mass spectrometry. *Anal. Chem.* **64,** 467R–502R, and references cited therein.
337. Chait, B. T. and Kent, S. B. H. (1992) Weighing naked proteins: practical, high-accuracy mass measurement of peptides and proteins. *Science* **257,** 1885–1894, and references cited therein.

338. Yazdanparast, R., Andrews, P. C., Smith, D. L., and Dixon, J. E. (1987) Assignment of disulfide bonds in proteins by fast atom bombardment mass spectrometry. *J. Biol. Chem.* **262,** 2507–2513.
339. Raschdorf, F., Dahinden, R., Maerki, W., Richter, W. J., and Merryweather, J. P. (1988) Location of disulphide bonds in human insulin-like growth factor (IGFs) synthesized by recombinant DNA technology. *Biomed. Environ. Mass Spectrom.* **16,** 3–8.
340. Griffin, P. R., Shabanowitz, J., Yates, III, J. R., Zhu, N. Z., and Hunt, D. F. (1989) Laser photodissociation Fourier transform mass spectrometry: new methodology for sequence analysis of oligopeptides and location of disulfide bonds, in *Techniques in Protein Chemistry* (Hugli, T. E., ed.), Academic, San Diego, pp. 160–167.
341. Sørenson, H. H., Thomsen, J., Bayne, S., Højrup, P., and Roepstorff, P. (1990) Strategies for determination of disulphide bridges in proteins using plasma desorption mass spectrometry. *Biomed. Env. Mass Spectrom.* **19,** 713–720.
342. Despeyroux, D., Bordas-Nagy, J., and Jennings, K. R. (1991) Determination of the amino acid sequence of cystine-containing peptides by tandem mass spectrometry. *Rapid Commun. Mass Spectrom.* **5,** 156–159.
343. Stewart, A. E., Raffioni, S., Chaudhary, T., Chait, B. T., Luporini, P., and Bradshaw, R. A. (1992) The disulfide bond pairing of the pheromones Er-1 and Er-2 of the ciliated protozoan *Euplotes raikovi. Protein Sci.* **1,** 777–785.
344. Mancini, M. L. (1989) Enhancing the stability of disulfide-bond containing peptides under fast-atom bombardment conditions. *Biomed. Environ. Mass Spectrom.* **18,** 1102–1104.
345. Visentini, J., Gauthier, J., and Bertrand, M. J. (1989) Effect of trifluoroacetic acid on the reduction of disulfide bridges in peptides analyzed by fast-atom bombardment mass spectrometry. *Rapid Commun. Mass Spectrom.* **3,** 390–395.
346. Bolgar, M. S. and DiDonato, G. C. (1992) Disulfide bond assignment of endothelin-1 and analogues by ionspray mass spectrometry, in *Proceedings of the 40th ASMS Conference on Mass Spectrometry and Allied Topics,* pp. 1799–1800.
347. Feng, R., Bell, A., Dumas, F., and Konishi, Y. (1992) A fast and simple method for accurate counting of cysteines, disulfide bridges and free SH groups in proteins using ionspray™ mass spectrometry, in *Biotechnology International* (North, K., ed.), Century, London, pp. 155–158.
348. Sun, Y. and Smith, D. L. (1988) Identification of disulfide-containing peptides by performic acid oxidation and mass spectrometry. *Anal. Biochem.* **172,** 130–138 and references cited therein.
349. Morris, H. R. and Pucci, P. (1985) A new method for rapid assignment of S–S bridges in proteins. *Biochem. Biophys. Res. Commun.* **126,** 1122–1128.
350. Savoy, L.-A., Greer, F. M., and Morris, H. R. (1993) Peptide and protein analysis by electrospray mass spectrometry, in *Peptides 1992: Proceedings of the Twenty-Second European Peptide Symposium* (Schneider, C. H. and Eberle, A. N., eds.), Escom, Leiden, The Netherlands, pp. 441–442.
351. Rodriguez, H., Nevins, B., and Chakel, J. (1989) Evaluation of methods for the analysis of disulfide containing peptides by fast atom bombardment mass spec-

trometry, in *Techniques in Protein Chemistry* (Hugli, T. E., ed.), Academic, San Diego, pp. 186–194.

352. Bean, M. F. and Carr, S. A. (1992) Characterization of disulfide bond position in proteins and sequence analysis of cystine-bridged peptides by tandem mass spectrometry. *Anal. Biochem.* **201,** 216–226, and references cited therein.
353. Hidaka, Y. and Schimonishi, Y. (1989) A new method for determination of disulfide pairing in peptides. *Bull. Chem. Soc. Jpn.* **62,** 1986–1994.
354. Hidaka, Y., Sato, K., Nakamura, H., Kobayashi, J., Ohizumi, Y., and Shimonishi, Y. (1990) Disulfide pairings in geographutoxin I, a peptide neurotoxin from *Conus geographus. FEBS Lett.* **264,** 29–32.
355. Grant, G. A. and Chiappinelli, V. A. (1985) κ-Bungarotoxin: complete amino acid sequence of a neuronal nicotinic receptor probe. *Biochemistry* **24,** 1532–1537.
356. Kopeyan, C., Martinez, G., and Rochat, H. (1985) Primary structure of toxin IV of *Leiurus quinquestriatus quinquestriatus.* Characterization of a new group of scorpion toxins. *FEBS Lett.* **181,** 211–217.
357. Ryan, R. J., Charlesworth, M. C., McCormick, D. J., Milius, R. P., and Keutmann, H. T. (1988) The glycoprotein hormones: recent studies of structure-function relationships. *FASEB J.* **2,** 2661–2669.
358. Nakaya, K., Omata, K., Okahashi, I., Nakamura, Y., Kolkenbrock, H., and Ulbrich, N. (1990) Amino acid sequence and disulfide bridges of an antifungal protein isolated from *Aspergillus giganteus. Eur. J. Biochem.* **193,** 31–38.

CHAPTER 8

Site-Specific Chemical Modification Procedures

Michael W. Pennington

1. Introduction

Identification of important or essential residues in proteins and peptides is an overall goal in understanding the mechanism of an enzyme or the binding affinity of a peptide ligand to its receptor. Most early attempts to accomplish this relied on chemical modification of amino acid residues in the native peptide or protein *(1)*. Selectivity of chemical modification reagents has always been the largest problem to overcome when this approach is taken. Isolation of the desired product from the heterogeneous mixture of components and determination of the modified residue or residues become very labor-intensive processes.

Peptide synthesis offered a viable alternative to this process by utilizing orthogonal deprotection strategies to position the specific residue or residues to be modified. As solid-phase synthetic procedures have improved, this process has become even easier by minimizing the more labor-intensive solution procedures. Advancement of the automation procedures *(2,3)* now permits routine synthesis of peptides and mini-proteins, such as HIV protease *(4,5)* and interleukin-3 *(6)*. Site-directed analogs can be easily synthesized by either incorporation of an orthogonal deprotection strategy, or by simply removing resin aliquots during the total synthesis of the native sequence and completing synthesis of the molecule following replacement of one or more amino acids in the sequence *(7,8)*.

In this chapter, several convenient methods for preparing specifically positioned chemically modified peptides will be described. These include

From: *Methods in Molecular Biology, Vol. 35: Peptide Synthesis Protocols*
Edited by: M. W. Pennington and B. M. Dunn

solid-phase and some solution procedures. Most of these methods are simple acylations on the N-terminus or another internal primary amine. Some of these procedures have been presented in detail in the literature, but many have not. Additionally, we describe the positioning of specifically modified side chains using a combination of Fmoc/*t*Bu-Bzl strategy. Lastly, a description of several carboxylic acid amidation reactions that have been useful in the preparation of cyclic and fluorescent peptides is presented.

2. Materials

1. All solvents described are purchased from commercial sources, such as Fisher (Fair Lawn, NJ) or Aldrich (Milwaukee, WI), and used as such.
2. 2,4 Dinitrofluorobenzene (DNFB), *N*-hydroxysuccinimide, pyridine–sulfurtrioxide complex, biotin, acetic anhydride, dansyl chloride, *p*-benzoylbenzoic acid (BBA), 7-methoxycoumarin-4-acetic acid, palmitic acid, myristic acid, and stearic acid may be obtained from Aldrich. Biotin *p*-nitrophenyl ester, 1-ethyl 3-(3-dimethylaminopropyl)carbodiimide HCl (EDAC) and 5-(2-aminoethylamino)-naphthalene sulfonic acid (Na salt) (1,5 EDANS) may be purchased from Sigma (St. Louis, MO). 4-Dimethylamino-azobenzene 4'-carboxylic acid (DABCYL) may be obtained from TCI America.
3. Specialized amino acid derivatives are commercially available from Bachem Bioscience (King of Prussia, PA). These derivatives include Fmoc-Bzl and Boc-Fmoc or Boc-(OFm)-protected residues.

3. Methods

The general strategy presented in this chapter is to exploit the solid-phase principle whenever possible. This greatly reduces difficult extractions and recrystallizations required when solution procedures are carried out. It is the goal of this chapter to present the most common types of chemical modifications of synthetic peptides. Many of these modifications are very similar, but very few procedures have been presented in the literature. Additionally, this chapter will present approaches for the synthesis of branched peptide structures and specifically positioned benzyl-type analogs requiring no special chemical modification protocols.

3.1. Simple N-Terminal Extensions

These modifications involve reaction of the free N-terminus of the completed peptide chain with an alkylating or acylating species. In all cases listed below, the N-terminal-protecting group has been removed. In the case of Boc synthesis, it is necessary also to perform a base wash with 10% DIEA in DCM to neutralize the TFA salt. All of these modifications will result in a blocked N-terminus, which is refractory to Edman degradation *(9,10)*. As a necessary control, a Kaiser test should be performed as described below in order to ensure coupling of the derivative.

3.1.1. Ninhydrin Analysis (Kaiser Test)

This test, developed by Kaiser et al. *(11)*, provides a clear indication of the amount of free amine present. This test is routinely performed before and after a coupling procedure to ensure completeness.

1. Dissolve 80 g phenol in 20 mL absolute ethanol, and keep in a brown bottle marked A.
2. Dilute 2 mL of 1 m*M* aqueous solution of KCN up to 100 mL with pyridine. Keep this in a brown bottle marked B.
3. Dissolve 500 mg of ninhydrin in 10 mL of absolute ethanol, and keep in a brown glass bottle marked C.
4. Take a small aliquot of peptide resin (approx 5–10 mg) and place in a 10 × 75 mm borosilicate test tube.
5. Add two drops of reagent A followed by two drops of reagent B and, finally, add two drops of reagent C.
6. Place this test tube in a heating block that is maintained at 100°C for 2 min.
7. Remove the test tube, and check the color of the solution. A dark blue result indicates a strong positive reaction for primary amines. A clear yellow solution indicates a strong negative reaction for primary amines. Lighter shades of blue and blue-green indicate an incomplete coupling reaction.
8. Certain amino acid residues do not react strongly with ninhydrin. Proline gives a resin bead with a red color for a strong positive. Asp, Asn, Glu, Gln, His, and Cys do not always react well with ninhydrin and often give a red to red brown color as a positive reaction.

3.1.2. Acetylation

Acetylation places an acetyl group on the N-terminus of the peptide. This will increase the mol wt of the peptide by 42 mass units.

1. Take the deblocked peptide resin, and add a solution of 20% acetic anhydride dissolved in DMF at a ratio of 7 mL/g of resin. Ninhydrin analysis of the deblocked peptide resin should be positive prior to this reaction. Addition of 1.5 Eq of DIEA after 5 min neutralizes the protons that have been generated.
2. Allow this acetylation to proceed for 30 min at room temperature. This is a fairly rapid reaction and is usually complete within 10 min.

3. Wash the resin with two washes of DMF, two washes of DCM, and two washes of MeOH.
4. Ninhydrin analysis at this point should be a strong negative.
5. No special cleavage procedures need to be followed in either Boc or Fmoc chemistries.
6. This procedure can be used to radiolabel the peptide by incorporation of radiolabeled acetic anhydride, which is commercially available.

3.1.3. Biotinylation

For a number of immunological procedures (*see* Chapter 10, *PAP*) as well as for structure–function studies, biotin has been a very useful probe *(12,13)*. For solid-phase biotinylation, we follow the procedure of Lobl et al. *(14)*.

1. Biotin is only moderately soluble in solvents compatible with peptide synthesis. We have found that biotin-*p*-nitrophenylester and biotin-*N*-hydroxysuccinimide ester are very soluble in DMF. We routinely both of these compounds.
2. Dissolve 3 Eq of either biotin-NHS or biotin-*p*-nitrophenylester in 8–10 mL of DMF/g of resin. Stir until all of the biotin is dissolved (usually 2–5 min).
3. Add this solution to the deblocked peptide resin. We have found that this reaction is fairly slow, and requires several hours or overnight reaction. The temperature may be increased up to 40°C to speed up the reaction. Ninhydrin analysis should again be positive prior to addition of the biotin and negative following the successful coupling.
4. No special cleavage procedures are required for either Boc or Fmoc peptides.
5. Biotin is fairly hydrophobic and usually will result in a later retention time for the biotinylated peptide on RP-HPLC (*see* Chapter 3, *PAP*).
6. The mol wt of the final product will be increased by 226 mass units (*see* Chapter 7, *PAP*).

3.1.4. Dinitrophenylation

Addition of the dinitrophenyl group turns the peptide a bright yellow color and increases the mol wt by 166 mass units.

1. The deblocked peptide resin is treated with 4 Eq of dinitrofluorobenzene and 4 Eq of DIEA in DMF for 3 h. We have found that the reaction is complete within 3 h for most peptides. However, some peptides may be less reactive and require a second treatment overnight.
2. The DNP-peptide resin is then thoroughly washed with DMF, followed by DCM, and finally, MeOH.

3. Ninhydrin analysis of the resin should now be a strong negative. If this is not the case, dinitrophenylation is not complete and must be repeated.
4. No special cleavage procedures are required for the peptide resins. Protection from light is recommended for storage of the resin as well as the final product.
5. The DNP-amino acid derivative is stable to acid hydrolysis and will not be detected by standard amino acid analysis methods.

3.1.5. Dansylation

Attachment of the dansyl group to the peptide will give the peptide a pale yellow-green color. This substituent will increase the mol wt by 233.3 mass units.

1. Dissolve 5 Eq of dansyl chloride in 10 mL of DMF/g of deblocked peptide resin. Add 2 Eq of DIEA to this solution to maintain the pH of the reaction.
2. Add this solution to the deblocked peptide resin and allow to react overnight at room temperature or increase the temperature of the reaction up to 35°C for 3–4 h. We have found that elevating the temperature slightly helps speed the reaction to completion.
3. Terminate reaction by draining the reaction vessel and washing thoroughly with two washes of DMF, followed by two washes of DCM, and finally, two washes of MeOH. Ninhydrin analysis of the final resin should now indicate a negative result for free amines.
4. No special cleavage protocols are required for dansylated peptides. These peptides and peptide resin should be stored in the dark.
5. The dansyl-amino acid derivative is stable to standard hydrolysis procedures and will not be detected by standard amino acid analysis.

3.1.6. Dabcylation and Dabsylation

These derivatives will turn the peptide a bright red color. The carboxylic acid form (dabcyl) and the sulfonyl chloride form (dabsyl) both react with the N-terminal amino acid. However the dabsyl derivative will be stable to normal amino acid analysis hydrolysis procedures.

1. For the dabcyl derivative, we routinely use NMP as the solvent of choice. Dabcyl is fairly soluble in DMF, but has much greater solubility in NMP. Dissolve 4 Eq dabcyl in 10 mL NMP/g of resin.
2. Activate the carboxyl group by adding 4.4 Eq of diisopropyl carbodiimide and 8 Eq of hydroxybenzotriazole (HOBt).
3. Add this solution to the deblocked peptide resin. Allow this coupling reaction to continue overnight. We have observed this to be a fairly slow coupling derivative.

4. Terminate the coupling process by washing the peptide resin extensively with NMP until no more red color is seen in the wash. Next, wash the resin with three washes of DCM followed by three washes of MeOH. Ninhydrin analysis should now be negative. If it is not negative, then recouple by performing a base wash of the resin and starting back at step 1.
5. The dabsyl derivative can be coupled directly by dissolving 4 Eq in NMP with 4 Eq of DIEA and coupling overnight at 35°C. Perform step 4, and check for completeness of reaction.
6. These derivatives are stable to both TFA- and HF-deprotection procedures. Addition of these derivatives will increase the mol wt by 251 mass units for the dabcyl and 288 for the dabsyl derivative.

3.1.7. Fatty Acid Acylation

Attachment of a fatty acid, such as myristic, palmitic, or stearic acid, is easily accomplished using standard coupling procedures (*see* Note 1).

1. Dissolve 4 Eq of the fatty acid in 10 mL of DCM or DMF, or mixtures of these two solvents/g of deblocked peptide resin.
2. Activate the fatty acid carboxyl group by addition of 4.4 Eq of diisopropylcarbodiimide and 8 Eq of hydroxybenzotriazole to this solution.
3. Add this solution to the peptide resin, and allow to react over night at 35°C. This acylation is rather slow and may require additional couplings to obtain complete acylation.
4. Terminate the reaction by draining the reaction vessel, and washing with three washes of DCM followed by two washes of DMF, followed by two washes of DCM, and lastly, two washes of MeOH. Ninhydrin analysis should be negative.
5. No special precautions must be followed in the cleavage of these peptides. Solubilization of the fatty acyl-peptide may be hampered by these fatty acid groups and may require slightly more drastic solvents, such as glacial acetic acid or even neat TFA. Molecular weight increases by 209, 238, and 267 mass units for myristic, palmitic, and stearic acids, respectively (*see* Note 2).

3.1.8. Acylation with 7-Methoxycoumarin 4-Acetic Acid (MCA)

Coumarin derivatives are **not stable to HF cleavage** and thus should only be used when employing an Fmoc strategy with a TFA cleavage.

1. Dissolve 4 Eq of the coumarin derivative in 10 mL of DMF/g of deblocked Fmoc-peptide resin.
2. Activate the carboxyl group by adding 4.4 Eq of diisopropylcarbodiimide and 8 Eq of hydroxybenzotriazole to this solution.
3. Add this solution to the deblocked peptide resin, and allow to couple for 4 h.

4. Terminate the reaction by draining the reaction vessel and washing with two washes of DMF, followed by two washes of DCM, followed by two washes of MeOH. Ninhydrin analysis of the resin should be negative. If it is not negative, perform a base wash procedure and recouple from step 1.
5. As mentioned above, these derivatives are not stable to HF. They should only be coupled when an Fmoc strategy has been used. Addition of 7-methoxy-coumarin 4-acetic acid will increase the mol wt of the peptide by 216 mass units.

3.1.9. Reductive Alkylation

This procedure can be used to place an alkyl or aromatic substituent at the N-terminus of the peptide. This procedure is a modification of the Jentoft and Dearborn method *(15)*. A novel method of limiting the number of small alkyl substituents was recently reported at the 22nd European Peptide Symposium where the N-terminal amine was protected with the Dod-protecting group prior to the reductive alkylation *(16)*. This group is then removed by treatment with TFA, and the peptide can be elongated or cleaved.

1. Dissolve 1–4 Eq of the appropriate aldehyde (i.e., benzaldehyde for a benzyl group, acetaldehyde for an ethyl group, and so on) in 10 mL of DMF containing 1% acetic acid/g of deblocked peptide resin. The greater the number of Eq of aldehyde used, the greater the potential to generate disubstituted products.
2. Add this solution to the deblocked peptide resin, and allow to mix for 30 min.
3. Initiate the reduction process by addition of 1–4 Eq of sodium cyanoborohydride (depending on the number of Eq of aldehyde used) to the reaction mixture. Allow reaction to mix for an additional 2 h.
4. Terminate the reaction by draining the reaction vessel and washing the resin as described above.
5. Reductive alkylation with smaller substituents, such as formaldehyde and acetaldehyde, usually results in the formation of dialkyl products.
6. This procedure can be used to incorporate a radiolabel into the N-terminus of the peptide by using a radiolabeled aldehyde.

3.1.10. Anthranilylation

Attachment of this amino acid derivative results in a fluorescent peptide derivative. This fluorescence can be internally quenched by *p*-nitro-Phe and nitro-Tyr derivatives, resulting in a convenient method of preparing internally quenched proteolytic substrates *(17,18)*.

1. Dissolve 4 Eq Boc-anthranilic acid in NMP with 4.4 Eq of DCC and 8 Eq HOBT.
2. We have found that incorporating potassium isothiocyanate to a final concentration of 0.4*M* in this coupling mixture improves this coupling.
3. Add this preactivated solution to the deblocked peptide resin, and allow to couple from 4 h to overnight.
4. Following coupling of the derivative, the Kaiser test will be negative.
5. Remove the final Boc group with 50% TFA in DCM prior to HF cleavage.
6. Addition of this residue will increase the mol wt of the peptide by 121 mass units.

3.1.11. Photoprobe Coupling (Addition of Benzoylbenzoic Acid)

Addition of benzoylbenzoic acid (BBA) results in a peptide that can be conveniently used as a photoprobe and may be covalently linked to its target by UV irradiation *(19,20)*. This derivative results in an addition of 208 mass units to the peptide. Additionally, incorporation of a photoprobe into the side chain of Phe has been reported *(21)*. This protected derivative Boc-Bpa-OH (*p*-benzoyl-phenylalanine) is now commercially available from Bachem Bioscience Inc.

1. Dissolve 4 Eq of BBA in NMP. Add 4.4 Eq of DCC and 8 Eq HOBt to this solution, and allow to preactivate for 30 min.
2. Add this solution to the deblocked peptide resin, and allow to couple for 2–4 h.
3. Following addition of this derivative, Kaiser analysis will be negative.
4. Cleavage of the peptide should proceed as a normal HF cleavage with **no thiol scavengers.**

3.2. Specialized Uses of Boc-Lys(Fmoc)-OH and Boc-Orn(Fmoc)-OH

These amino acid derivatives have the unusual property of being differentially labile while still attached to the solid-phase support. This property allows both Boc and Fmoc chemistries to be performed in an orthogonal deprotection scheme. This allows the chemist to create site-specific modifications and create branched peptide structures very easily.

3.2.1. Placement Specificity for Chemical Modification

The following procedure will refer to all of the reactions described above by incorporation of specialized amino acid derivative Boc-Lys(Fmoc)-OH.

1. This derivative can be positioned during a Boc strategy synthesis so as to incorporate a differentially labile protecting group stable to the TFA-deprotection steps of the Boc synthesis.
2. On completion of the solid-phase assembly of the peptide chain, the final Boc group is not removed. Instead, the Fmoc group on this internal Lys residue is removed by treatment with 20% piperidine in DMF.
3. This results in an internally positioned primary amine that can then be reacted with any one of the abovementioned acylation reactions stable to HF.
4. The resulting peptide resin is then treated to remove the Boc group. Following Boc removal, this resin can then be further reacted with another acylating reagent if desired. The resulting product is subsequently deprotected using standard HF deprotection procedures yielding the monospecific or dispecific adduct.
5. By incorporating multiple residues of the Boc-Lys(Fmoc)-OH sequentially, these can be modified prior to elongating the peptide chain to create multiple derivatization sites. For example, a fatty acid may be linked near the C-terminus, an internal photoprobe can be positioned in the middle of the molecule, and the N-terminus may be biotinylated. This could be easily accomplished by incorporating two Boc-Lys(Fmoc)-OH derivatives and N-terminal acylation. However, synthesis would proceed with coupling the first derivative and immediate removal of the Fmoc group prior to elongation of the peptide chain. The side chain is acylated with the appropriate fatty acid derivative. Extension of the primary chain then resumes by removal of the Boc group and addition of the next portion of the sequence until the second derivatization point is encountered. At this point, the Fmoc group would be removed, and the 4-benzoylbenzoic acid coupling reaction is performed. Following this coupling, the remainder of the sequence would proceed after removal of the Boc group, and lastly, the N-terminus would be derivatized with biotin using biotin nitrophenylester.

3.2.2. Synthesis of Branched Peptide Structures

By incorporating the Boc-Lys(Fmoc)-OH at the appropriate position in the peptide sequence, a branched peptide can be synthesized one of two ways.

1. Following incorporation of this derivative, synthesis continues on the main peptide sequence until it is complete. The main peptide chain is left either protected with the Boc group intact, or the Boc group may be removed and the N-terminus acetylated. Following completion of the main peptide chain, the branched peptide chain can be assembled by deblocking the Fmoc group with 20% piperidine in DMF. Coupling of the branched peptide can either proceed using an Fmoc strategy if the final Boc group was

not removed or with either a Boc or Fmoc strategy if the N-terminus has been acetylated.

2. This method requires the use of Fmoc-Bzl amino acid derivatives (these derivatives are commercially available). If one desires to synthesize the branched peptide chain first, this is accomplished by stopping the Boc strategy immediately following the coupling of the Boc-Lys(Fmoc)-OH. At this point, an Fmoc strategy may be employed, but the Bzl side-chain-protection strategy must be used on all of the side chains of the branched portion. Otherwise the *t*Bu-protecting groups added during the Fmoc portion of the synthesis would be removed when the Boc strategy resumes. On completion of the branched portion of the molecule, the Fmoc can be left on or the branched N-terminus must be acetylated. Now, the Boc synthesis of the main peptide chain resumes until completed. If the Fmoc group has not been removed, deblock with 20% piperidine **prior to HF cleavage.**
3. When following either of these strategies, use of a scavenger, such as DTT and 4-methylindole, in the TFA-deblocking solution during the Boc procedure is strongly recommended if Trp and Met are present. It is more important, especially in the case of Scheme 2, that the Trp is present without the formyl-protecting group in Fmoc synthesis. On repetitive cycles of TFA exposure during the ensuing Boc section following the Fmoc portion of synthesis, the Trp indole side chain will be destroyed if a scavanger is not incorporated.

3.3. Positioning of Benzyl-Protected Side Chains

When an Fmoc strategy is being utilized, specific analogs may be generated by utilizing an Fmoc-Bzl-protected derivative instead of the standard Fmoc-*t*Bu derivative at selected positions. The benzyl group is not cleaved during standard TFA cleavage schemes for Fmoc peptides. This will result in a specifically benzylated derivative following cleavage with TFA. This type of strategy can be used to increase the hydrophobicity of a peptide. This strategy was employed to make the dibenzylated CD_4 receptor fragment, which inhibited HIV-induced cell fusion and infection in vitro *(22)*.

3.4. Carboxylic Acid Modifications

Up to this point, all of the reactions described have been simple manipulations of general coupling protocols using the solid-phase approach. Reactions at carboxyl groups are generally carried out using solution procedures. Recently, with the synthesis of Boc-Glu(OFm)-OH and Boc-Asp(OFm)-OH, some solid-phase carboxyl modifications are now possible. These include cyclization through a primary amine to the deblocked OFm derivative *(23)* and preparation of a fluorescent Boc-Glu-(EDANS)-OH derivative from the Boc-Glu(OFm)-OH derivative while attached to the solid-phase support (M. W. P., personal communication). This section will describe procedures for the preparation of the EDANS derivative using a solution procedure. For peptides containing only one carboxyl group, where either the C-terminal carboxyl group or the C-terminus is amidated, and only one internal side chain carboxyl exists, a Boc strategy may be used. However, if more than one carboxyl group is present, we recommend using an Fmoc strategy employing Sasrin™ resin (available from Bachem Bioscience Inc.) to prepare the free acid of the protected peptide *(24)* and subsequent solution reaction on this protected peptide to synthesize the C-terminal adduct (*see* Note 3).

3.4.1. Attachment of EDANS to Carboxyl Group

Dabcyl and EDANS are two fluorophores with excellent donor-acceptor qualities for fluorescence energy transfer (FRET) *(25)*. Incorporating them into opposite sides of proteolytic cleavage site on peptide substrates has been exploited to create a continuous fluorescence assay for HIV protease *(26)*. EDANS can be attached to either the C-terminal carboxyl or the α- or β-carboxyls of Asp and Glu, respectively. If the peptide contains only one carboxyl group, a standard Boc synthesis is recommended. The N-terminus of the peptide must be blocked (acetylated, dabcylated, and so forth).

1. Dissolve 0.25 mmol of the cleaved, N-terminally blocked peptide in 100 mL of DMF (*see* Note 4).
2. Add 2.2 Eq of EDAC (1-ethyl 3-[3-dimethylaminopropyl] carbodiimide) HCl and 2.2 Eq of hydroxybenzotriazole to initiate the reaction.
3. Add 0.5 mmol of 1,5-EDANS (sodium salt) to this solution, and mix until fully dissolved.
4. Allow the reaction to proceed for 3 h at room temperature. Follow the reaction by RP-HPLC. Inject a small aliquot (5 µL diluted with three parts water) directly onto an RP-HPLC column. A massive breakthrough peak will be observed from the DMF.
5. We have routinely observed the retention time to shift to earlier values on successful coupling.
6. Terminate the reaction by diluting with five to eight parts of water and load onto preparative RP-HPLC (*see* Chapter 4, *PAP*). (Do not load this solution onto the HPLC if the peptide begins to precipitate. This solution

must be filtered to prevent damage to the HPLC column.) The DMF, all of the coupling agents, and the unreacted EDANS will wash through on a normal C_{18}-HPLC column. More hydrophobic peptides will be retained and can be isolated by a normal gradient elution.

7. This substituent will increase the mol wt by 247 mass units.

3.4.2. Solid-Phase Cyclization Reactions

Cyclization of protected peptides has been accomplished by incorporating specially derivatized amino acid residues, such as Boc-Glu(OFm)-OH and Boc-Asp(OFm)-OH, in combination with Boc-Lys(Fmoc)-OH *(23)*. When these derivatives are used in combination with one another, it is possible to form a cyclized peptide through the amide bond generated by coupling the side-chain amino group of the Lys or the α-amino group to the side-chain carboxyl group of the Glu or Asp. Other cyclic peptides have also exploited this solid-phase cyclization by incorporating a combination of Fmoc-*t*Bu and Fmoc-Bzl amino acids in the preparation of $Asu^{(1,7)}$-calcitonin *(27)*. More recently, a cyclized derivative of endothelin was prepared that resulted in a potent endothelin antagonist *(28)*. Both of these molecules utilized this amide linkage as a disulfide bond surrogate.

1. During synthesis, the positioning of the differentially labile derivatives is crucial. Information about secondary structure can help determine the position for the complementary pair of Boc-Lys(Fmoc)-OH and the Boc-Glu(OFm)-OH or Boc-Asp(OFm)-OH.
2. Once the two derivatives have been incorporated, the peptide resin N-terminal Boc is not removed. Instead, the internal side-chain blocking groups (the Fmoc group and the Fm group) are removed by treatment with 20% piperidine in DMF for 30 min at room temperature.
3. The internally deblocked peptide resin is then sequentially washed with DMF, iPOH, and DCM.
4. The amide bridge is formed by adding 6 Eq of BOP to the peptide resin in 6 mL of NMP/g of resin in the presence of 6 Eq of triethylamine. This coupling is allowed to proceed for 2 h at room temperature with constant mixing (longer reaction time may be required).
5. Following the cyclization, the Kaiser test should be negative, indicating quantitative cyclization of the internal primary amine.
6. Synthesis on the remainder of the molecule by a normal Boc-based procedure can resume following the cyclization. This internal cyclization will result in the loss of the mass of a water molecule (18 mass units) from the final mol wt of the deprotected peptide.

7. Cleavage of the peptide requires no special precautions other than those required by the amino acids present in the entire peptide sequence.
8. It is also possible to cyclize the molecule by positioning only the Boc-Glu(Fm)-OH or Boc-Asp(Fm)-OH in the molecule, and removing the N-terminal Boc group. Treatment of the resin with piperidine to remove the Fm group, followed by BOP cyclization as described above will result in the formation of cyclic molecule involving the α-amino group of the molecule and the internal side-chain carboxyl group. This will result in a peptide refractory to Edman degradation or further elongation.
9. It is also possible to substitute alternative differentially protected amino acids, such as Boc-Asu(Fm)-OH (Asu = aminosuberic acid), instead of the Asp and Glu derivatives.
10. Head-to-tail cyclization is also possible using a special resin where the side chain of Glu or Asp is linked to the resin and the C-terminal carboxyl group is protected with the Fm group. Using the same procedure as described above will result in either a head-to-tail arrangement or a side chain-to-tail orientation.

4. Notes

1. Chemical modification reactions involving fatty acids often result in peptides with solubility problems. Aggregation may create a major solubilization problem that may require more drastic measures, such as neat TFA or DMSO, to achieve satisfactory solubilization.
2. Peptides containing the sulfated Tyr ester may be conveniently prepared by Fmoc strategy using the Fmoc-Tyr(SO_3 Ba salt)-OH available from Bachem Bioscience. Cleavage of the final peptide is carried out in 50% TFA solution in DCM containing 5% 1,2 ethanedithiol and other scavengers as required by the other amino acid residues.
3. Peptides containing the amino acid γ-carboxyglutamic acid can also be prepared by an Fmoc strategy. This derivative Fmoc-Gla$(OtBu)_2$-OH is available from Bachem Bioscience. Cleavage of the peptide from the resin is accomplished in 40–50% TFA in DCM containing 5% 1,2 ethanedithiol and other appropriate scavengers *(29,30)* as required by the other amino acids present.
4. Dabcyl-EDANS substrates are very insoluble in most aqueous buffers. We routinely solubilize these compounds in DMF and dilute this into the aqueous buffer we are using.

References

1. Means, G. E. and Feeny, R. E. (1971) *Chemical Modification of Proteins*, Holden-Day, San Francisco, CA.
2. Merrifield, R. B. (1986) Solid-phase synthesis. *Science* **232,** 341–347.

3. Fields, G. B. and Noble, R. L. (1990) Solid-phase peptide synthesis utilizing 9-fluorenylmethoxycarbonyl amino acids. *Int. J. Peptide Prot. Res.* **35,** 161–214.
4. Nutt, R. F., Brady, S. F., Darke, P. L., Ciccarone, T. M., Colton, C. D., Nutt, E. M., Rodkey, J. A., Bennet, C. D., Waxman, L. H., Sigal, I. S., Anderson, P. S., and Veber, D. F. (1988) Chemical synthesis and enzymatic activity of a 99-residue peptide with a sequence proposed for the human immunodeficiency virus protease. *Proc. Natl. Acad. Sci. USA* **85,** 7129–7133.
5. Schneider, J. and Kent, S. B. H. (1988) Enzymatic activity of a synthetic 99 residue protein corresponding to the putative HIV-1 protease. *Cell* **54,** 363–368.
6. Lewis, I. C., Aebersold, R., Ziltener, H., Schrader, J. W., Hood, L., and Kent, S. B. H. (1986) Automated chemical synthesis of a protein growth factor for hemopoetic cells, Interleukin-3. *Science* **231,** 134–1139.
7. Garsky, V. M., Lumma, P. K., Freidinger, R. M., Pitzenberger, S. M., Randall, W. C., Veber, D. F., Gould, R. J., and Friedman, P. A. (1989) Chemical syntheis of echistatin, a potent inhibitor of platelet aggregation from *Echis carinatus*: synthesis and biological activity of selected analogs. *Proc. Natl. Acad. Sci. USA* **86,** 4022–4026.
8. Pennington, M. W., Kem, W. R., and Dunn, B. M. (1990) Synthesis and biological activity of six monosubstituted analogs of a sea anemone polypeptide neurotoxin. *Peptide Res.* **3,** 228–232.
9. Edman, P. (1950) Method for determination of amino acid sequence in peptides. *Acta Chem. Scand.* **4,** 283–293.
10. Edman, P. (1956) On the mechanism of phenyl isothiocyanate degradation of peptides. *Acta Chem. Scand.* **10,** 761–768.
11. Kaiser, E., Colescott, R. L., Bossinger, C. D., and Cook, P. I. (1970) Color test for detection of free terminal amino groups in the solid phase synthesis of peptides. *Anal. Biochem.* **34,** 595–598.
12. Finn, F. M., Titus, G., and Hofmann, K. (1984) Synthesis of biotinylated and dethiobiotinylated insulin. *Biochemistry* **23,** 2547–2553.
13. Scott, D., Nitecki, D. E., Kindler, H., and Goodman, J. W. (1984) Immunogenicity of biotinylated hapten-avidin complexes. *Mol. Immunol.* **21,** 1055–1060.
14. Lobl, T. J., Deibel, M. R., and Wu, A. W. (1988) On-resin biotinylation of chemically synthesized proteins for one-step purification. *Anal. Biochem.* **170,** 502–511.
15. Jentoft, N. and Dearborn, D. G. (1983) Protein labeling by reductive alkylation. *Methods Enzymol.* **91,** 570–579.
16. Kaljuste, K. and Unden, A. (1992) *N*-monomethylation of peptides on solid phase, in *Peptides 1992* (Schnieder, C. H. and Eberle, A. N., eds.), Escom, Leiden, Netherlands, in press.
17. Meldal, M. and Breddam, K. (1991) Anthranilamide and nitrotyrosine as a donor-acceptor pair in internally quenched fluorescent substrates for endoproteases: multicolumn synthesis of enzyme substrates for subtilisn Carlsberg and pepsin. *Anal. Biochem.* **195,** 141–147.
18. Toth, M. V. and Marshall, G. R. (1990) A simple continuous fluorometric assay for HIV protease. *Int. J. Peptide Prot. Res.* **36,** 544–550.
19. Parker, J. M. R. and Hodges, R. S. (1985) Photoaffinity probes provide a general method to prepare synthetic peptide-conjugates. *J. Prot. Chem.* **3,** 465–478.

20. Gorka, J., McCourt, D. W., and Schwartz, B. D. (1989) Automated synthesis of a C-terminal photoprobe using combined Fmoc and *t*-Boc synthesis strategies on a single automated peptide synthesizer. *Peptide Res.* **2,** 376–380.
21. Kauer, J. C., Erickson-Viitanen, S., Wolfe, H. R., Jr., and DeGrado, W. F. (1986) *p*-Benzoyl-phenylalanine, a new photoreactive amino acid: photoaffinity labeling of calmodulin with a synthetic calmodulin-binding peptide. *J. Biol. Chem.* **261,** 10,695–10,700.
22. Lifson, J. D., Hwang, K. M., Nara, P. L., Fraser, B., Padgett, M., Dunlop, N. M., and Eiden, L. E. (1988) Synthetic CD4 peptide derivatives that inhibit HIV infection and cytopathicity. *Science* **241,** 712–716.
23. Felix, A. M., Wang, C. T., Heimer, E. P., and Fournier, A. (1987) Applications of BOP reagent in solid-phase peptide synthesis: solid-phase side-chain to side-chain cyclizations using BOP reagent. *Int. J. Peptide Prot. Res.* **31,** 231–238.
24. Mergler, M., Tanner, R., Gostelli, J., and Grogg, P. (1988) Peptide synthesis by a combination of solid-phase and solution methods I: a new very acid labile anchor group for the solid phase synthesis of fully protected peptides. *Tetrahedron Lett.* **29,** 4005–4008.
25. Wang, G. T., Matayoshi, E., Huffaker, H. J., and Kraft, G. A. (1990) Design and synthesis of new fluorogenic HIV protease substrates on resonance energy transfer. *Tetrahedron Lett.* **31,** 6493–6496.
26. Matayoshi, E., Wang, G. T., Kraft, G. A., and Erickson, J. (1990) Novel fluorogenic substrates for assaying retroviral proteases by resonance energy transfer. *Science* **247,** 954–958.
27. Ho, P., Slavazza, D., Chang, D., Bassi, K., and Chang, K. (1990) Synthesis of cyclic ($Asu^{1,7}$)-eel calcitonin by segment condensation on solid support, in *Peptides: Chemistry, Structure and Biology* (Rivier, J. and Marshall, G., eds.), Escom, Leiden, Netherlands, pp. 993–995.
28. Spinella, M. J., Malik, A. B., Everitt, J., and Anderson, T. T. (1991) Design and synthesis of a specific endothelin 1 antagonist: effects on pulmonary vasoconstriction. *Proc. Natl. Acad. Sci. USA* **88,** 7443–7446.
29. King, D. S., Fields, C. G., and Fields, G. B. (1990) A cleavage method which minimizes side reactions following Fmoc solid phase peptide synthesis. *Int. J. Peptide Prot. Res.* **36,** 255–266.
30. Rivier, J., Galyean, R., Simon, L., Cruz, L. J., Olivera, B. M., and Gray, W. R. (1987) Total synthesis and further characterization of the gamma-carboxyglutamate-containing sleeper peptide from *Conus geographus*. *Biochemistry* **26,** 8508–8512.

[illegible] and Scott, [illegible] (1988) [illegible] system using conditioned [illegible]

[illegible]

[illegible] and [illegible] (1985) [illegible]

[illegible] and [illegible] (19[illegible]) [illegible]

CHAPTER 9

Synthesis of Phosphopeptides Containing *O*-Phosphoserine and *O*-Phosphothreonine

Anatol Arendt and Paul A. Hargrave

1. Introduction

Phosphorylation and dephosphorylation of proteins represents one of the most widespread and important reactions in the regulation of cellular processes. Specific serine, threonine, or tyrosine residues in substrate proteins become phosphorylated by the action of protein kinases that catalyze the transfer of phosphate from high-energy nucleoside triphosphate. Major proteins in bones, teeth, eggs, and milk are highly phosphorylated. Preparation of phosphopeptides related to sequences of phosphoproteins is important for the study of their properties. Enzymatic methods for the synthesis of phosphopeptides can be useful, but are limited to very specific peptide sequences.

Chemical phosphorylation was reviewed by A. W. Frank in 1984 *(1)*, but at that time, only phospho-amino acids and phosphodipeptides were prepared. In the same year, Alewood et al. *(2)* introduced the use of the dibenzyl-blocking group to protect phosphoserine during peptide synthesis. Unfortunately, the dibenzyl group proved to be insufficiently stable during typical conditions of solid-phase peptide synthesis *(3)*. Arendt and Hargrave *(4)* used the more stable diphenyl protection of phosphoserine and phosphothreonine that made possible synthesis of phosphopeptides by solid phase. Diphenyl triesters were deprotected by

From: *Methods in Molecular Biology, Vol. 35: Peptide Synthesis Protocols*
Edited by: M. W. Pennington and B. M. Dunn Copyright ©1994 Humana Press Inc., Totowa, NJ

tetrabutylammonium fluoride (TBAF), and then the peptide was cleaved from the resin using the standard HF method (*see* Chapter 4). Penta and nonapeptides containing both phosphoserine and phosphothreonine were prepared using this method. However, this procedure is sometimes difficult to control, and some nonphosphorylated peptides are obtained when contact with TBAF is prolonged. Catalytical hydrogenolysis of diphenyl phosphopeptides yielded good results and eliminated this inconvenience *(5)*. This also made it possible to deprotect the diphenyl phosphopeptides following HF cleavage of the peptide from the resin. This modification was used to synthesize two heptaphosphopeptides, $PSer^5$ and $PThr^5$ kemptide *(6)*. Their identity was verified by enzymatic synthesis and mass spectrometry (*see* Chapter 7, *PAP*). Four undecaphosphopeptides (two with PSer and two with PThr) were synthesized by the same procedure and used as substrates for casein kinase II *(7)*. Seven different monophosphorylated peptides from the sequence of bovine rhodopsin have been made and tested as substrates for rhodopsin kinase *(8)*. Although this method has been very useful, we have also found that it has some limitations. When we synthesized the myelin peptide AcAlaSerAlaGln LysArgPro(P)SerGlnArgSerLysTyrNH_2 and then removed the diphenyl-blocking group by catalytic hydrogenolysis, we observed that the tyrosine phenyl group was also catalytically reduced (Arendt, A. unpublished results). Also, the hydrogenolysis of longer (>15) and hydrophobic peptides is more difficult, slower and synthetic yields are much smaller than with shorter ones. This chapter describes the synthetic method we have developed for preparation of *N*-(*t*-butoxycarbonyl)-*O*-(diphenylphosphono)-L-serine and *N*-(*t*-butoxycarbonyl)-*O*-(diphenylphosphono)-L-threonine, and the application of these substrates in solid-phase synthesis of peptides. Additional methods for synthesis of phosphopeptides are described elsewhere *(6,9)*.

2. Materials

t-Butoxycarbonyl (Boc) serine and threonine may be obtained from Bachem Bioscience Inc. (Philadelphia, PA); trifluoroacetic acid (TFA), pyridine, and triethylamine (NEt_3) from Fisher Scientific (Pittsburgh, PA); hydrogen fluoride (HF) from Matheson (Secaucus, NJ); and benzyl chloride, diphenyl chlorophosphate, dicyclohexylamine (DCHA), 10% palladium on charcoal (Pd/C), and platinum oxide from Aldrich (Milwaukee, WI).

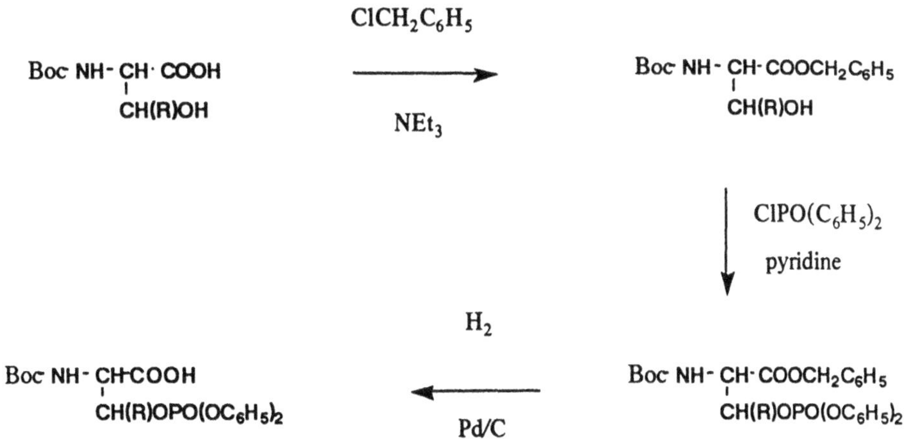

R = H (Serine) or R = $-CH_3$ (Threonine)

Scheme 1.

3. Methods

3.1. Preparation of the Protected Phosphoserine and Phosphothreonine Derivatives

This is a three-step synthesis involving:

1. Temporary protection of the reactive carboxyl group of BocSer or BocThr;
2. Phosphorylation of the free hydroxyl group; and
3. Selective deprotection of the carboxyl group (Scheme 1).

3.1.1. Synthesis of N-(t-Butoxycarbonyl)-O-(Diphenylphosphono)-L-Serine Benzyl Ester

1. Dissolve 100 g (0.339 mol) of Boc-L-SerOBzl in 750 mL of pyridine.
2. Cool the mixture in an ice bath, and add 100 g (0.373 mol) of diphenylphosphochloridate slowly with stirring.
3. Remove the cooling bath, and stir mixture overnight at room temperature.
4. Transfer the mixture to round-bottom flask, and evaporate to solid on rotatory evaporator.
5. Dissolve the residue in 800 mL of chloroform, wash 2× water, 2× 1*M* HCl, 2× water (2 L each), and evaporate organic layer on rotatory evaporator (without drying).
6. Dissolve crystalline mass in 400 mL of warm ethyl acetate, and add equivalent amount of petroleum ether. Let crystallize.

7. Remove crystals by filtration, wash them with a small amount of petroleum ether, and dry. You can obtain 166–170 g of product containing only a small amount (<1%) of substrate.
8. For a better quality product, recrystallize from isopropanol (74 g/L). After this, 159–163 g (89–91%) pure product can be obtained.

3.1.2. Synthesis of N-(t-*Butoxycarbonyl)*-O-*(Diphenylphosphono)*-L-*Serine*

1. Suspend 15 g (28.4 mmol) of *N*-(*t*-butoxycarbonyl)-*O*-(diphenylphosphono)-L-serine benzyl ester and 850 mg 10% Pd/C in the mixture of 50 mL ethyl acetate and 50 mL of methanol in pressurized bottle.
2. Pressurize the bottle with hydrogen gas at 4.05 bar (4 atm), and mix the residue magnetically. When the mixture becomes homogeneous, the hydrogenolysis is complete.
3. Filter the solution after ~3 h of reaction.
4. Evaporate solvent *in vacuo* by rotatory evaporator.
5. Dissolve the resulting oil in 40 mL of warm isopropanol. Add water until the solution becomes slightly cloudy. Keep mixture in refrigerator (at about 5°C) overnight.
6. Filter crystals and wash them with water. You can obtain 11.5–12 g (92–96%) of crystalline product.

Hydrogenolysis can be performed with good results also under atmospheric pressure, but more solvent must be used and the reaction is slower. For details, *see* ref. *6*. For longer storage, protected phosphoserine can be converted to the stable dicyclohexylammonium (DCHA) salt *(6)*. Boc-Ser[PO(OPh)$_2$]-OH·DCHA is commercially available from AminoTech Inc. (Canada).

3.1.3. Synthesis of N-(t-*Butoxycarbonyl)*-L-*Threonine Benzyl Ester*

1. Using conditions similar to that for the serine analog, from 38.9 g (0.177 mol) of BocThr, 43.4 mL (0.311 mol) triethylamine and 35.1 mL (0.305 mol) of benzyl chloride in 273 mL of ethyl acetate 47.7–50 g (87–91%) of product can be obtained.
2. The syrup of BocThrOBzl crystallizes very slowly and can be used for the next step without further purification. Ethyl ether/hexane can be used for crystallization of product.

3.1.4. Synthesis of N-(t-*Butoxycarbonyl)*-O-*(Diphenylphosphono)*-L-*Threonine Benzyl Ester*

From 46.4 g (0.15 mol) BocThrOBzl and 44.3 g (0.165 mol) diphenylphosphochloridate in 332 mL of pyridine, using the procedure for the serine analog, 75–77 g (92–95%) of noncrystalline, but chromatographically homogeneous product can be obtained.

3.1.5. Synthesis of DCHA Salt of N-*(*t-*Butoxycarbonyl)*-O-*(Diphenylphosphono)*-L-*Threonine*

1. 15.4 g (28.4 mmol) of noncrystalline protected phosphothreonine compound from the previous synthesis can be hydrogenolyzed under similar conditions to that for the serine analog. After evaporation of the solution, the resulting heavy oil can be converted to a crystalline dicyclohexylammonium salt.
2. Dissolve product in 350 mL ethyl ether/hexane (1:1 v/v), and add 5.6 mL (28 mmol) of dicyclohexylamine.
3. Let the solution crystallize at 0°C (~12 h).
4. Product can be recrystallized from a mixture of isopropanol/ethyl ether. Yield is 15–16 g (85–90%).

3.1.6. Conversion of DCHA Salt of N-*(*t-*Butoxycarbonyl)*-O-*(Diphenylphosphono)*-L-*Serine or* N-*(*t-*Butoxycarbonyl)*-O-*(Diphenylphosphono)*-L-*Threonine to Free Acid*

Before peptide synthesis, the necessary amount of Boc-Ser[PO(OPh)$_2$]-OH or Boc-Thr[PO(OPh)$_2$]-OH is first liberated from the salt:

1. Suspend the calculated amount of DCHA salt of phosphoserine or phosphothreonine derivative in ethyl acetate in a separatory funnel.
2. Wash this with 1*M* sulfuric acid solution and separate layers.
3. Wash the organic phase 2× with water, dry with $MgSO_4$, and evaporate *in vacuo*.
4. Dissolve in calculated volume of a suitable solvent and use for synthesis of phosphopeptide.

3.2. Synthesis of Phosphopeptides

Synthesis of phosphopeptides on solid phase can be performed manually, on an automated synthesizer, or on a multiple-peptide synthesizer, using any procedure suitable for Boc amino acids. Incorporation of Boc-Ser[OP(OPh)$_2$] and Boc-Thr[OP(OPh)$_2$] into the peptide chain proceeds in the same manner as nonphosphorylated amino acids.

3.3. Cleavage of Phosphopeptides from Resin

Peptides can be cleaved from the resin using the standard water-free HF (*see* Chapter 4) or TFMSA (*see* Chapter 5) procedure. Phenyl protection of phosphoserine or phosphothreonine is usually stable under these conditions. In some cases, 10–20% of nonphosphorylated peptide and monophenyl phosphopeptide is present (calculated from HPLC chromatogram; *see* Chapter 3, *PAP*). Product can be easily separated from the substrate at this step because of the great difference in polarity of these compounds.

3.4. Cleavage of Phenyl Group from Phosphopeptides

1. Dissolve peptide with phenyl-protected phospho group in a mixture of 40% trifluoroacetic acid in acetic acid (optimal concentration 10–40 mL/mmol).
2. Add 482 mg of amorphous PtO_2/phospho-group/mmol (1 Eq).
3. Perform hydrogenolysis for 24 h at room temperature under 4.05 bar (4 atm) of hydrogen pressure. Under atmospheric pressure, the process is too slow.
4. Evaporate the mixture to dryness under vacuum, suspend in water, and remove the catalyst by filtration.

3.5. Phosphopeptide Purification

Phosphopeptide can be purified by preparative HPLC on a reverse-phase or ion-exchange column using a mixture of volatile solvents. A good method for separation of phosphopeptides from nonphosphorylated impurities is affinity chromatography on Fe^{2+} Chelex gel (iminodiacetic acid epoxy activated Sepharose 6B, Sigma) *(6)*.

Free phosphopeptide cannot be stored dry, even in the freezer, for a long time. However, storing the DCHA salt of the phosphopeptide resulted in no observable dephosphorylation for at least 1 yr of storage at –20°C.

Acknowledgments

This work was supported in part by research grants EY06225 and EY06226 from the National Eye Institute of the National Institutes of Health, an unrestricted departmental award from Research to Prevent Blindness, Inc., and an International Human Frontier Science Program award. P. A. H. is Francis N. Bullard Professor of Ophthalmology.

References

1. Frank, A. W. (1984) Synthesis and properties of *N*-, *O*-, and *S*-phospho derivatives of amino acids, peptides, and proteins. *CRC Crit. Rev. Biochem.* **16,** 51–101.
2. Alewood, P. F., Perich, J. W., and Johns, R. B. (1984) Preparation of *N*-(*t*-Butoxycarbonyl)-*O*-(dibenzylphosphono)-L-serine. *Aust. J. Chem.* **37,** 429–433.
3. Alewood, P. F., Perich, J. W., and Johns, R. B. (1984) A novel approach to phosphopeptide synthesis—preparation of Glu-PSer-Leu. *Tetrahedron Lett.* **25,** 987–990.
4. Arendt, A. and Hargrave, P. A. (1985) Solid-phase synthesis of phosphopeptides: synthesis of phosphopeptides from the carboxyl-terminus of rhodopsin, in *Pep-*

tides: Structure and Function (Deber, C. M., Hruby, V. I., and Kopple, K. P., eds.), Pierce Chemical Co., Rockford, IL, pp. 237–240.

5. Perich, J. W., Valerio, R. M., and Johns, R. B. (1986) Solution-phase synthesis of an *O*-phosphoseryl-containing peptide using phenyl phosphorotriester protection. *Tetrahedron Lett.* **27,** 1373–1376.
6. Arendt, A., Palczewski, K., Moore, W. T., Caprioli, R. M., McDowell, J. H., and Hargrave P. A. (1989) Synthesis of phosphopeptides containing *O*-phosphoserine or *O*-phosphothreonine. *Int. J. Peptide Protein Res.* **33,** 468–476.
7. Litchfield, D. W., Arendt, A., Lozeman, F. J., Krebs, E. G., Hargrave, P. A., and Palczewski K. (1990) Synthetic phosphopeptides are substrates for casein kinase II. *FEBS Lett.* **261,** 117–120.
8. Adamus, G., Arendt, A., Hargrave, P. A., Heyduk, T., and Palczewski, K. (1993) The kinetics of multi-phosphorylation of rhodopsin. *Arch. Biochem. Biophys.* **304,** 443–447.
9. Perich, J. W. (1991) Synthesis of *O*-Phosphoserine- and *O*-Phosphothreonine-containing peptides. *Methods in Enzymology* **201,** 225–233.

CHAPTER 10

Solid-Phase Synthesis of Phosphorylated Tyr-Peptides by "Phosphite Triester" Phosphorylation

Michael W. Pennington

1. Introduction

Phosphorylation of tyrosine, serine, and threonine residues by protein kinases has been shown to be a key step in the regulation of many cellular events *(1)*. A better understanding of the molecular basis of this regulatory step is needed in order to gain insight into the specificity of the kinases that control these events.

Chemical methodology for preparing the phosphate esters of Tyr, Ser, and Thr has recently improved making it possible to synthesize peptides containing one or more of these residues routinely *(2–6)*. Some of these methods require the preparation of diphenyl-phosphono esters of Thr and Ser, which are reported to be stable to HF treatment (*see* Chapter 4) making a Boc-synthesis strategy possible. These derivatives are ultimately deblocked by hydrogenation *(7)*. Use of this strategy for preparing Tyr-P peptides is hampered by reduction of the aromatic ring of Tyr. However, a newer strategy has recently been developed that allows the convenient synthesis of phosphotyrosine. This method, developed by Perich and coworkers *(8)*, utilizes an Fmoc strategy where the Tyr residue to be phosphorylated is incorporated with an unprotected side chain. Following completion of synthesis of the remaining sequence, the peptide resin is subjected to phosphitylation and subsequent oxidation. Early reports have shown that Ser, Thr, and Tyr could be successfully incorporated

From: *Methods in Molecular Biology, Vol. 35: Peptide Synthesis Protocols*
Edited by: M. W. Pennington and B. M. Dunn

with unprotected side chains into the peptide in Fmoc strategies either by carbodiimide/HOBt or BOP reagent *(9–11)*. Automated syntheses of phosphopeptides using this type of procedure have also recently been reported *(12,13)*. Additionally, peptides containing multiple phosphorylated residues have been prepared using this method as a "global" phosphorylation method *(14)*. Although these procedures were all developed in other laboratories and are all well documented in the literature, since we routinely prepare our phosphopeptides by this method, it is our desire to include a chapter describing this methodology to make our volume as complete and up to date as possible.

2. Materials

1. An Fmoc strategy is necessary to use this procedure. All Fmoc derivatives are available from Bachem Bioscience (King of Prussia, PA). The special Fmoc amino acids with unprotected side chains (Fmoc-Ser-OH, Fmoc-Thr-OH and Fmoc-Tyr-OH) are also available from the same vendor.
2. The phosphitylation reagents di-*tert*-butyl-*N,N* diethyl phosphoramidite and 1H-tetrazole are available from Aldrich (Milwaukee, WI). These reagents must be kept frozen and dry.
3. The oxidation reagents used to convert the phosphite triester to the phosphate most commonly used are *m*-chloroperbenzoic acid and *t*-butyl hydroperoxide. These reagents are also available from Aldrich.
4. All solvents required from peptide synthesis are available from standard commercial sources.
5. Commercially prepared protected Boc-diphenylphosphonate derivatives of Ser, Thr, and dimethylphosphonate-Tyr are available from Bachem Bioscience. Additionally, the Fmoc-Tyr dimethylphosphonate derivative is also available.

3. Methods

This short section describes the standard procedure for preparing phosphorylated peptides containing Tyr using an Fmoc-based synthesis strategy. An excellent review of Fmoc-strategy solid-phase synthesis may be found in Atherton and Sheppard *(15)*. This phosphorylation procedure will work well on peptides that do not contain oxidation-prone residues, such as Met, Cys, and Trp. In cases where these residues are present, a Boc strategy, such as that described by Kitas et al. *(5)*, may be more appropriate. The incorporation of the Boc-protected dimethylphospho derivative instead of the derivative containing the unprotected side-chain hydroxyl group is suggested. This will eliminate the postphosphitylation oxidation step and help to preserve these susceptible residues.

3.1. Preparation of the Protected Phosphorylated Peptide

Initially, the phosphite triester is formed on the unprotected phenolic hydroxyl group of Tyr. Subsequently, the phosphitylated derivative is oxidized to the phosphonate derivative by treatment with an oxidant.

1. Synthesis of the desired peptide sequence is performed using a standard Fmoc strategy, except that at the position where the phosphorylated residue is located, an Fmoc amino acid derivative of Tyr is incorporated with the side chain unprotected (*see* Note 1). This coupling may be mediated by either standard carbodiimide/HOBT methods *(16)* or with BOP or similar uronium-type reagents *(17)*.
2. Synthesis then resumes until the entire peptide chain is assembled. We recommend incorporating the final amino acid as the appropriate Boc-*t*Bu derivative. These groups are cleaved when the entire peptide resin is deprotected and cleaved. Maintaining the blocked N-terminus prevents undesired reactions at this position by incorporating acid-labile-protecting groups. This eliminates exposure to a final treatment of piperidine to remove the N-terminal Fmoc group.
3. Prepare a fresh solution of di-*tert*-Butyl *N,N* diethyl-phosphoramidite in fourfold excess over the total theoretical peptide amount to be phosphorylated, and dilute this into 1–2 mL/g of resin of dry DCM. The amount of reagent to be prepared is determined by the relative substitution of the solid-phase resin times the amount of resin used for the synthesis. For example, 2 g of resin with an initial substitution of 0.65 mmol/g would result in 1.3 mmol of peptide, thus requiring 5.2 mmol of the phosphoramidite. **Always remove a portion of resin in case a problem occurs during the phosphorylation procedure or cleavage.**
4. This solution is added to the peptide resin preswollen in dry DCM.
5. 1H-Tetrazole is added directly to the reaction vessel in a tenfold excess over the total amount of the phosphoramidite.
6. The reaction vessel is mixed for 1–1.5 h at room temperature. Following this reaction time, the vessel is drained and the peptide resin washed with three DCM washes.
7. To complete the phosphorylation procedure:
   a. The phosphitylated derivative is oxidized *in situ* by treatment with a solution of 85% *m*-chloroperbenzoic acid (3–5 Eq based on the amount of peptide) in 5 mL of DCM/g of resin for 10–30 min at room temperature.

b. Alternatively, oxidation can be performed with a solution of *t*-butyl hydroperoxide in DMF at a ratio of 6–20 Eq/Eq of peptide resin for 30–60 min at room temperature (*see* Note 2). This step may be used successfully with Met peptides if the temperature is lowered to 4°C.

8. Following the oxidation, the reaction vessel is drained and washed with 5 × 5 mL/g DCM, and dried.

3.2. Cleavage of the Final Phosphopeptide

Following the completion of the phosphorylation reaction, the peptide resin is ready for final cleavage and deprotection. As with all Fmoc-based peptides, the scavenger cocktail is dictated by the types of protecting groups present. We have found the "Reagent K" cocktail to be the most satisfactory for phosphopeptides *(18)*. The cleavage of Fmoc-synthesized peptides is covered in greater detail elsewhere in this volume (*see* Chapter 5).

1. Prepare a solution of 82.5% trifluoroacetic acid, 5% thioanisole, 5% phenol, 5% H_2O, and 2.5% 1,2-ethanedithiol at a ratio of 10 mL of cleavage cocktail/g of resin.
2. Cool this solution on ice for 1 h.
3. Initiate the cleavage by adding this cold solution to a flask containing the peptide resin and a magnetic stir bar. Actuate the stir bar, and allow the cleavage vessel to warm up to room temperature.
4. Stir this solution for 1–2.5 h depending on the number of Arg(Pmc) residues present. Longer cleavage time is sometimes necessary for peptides containing more than three or four Arg residues.
5. Filter the peptide through a fritted funnel, and wash the resin fines with TFA.
6. Precipitate the peptide into ice-cold ether.
7. Analyze the crude peptide by RP-HPLC (*see* Chapter 3, *PAP*), and compare the retention time vs the unphosphorylated form. Phosphopeptides generally elute significantly earlier than the nonphosphorylated peptide on a C_{18}-column.
8. FAB/MS analysis (*see* Chapter 7, *PAP*) of the product (*see* Notes 3 and 4) should result in a mol-wt increase of 79.5 mass units/phosphate incorporated. If the mass is 16 mass units low, the oxidation reaction was unsuccessful and should be repeated for a longer period with a greater excess of *m*-chloroperbenzoic acid or *t*-butyl hydroperoxide (*see* Note 5).

4. Notes

1. An Fmoc strategy must be employed when using this methodology.
2. It has been reported that use of *t*-butyl hydroperoxide was successfully employed in peptides containing Met and Cys *(19)*. In this scenario, the

Cys(Trt) was resistant to oxidation, and the Met was only partially oxidized.

3. FAB/MS analysis is required for all peptides containing Tyr-P. Also, the UV spectrum of Tyr-P is shifted with a λ_{max} at 266 nm compared to Tyr with a λ_{max} at 275 nm *(20)*.
4. Standard amino acid hydrolysis procedures destroy the phospho-ester linkage, and the amino acid derivative will be detected as the standard underivatized form with the minor amount of destruction associated with hydrolysis of these hydroxy residues.
5. Peptides containing any of oxidation-sensitive residues, such as Met, Cys, and Trp, may be synthesized with a Boc strategy using the appropriate protected diphenylphosphonate derivative of Ser or Thr and dimethylphosphonate for Tyr. These derivatives are commercially available. The phenyl-protecting groups are stable to the cleavage reaction and removed by catalytic hydrogenation as described in the preceding chapter. The Tyr derivative may be incorporated as the dimethylphosphonate, and the protecting groups for the entire peptide as well as the dimethylphosphonate-Tyr are cleaved by TFMSA/TFA/thioanisole *(5)*. Peptides containing phosphoserine or phosphothreonine can also be prepared with this type of strategy, but the phosphorylation proceeds much slower than with the phosphotyrosine residue. Peptides containing these residues may be best prepared by the method described in Chapter 9.

References

1. Cohen, P. (1982) The role of phosphorylation in neural and hormonal control of cellular activity. *Nature* **296,** 613–620.
2. De Bont, H. B. A., Veeneman, G. H., Van Boom, J. H., and Liskamp, R. M. (1987) Synthesis of phosphopeptides: a simple procedure for serine and threonine. *Recl. Trav. Chim. Pays Bas* **106,** 641, 642.
3. Bannwarth, W. and Trezeciak, A. (1987) A simple and effective chemical phosphorylation procedure for biomolecules. *Helv. Chim. Acta* **70,** 175–186.
4. Perich, J. W. and Johns, R. B. (1988) Di-*t*-Butyl *N,N*-diethylphosphoramidite and dibenzyl *N,N*-diethylphosphoramidite. Highly reactive reagents for the "phosphite-triester" phosphorylation of serine-containing peptides. *Tetrahedron Lett.* **29,** 2369–2372.
5. Kitas, E. A., Perich, J. W., Tregear, G. W., and Johns, R. B. (1990) Synthesis of *O*-phosphotyrosine-containing peptides. 3. Synthesis of H-Pro-Tyr(P)-Val-OH via dimethyl phosphate protection and use of improved deprotection procedures. *J. Org. Chem.* **55,** 4181–4187.
6. Valerio, R. M., Alewood, P. F., Johns, R. B., and Kemp, B. E. (1989) Synthesis of *O*-phosphonotyrosyl peptides. *Int. J. Peptide Protein Res.* **33,** 428–438.
7. Arendt, A., Palczewski, K., Moore, W. T., Caprioli, R. M., McDowell, J. H., and Hargrave, P. A. (1989) Synthesis of phosphopeptides containing *O*-phosphoserine and *O*-phosphothreonine. *Int. J. Peptide Protein Res.* **33,** 468–476.

8. Perich, J. W., Le Nguyen, D., and Reynolds, E. C. (1991) The facile synthesis of Ala-Glu-Tyr(P)-Ser-Ala by global Di-*t*-Butyl *N,N* Diethylphosphoramidite "phosphite-triester" phosphorylation of a resin bound peptide. *Tetrahedron Lett.* **32,** 4033–4034.
9. Fields, C. G., Fields, B., Noble, R. L., and Cross, T. A. (1989) Solid phase peptide synthesis of N^{15}-gramacidins A, B, and C and high performance liquid chromatographic purification. *Int. J. Peptide Protein Res.* **33,** 298–303.
10. Fournier, A., Wang, C. T., and Felix, A. M. (1988) Applications of BOP reagent in solid phase synthesis: advantages of BOP reagent for difficult couplings exemplified by a synthesis of [Ala^{15}]-GRF-1-29-NH_2. *Int. J. Peptide Protein Res.* **31,** 86–97.
11. Fournier, A., Danho, W., and Felix, A. M. (1989) Applications of BOP reagent in solid phase peptide synthesis: III. Solid-phase peptide synthesis with unprotected aliphatic and aromatic hydroxy amino acids using BOP reagent. *Int. J. Peptide Protein Res.* **33,** 133–139.
12. De Bont, H. B. A., Von Boom, J. H., and Liskamp, R. M. J. (1990) Automatic synthesis of phosphopeptides by phosphorylation on the solid phase. *Tetrahedron Lett.* **31,** 2497–2500.
13. Andrews, D. M., Kitchin, J., and Seale, P. W. (1992) The solid-phase synthesis of a range of *O*-phosphorylated peptides by post-assembly phosphitylation and oxidation, in *Peptides: Chemistry and Biology* (Smith, J. A. and Rivier, J. E., eds.), Escom, Leiden, Netherlands, pp. 619–620.
14. Perich, J. W. (1992) Efficient solid phase synthesis of mixed Thr(P)-, Ser(P)- and Tyr(P)-containing phosphopeptides by global "phosphite-triester" phosphorylation. *Int. J. Peptide Protein Res.* **40,** 134–140.
15. Atherton, E. and Sheppard, R. C. (1989) *Solid Phase Peptide Synthesis: A Practical Approach.* IRL, New York.
16. Konig, W. and Geiger, R. (1973) New catalysts in peptide synthesis, in *Chemistry and Biology of Peptides* (Meienhofer, J., ed.), Ann Arbor Science Publishers, Ann Arbor, MI, pp. 343–350.
17. Coste, J., Le Nguyen, D., and Castro, B. (1990) PyBOP: a new coupling reagent devoid of toxic byproducts. *Tetrahedron Lett.* **31,** 205–208.
18. King, D. S., Fields, C. G., and Fields, G. B. (1990) A cleavage method which minimizes side reactions following Fmoc solid phase synthesis. *Int. J. Peptide Protein Res.* **36,** 255–266.
19. Andrews, D. M., Kitchin, J., and Seale, P. W. (1991) Solid phase synthesis of a range of *O*-phosphorylated peptides by post-assembly phosphitylation and oxidation. *Int. J. Peptide Protein Res.* **38,** 469–475.
20. Turck, C. W. (1992) Identification of phosphotyrosine residues in peptides by high performance liquid chromatograhy on-line derivative spectroscopy. *Peptide Res.* **5,** 156–160.

CHAPTER 11

Design of Novel Synthetic Peptides Including Cyclic Conformationally and Topgraphically Constrained Analogs

Victor J. Hruby and G. Gregg Bonner

1. Introduction

The properties of a peptide in biological systems are dependent on its structure. Thus, our ability to use rational design for the generation of useful peptides is dependent on our commensurate ability to determine the specific relationships of molecular structure to biological activity. The ability to recognize these relationships is complicated by a variety of uncertainties, both in the assay systems and in the interpretation of data. One major complicating factor is the difficulty in ascertaining the three-dimensional structure of the peptide itself. Most peptides are inherently flexible and, thus, assume many conformations in solution. It is difficult to determine which of these conformations is responsible for the observed activity of the peptide, and many peptides may be active in more than one conformation. The use of conformational constraints has been helpful in elucidating these structure–function relationships. The logic is that if the peptide is restricted to a particular conformation or closely related family of conformations, then the activity measured will reflect that structure. Although a perfectly rigid molecule is impossible, by creating analogs with prescribed structural motifs, one can begin to ascribe certain biological activities to their causal structures. In favorable cases, knowledge of requirements for peptide activity at receptor- and enzyme-active sites can be obtained. From this knowledge, one can design peptides and

From: *Methods in Molecular Biology, Vol. 35: Peptide Synthesis Protocols*
Edited by: M. W. Pennington and B. M. Dunn

peptide mimetics of very high activity and specificity. Moreover, one can create molecules that have the elements for binding, but not for signal transduction, and therefore are antagonists. Also, one can create molecules that have the ability to distinguish between receptor types and subtypes. In short, the utilization of conformational constraints can be helpful in developing peptides with high affinity for their receptor, with antagonist activity, and with high selectivity. Indeed, this approach also can play a role in the properties of biodistribution, efficacy, and metabolic stability, among others.

Conformational constraints are basically of two types: (1) global: the alteration of the three-dimensional structure accessible over a large number of amino acid residues (conformational change); and (2) local: usually the alteration of the amino acid side-chain conformations (topographical change) or fixing the backbone conformation of one or two amino acid residues. Global constraints often consist of covalent peptide cyclization and pseudocyclization, or of stabilization of secondary hydrogen-bonded structures (such as α-helices and β-turns) *(1,2)*. Local constraints consist of short-range, usually covalent, cyclizations (within a single residue or between residues that are adjacent in the primary structure), rotamer bias, and peptide bond replacements, to name a few. Clearly, there can be some overlap between these two types of constraints.

Many methods for the introduction of conformation constraints exist. This chapter is intended to describe the basis for some of these techniques and to give examples of peptides systems to which they have been applied.

2. Global Constraints

2.1. Cyclization and Modification of Cyclic Structure

The cyclization of a peptide necessarily makes it less flexible. Through proper design principles, these cyclizations have provided peptides that are "locked" into their biologically relevant conformation. Cyclization has been instrumental in:

1. Development of potent receptor ligands;
2. Conversion of agonistic molecules to antagonists;
3. Receptor selectivity enhancement; and
4. Separation of multiple activities, for some examples.

The following sections will relate selected successes using this approach.

2.1.1. *Oxytocin*

Oxytocin (OT) is a neurohypophyseal hormone that is involved in regulation of uterine contraction and milk ejection, and was the first peptide hormone to be synthesized in the laboratory.

OT, H-Cys[1]-Tyr[2]-Ile[3]-Gln[4]-Asn[5]-Cys[6]-Pro[7]-Leu[8]-Gly[9]-NH_2

as a result of its cyclic structure, is already relatively conformationally restricted. Yet, OT (and des-amino OT) retains some flexibility, as shown by X-ray analysis *(3)* and NMR *(4,5)*, despite the disulfide constraint. The X-ray data indicated that the residues in positions 4 and 8 might be brought in close proximity. By forcing these residues into close proximity via a lactam bridge between a glutamic acid in position 4 and a lysine at position 8, researchers could investigate the importance of this juxtaposition. This bicyclic OT analog cyclo(1-6,4-8)[Mpa[1],Glu[4],Lys[8]]OT (where Mpa = β-mercaptopropionic acid) was one of the most potent antagonists in the uterine receptor assay, whereas its monocyclic counterpart was a very weak agonist *(6)*. Cyclization has also played a role in the development of potent antagonists of other biologically active peptides, such as GnRH *(7)*, LHRH *(8)*, bombesin *(9)*, α-melanotropin *(10)*, somatostatin at the opioid receptor *(11)*, and angiotensin *(12)* among others. The results of these cyclizations indicate that the peptides are constrained to a conformation that facilitates binding, but is incompatible with signal transduction. The OT analog incorporates two of the most common global constraints, a lactam constraint and a disulfide constraint. The synthesis of this analog is outlined in Scheme 1, and is essentially that of Hill et al. *(6)*. Bicyclization is being used more commonly now, and some investigators are even using bicyclic model peptides in studies of ionophores, enzyme-active sites, and receptor active sites *(13)*. Table 1 lists some bicyclic peptides and their activities; others will be discussed below.

2.1.1.1. Synthesis of a Bicyclic Analog of Oxytocin (Scheme 1) *(6)* [β-Mpa[1],Glu[4],Cys[6],Lys[8]]oxytocin

The linear peptide was made starting with 0.93 g of *p*-MBHA resin (1.1 mmol/g titratable amine, 1.0 mmol). The N^{α}-Boc amino acids were coupled to the resin using a twofold excess of amino acid, DCC and HOBt in DCM/DMF. The *p*-nitrophenyl ester of asparagine was added in fourfold excess and allowed to couple in the presence of fourfold excess HOBt in DMF. Coupling was monitored via a ninhydrin test (*see* Chapter 8). This

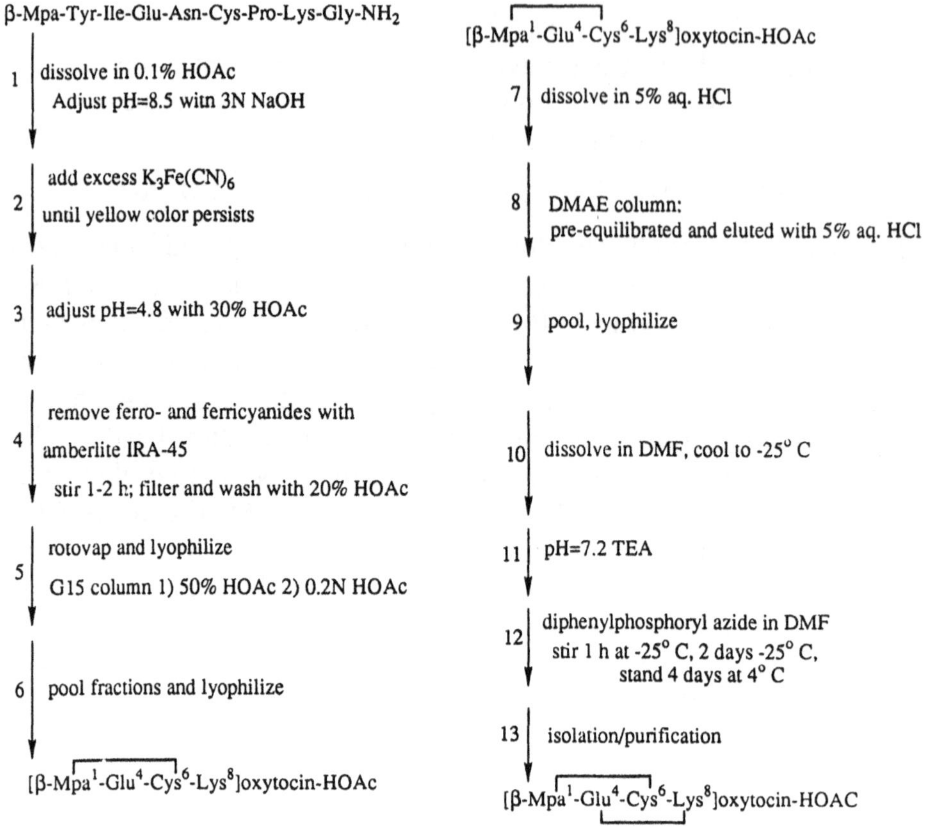

Scheme 1. Bicyclization of an oxytocin analog *(6)*.

generated the protected peptide resin β-Mpa(S-4-MeBzl)-Tyr(O-BrZ)-Asn-Cys(S-4-MeBzl)-Pro-Lys(2,4-Cl_2-Z)-Gly-MBHA resin (2.06 g). The peptide resin was washed with DCM, ETOH, and DCM, and dried *in vacuo*. 0.69g (0.33 mmol) was treated with anhydrous HF (10 mL) and anisole (3 mL) (45 min, 0°C) to deprotect and cleave the peptide (*see* Chapter 4) from the resin. After *in vacuo* removal of HF and anisole, the residue was washed with deaerated ethyl acetate (3 × 30 mL). The resin was then extracted with deaerated glacial acetic acid (50 mL) and three 30-mL portions of each of the following: 30% HOAc, 0.2*N* HOAc, and water. The aqueous extracts were combined and lyophilized to yield the crude linear peptide.

2.1.1.2. [β-MPA[1],GLU[4],CYS[6],LYS[8]]OXYTOCIN

Dissolve the white powder in 0.1% HOAc (1000 mL), which had been deaerated and bubbled with N_2 for 90 min. Adjust the pH to 8.5 with 3*N*

Table 1
Synthetic Bicyclic Peptides

Peptide	Activity	References
Cyclo(1-6,4-8)[Mpa^1,Glu^4,Lys^8]OT	Antagonist in uterine receptor; monocyclic was weak agonist	*6*
Cyclo(1-6,5-8)[Mpa^1,Phe^2,Asp^5]LVP	Potent antidiuretic; no pressor activity; monocyclics had pressor activity	*14*
Cyclo[des(Ala^1,Gly^2)(Cys^{5,12},D-Trp^8)-desamino(Cys^3)-descarboxy(Cys^{14})-dicarba^{3,14}]Somatostatin	High potency	*16*
Bicyclic Atriopeptin 103-125	Increased activity compared to the monocyclic parent	*15*
Endothelin	Outer cycle lactam analog is potent antagonist	*35*
Cyclo(Glu-Leu-Pro-Gly-Ser-Ile-Pro-ala)cyclo(1-γ-5-β)-Phe-Gly	Model peptide for study of several systems	*13*

NH_4OH, and oxidize the sulfhydryl groups with excess 0.01*N* $K_3Fe(CN)_6$ (40 mL) (as indicated by the persistence of the yellow color for 30 min). Stir the solution for 1 h, adjust the pH to 4.8 with 30% HOAc, and add anion-exchange resin (Amberlite IRA-45; 30-mL settled vol) to remove excess ferro- and ferricyanides. Stir the suspension for 2 h, and remove the resin by filtration and washing with 20% HOAc (3 × 25 mL). Reduce the volume by rotary evaporation and lyophilization to yield crude cyclic peptide

[β-Mpa^1,Glu^4,Cys^6,Lys^8]OT

The monocyclic peptide can be purified by sequential Sephadex G-15 columns (*see* Chapter 5, *PAP*) (2.65 × 100 cm). Use 50% HOAc as the eluent in the first column and 0.2*N* HOAc as the eluent in the second column. The fractions (3.5 mL) are monitored for peptide material (280 nm) and fractions with the major peak pooled and lyophilized to yield fluffy white powder (132.30 mg).

2.1.1.3. [β-Mpa^1,Glu^4,Cys^6,Lys^8]Oxytocin

Dissolve the monocyclic peptide-HOAc salt (100 mg, 0.1 mmol) in 5% aqueous HCl (2 mL), and apply to a dimethylaminoethylcellulose ion-exchange column (*see* Chapter 2, *PAP*) (previously equilibrated with 5% aqueous HCl). The column is then eluted with 5% HCl (aqueous), and the eluent monitored at 280 nm continuously. Pool the main peak and lyophilize. Cool a solution of the cyclic peptide hydrochloride salt (0.1 mmol) in DMF (100 mL) to –25°C with stirring before altering the pH to 7.2 (as determined with damp narrow-range pH paper) with TEA. Add diphenylphosphorylazide (DPPA) (26 µL, 0.12 mmol) in DMF (80 µL) and stir for 1 h at –25°C. The reaction vessel is then left to stand for 2 d at –25°C, and 4 d at 4°C. The pH can be kept in the range of 7–7.5 by TEA as needed throughout. TLC and ninhydrin tests are performed to determine if the reaction is complete. Add water (12 mL) and mixed-bed ion-exchange resin (AG 5018; 12-mL settled vol), and stir the mixture for 6 h. Filter the resin and remove the solvent under reduced pressure. Dissolve the residue in 3*M* aqueous HOAc (2 mL) before applying to a gel-filtration column (Sephadex G-25, 100 × 25 cm). Elute the column with 3*M* HOAc (aqueous) at 10 mL/h, and monitor the eluent continuously at 280 nm; 3.3 mL fractions are collected, and the product is found in fractions 79–90, which are pooled and lyophilized to yield 39 mg of bicyclic peptide. Purify the peptide further via HPLC with a linear gradient of 15–31% CH_3CN in aqueous 0.1% TFA over 16 min. This yields a fluffy white powder (13 mg) on lyophilization.

2.1.2. Vasopressin

Bicyclization was performed on vasopressin:

$$\text{H-}\overline{\text{Cys}^1\text{-Tyr}^2\text{-Phe}^3\text{-Gln}^4\text{-Asn}^5\text{-Cys}^6}\text{-Pro}^7\text{-(Lys or Arg)}^8\text{-Gly}^9\text{-NH}_2,$$

LVP or AVP

These nonapeptides are structurally related to OT, but have antidiuretic and vasoconstrictive properties. Positions 5 and 8 were cyclized to produce (1-6,5-8)[Mpa^1,Phe^2,Asp^5]LVP. It was suggested that such bicyclization would yield augmented activity because the putatively important interaction would always be present. The resultant LVP analog was a potent antidiuretic, yet was inactive as a pressor agent. This is in contrast to all the potent monocyclic antidiuretic LVP analogs in that the monocyclics all maintained pressor activity *(14)*. This cyclization apparently forces the peptide into a conformation that yields antidiuretic activity, yet does not allow alternative folding of the peptide into the shape required for pressor activity.

Another example of an increase in biological activity on bicyclization was seen with atriopeptin *(15)*. The bicyclic atriopeptin with a second ring closure was afforded by a disulfide bridge between residues 108 and

117. These compounds were more active than the parent compounds with EC_{50} from 0.05–3 μ*M* *(15)*.

2.1.3. Somatostatin

Somatostatin (SMS) is a tetradecapeptide that inhibits release of growth hormone and other factors. The bicyclic analog, des[Ala^1-Gly^2]-[$Cys^{5\text{-}12}$, D-Trp]desamino[Cys^3]-descarboxy[Cys^{14}]dicarba$^{3\text{-}14}$-somatostatin exhibited high activity *(16)*. Analogs of SMS also lead to the identification of a new class of highly potent μ opioid receptor antagonists. These peptides have the structure *(11)*

D-Phe^1-Cys^2-Tyr^3-D-Trp^4-Lys^5-Thr^6-Pen^7-Thr^8-NH_2 (CTP)

These peptides have a structure that fulfills the requirements for binding at the μ receptor as an antagonist, but only weakly at the δ-receptor. The constraint of the disulfide bond makes the alternative folding energetically impossible. These examples show that even in peptides that are naturally somewhat restricted in conformational mobility, the addition of further conformational constraints can lead to analogs that have novel and useful properties.

Except for an SMS analog above, the peptides that have been discussed have been cyclized via their side-chain moieties. The termini of the peptides are another synthetically useful location for cyclization. Cyclizations using direct amino terminal to carboxyl terminal bond formation have met with limited success (some SMS analogs, e.g., are exceptions), whereas cyclizations using one terminus and a side-chain group have yielded more rewarding results. The discussion below (and in Table 2) describes some successful products of monocyclization.

2.1.4. Enkephalins

Enkephalins are pentapeptides that bind to opioid receptors, and have the sequence Tyr^1-Gly^2-Gly^3-Phe^4-(Leu or Met)^5, [Leu^5]Enkephalin or [Met^5]Enkephalin. Binding studies have suggested that these peptides might have a compact structure when bound, and that cyclization might be a way to "freeze" the peptides into a compact bioactive structure. The cyclic analog described by Schiller and DiMaio (H-Tyr-c[N^γ-D-A_2bu-Gly-Phe-Leu]) had a MVD/GPI potency ratio of 5.8, as compared to about 1.6 for its linear counterpart *(17)*. The conclusion drawn from this information was that the cyclic molecule folded preferentially into the shape required for binding at one receptor type.

Table 2
Synthetic Cyclic Peptides

Cyclic peptides	Activity	References
Enkenphalins	Highly potent and selective	*17,18*
Somatostatin	Highly potent	*16,33*
Melanotropin	Superagonists	*21,33*
Leuteinizing hormone-releasing hormone	3X Increased antagonism over linear parent molecule	*8*
Cholecystokinin	Increased central receptor Selectivity compared to linear analogs	*24,25*
Hirudin	Greater antithrombin activity than linear parent	*30*
Parathyroid hormone-related protein	5X Increased agonism over linear parent molecule	*29*
Bombesin	Full agonist; antagonist also developed	*9*
Angiotensin II	Hcy^{3-5} disulfide; excellent binding and contractile activity Antagonists also prepared	*12*
Substance P	Selective ligands for eledoisin site	*32*
Neurokinin B	Selective cyclic agonist for neurokinin 3 receptor	*32*
Dermorphin	Tetrapeptides were highly μ opioid selective	*27*
Gonadotropin-releasing hormone	High potency, limited flexibility	*26*
Thymopentin	Fully active	*31*
Physalaemin	Model peptide, a good mimic of physalaemin salt bridge	*29*
Glucagon	New biological profile	*23*

Another important cyclic enkephalin analog is

DPDPE (H-Tyr-D-Pen-Gly-Phe-D-Pen-OH)

where Pen = Penicillamine or β,β'-dimethyl cysteine). Conformational studies have shown that DPDPE is highly constrained *(18–20)*. This cyclic peptide is highly potent and highly selective for the δ opioid receptor.

2.1.5. α-MSH and Other Hormones

One of the most striking examples of the effects of cyclization on peptide activity has been found in α-melanocyte-stimulating hormone (α-MSH) or α-melanotropin. Its natural sequence is Ac-Ser^1-Tyr^2-Ser^3-Met^4-Glu^5-His^6-Phe^7-Arg^8-Trp^9-Gly^{10}-Lys^{11}-Pro^{12}-Val^{13}-NH_2. A putative turn structure about Phe^7 was stabilized by converting Met^4 and Gly^{10} to cysteines and forming the disulfide bridge between them. This peptide was a superagonist in the frog skin assay *(21)*, but was only modestly potent in mammalian systems. The latter problem was overcome by the design of cyclic lactam analogs involving positions 5 and 10 *(22)*, suggestive of a role for constraint in receptor selectivity.

Interestingly, cyclic lactam analogs of glucagon have also been synthesized.

[Asp^9,Lys^{12},$Lys^{17,18}$,Glu^{21}]glucagon

has novel properties in that the conformational restriction leads to an analog with two distinct binding modes *(23)*.

LHRH analogs have been made in which the proximity of residues 5 and 8 are fixed by cyclization. The analog

Ac-D-Phe(p-Cl)-D-Phe(p-Cl)-D-Trp-Ser-Glu-D-Arg-Leu-Lys-
Pro-D-Ala-NH_2

had an ED_{50} value of 91.9 μg/kg in the inhibition of ovulation test. The linear counterpart was three times less potent *(8)*.

Cyclization via lactam formation of CCK_8 *(24,25)* has also yielded potent and selective compounds. For example,

Boc-γ-D-Glu-Tyr(SO_3H)-Nle-D-Lys-Trp-Nle-Asp-Phe-NH_2

had an inhibition constant of 0.56 n*M* and had a pancreas/brain selectivity ratio of 4464 *(24)*.

Highly potent analogs of GRF with in vivo activity have also been made via cyclization and subsequent substitutions thereof. One such analog is cyclo(Asp^8-Lys^{12}-(D-Ala^2,Asp^8, Ala^{15})-GRF(1-29)-NH_2) *(26)*.

Cyclic lactam analogs related to dermorphin have also been created. These molecules (Tyr-c(D-Orn-Phe-Asp)-NH_2 is an example) are highly μ opioid-specific, presumably because of the effects of the cyclic constraint forcing the Tyr and Phe side chains to be separated by a distance that is compatible with μ opioid receptor binding *(27)*.

Cyclic analogs or mimics of other biologically active peptides include Parathyroid-Related Protein *(28)*, Physalaemin *(29)*, Hirudin *(30)*, Thymopoetin *(31)*, Substance P, and Neurokinin B *(32)*, among others.

These examples have illustrated the effect of cyclization on selectivity, potency, development of antagonists, and separation of activities. Cyclization has also been shown to have other effects, such as prolongation of activity, and increased biological stability and efficacy. Table 2 lists some other peptides in which cyclization has led to novel properties.

The importance of type of cyclization should not be overlooked. For example, somatostatin

SMS (Ala-Gly-Cys-Lys-Asn-Phe-Phe-Trp-Lys-Thr-Phe-Thr-Cys-OH)

is relatively indifferent to the type of cyclization, as evidenced by the oral activity and potency of the cyclic analog c[Pro-Phe-D-Trp-Lys-Thr-Phe] *(33)*, but even it is sensitive to a change from disulfide to methylene bridging. Conotoxin GI, which is normally a $Cys^{1\text{-}6}$ disulfide, has greatly reduced paralytic activity when the disulfide is replaced with a lactam cycle *(34)*. Decreasing ring size (a simple and common technique for constraining peptides) also was a determinant in the latter case. Endothelin, a naturally occurring bicyclic disulfide, has been converted to a potent and specific antagonist by replacing the outer disulfide with a {Dpr^{1}-Asp^{15}} lactam cyclic linkage *(35)*.

The cyclizations described above are mostly disulfide or lactam cycles (except for SMS). Cyclizations include formation of carba analogs and thioethers *(36)*, bis-thioethers *(37)*, azopeptides *(38)*, and other cyclic structures, such as bridging structures *(24,39,40)*. Table 3 lists some of the types of cyclizations that have been utilized in peptides. Cyclization from backbone-to-backbone positions also is increasingly used.

Bridging is a special type of cyclization (Table 3, type B), where distant sites in a peptide are brought together with separate molecules or fragments. Bridging molecules have included succinic anhydride molecules *(24)*, and carboxymethylene fragments *(39)*. Bridging by metals could also be placed in this group *(40)*. CCK_8 analogs with Lys in positions 28 and 31 have been bridged via succinic acid, creating two amide bonds and the cyclic product. This case (*see* Scheme 2) is unusual in that the cyclic peptide is created by linking the peptide backbone after the side chain groups are connected. These CCK analogs are potent in brain membrane-binding inhibition, but have poor affinity for pancreatic acini and therefore are highly selective for central receptors *(24)*. Details of preparation for one of these analogs are given below.

Table 3
Cyclizations

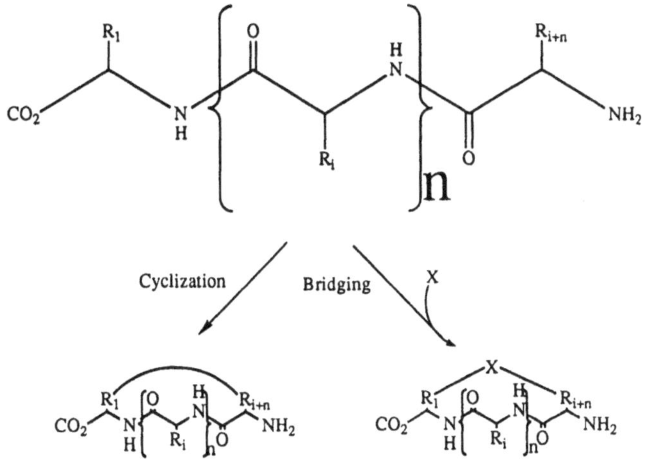

Cyclization		Bridging	
Cyclic group	Reference	Bridging group	Reference
—CH_2—S—S— (Disulfide)	*36*	Succinic acid	*24*
—$(CH_2)_3$— (Carba)	*36*	Carboxymethylene	*39*
—CH_2—S—CH_2—(Thioether)	*36*	Metals	*40*
—S—$(CH_2)_n$—S— (bisthioether)	*37*		
—N=N— (Azo)	*38*		

2.1.6. Synthesis of a Cyclic CCK Analog (JMV320) via Two Lysine Side-Chain Groups and Succinic Anhydride (Scheme 2) (24)

2.1.6.1. Boc-Lys-(Z)-Gly-OBz (I)

1. Treat Boc-Gly-OBz (5.6 g, 14 mmol) with TFA (30 mL) for 30 min at room temperature. Add ether to give the TFA salt of Gly-OBz as a white crystalline solid. Collect, wash with ether, and dry *in vacuo* over KOH pellets.
2. To a cold (–20°C) solution of Boc-Lys(Z) (5 g, 13.14 mmol) in DMF (30 mL), add successively NMM (1.47 mL, 13.14 mmol) and IBCF (1.78 mL, 13.14 mmol).
3. After 5 min of stirring, add the TFA salt of Gly-OBz, followed by DIEA (2.41 mL, 14 mmol), and stir the mixture for 1 h at room temperature.
4. Add ETOAc (200 mL), wash the solution with 1*N* potassium hydrogensulfate (3 × 200 mL), water, saturated $NaHCO_3$ (3 × 200 mL), and brine, and dry with $MgSO_4$, and concentrate under reduced pressure to yield a residue

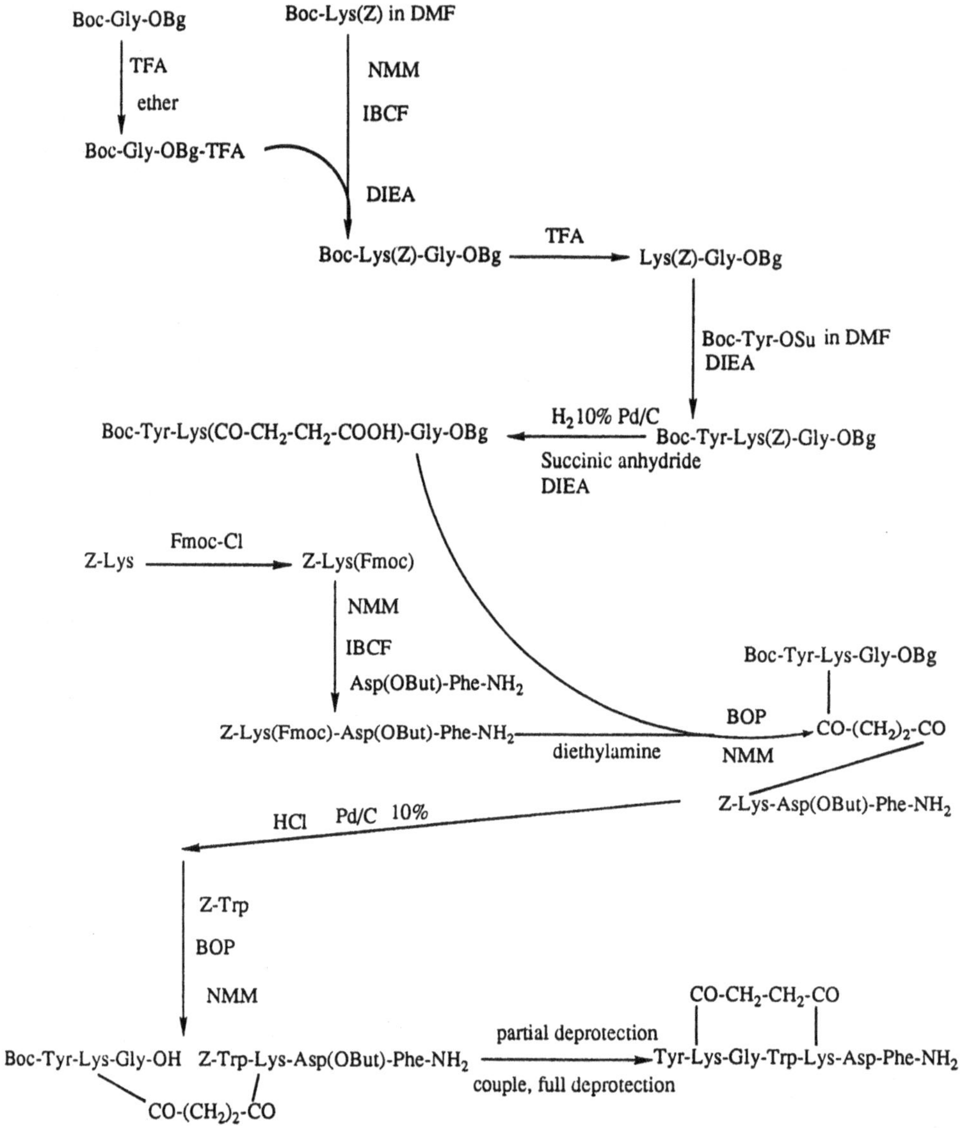

Scheme 2. Bridging of lysine side-chain groups via succinic anhydride *(24)*.

that crystallizes on trituration in hexane. Boc-Lys(Z)-Gly-OBz (8.3 g, 12.5 mmol) is then partially deprotected with TFA as before (step 1).

2.1.6.2. Boc-Tyr-Lys(Z)-Gly-OBz (II)

Add the partially deprotected material to a solution of Boc-Tyr-OSu (4.35 g, 11.5 mmol) in DMF (30 mL) followed by DIEA (2.16 mL, 12.56 mmol), and stir for 2 h at room temperature. Dilute the mixture, and treat with ethyl acetate as before. Evaporation of solvent and trituration with ether gives 8.7 g (92%) of the expected white solid.

2.1.6.3. Boc-Tyr-Lys(CO-CH_2-CH_2-COOH)-Gly-OBz (III)

1. Hydrogenate this compound (a total of 9.5 g, 11.53 mmol) overnight in DMF (100 mL) containing 1.1 mL of concentrated HCl, in the presence of 10% Pd/C catalyst.
2. Filter off the catalyst and concentrate the filtrate *in vacuo*. The residue solidifies on trituration in ether. Collect the solid, wash with ether, and dry *in vacuo* over KOH pellets.
3. Add succinic anhydride (1.15 g, 11.53 mmol) and DIEA (1.98 mL, 11.53 mmol) to a solution of this partially deprotected peptide in DMF (40 mL).
4. After 3 h stirring at room temperature, dilute the reaction with 500 mL of ETOAc, and wash with 1*N* aqueous potassium hydrogensulfate solution (3 × 200 mL), brine, dry over $MgSO_4$, and concentrate under reduced pressure to give a residue that crystallizes on trituration in ether (7.66 g, 84%).

2.1.6.4. Lys(Fmoc) (IV)

1. Dissolve Z-Lys (2.8 g, 10 mmol) in a 1*M* aqueous $NaHCO_3$ solution (25 mL), and cool the mixture to 4°C.
2. Add 1,4-Dioxane (15 mL) followed by Fmoc-Cl (2.59 g, 10 mmol).
3. Stir 2 h at 4°C, then add a 0.5*N* aqueous potassium hydrogensulfate solution (3 × 200 mL), and stir the suspension for 30 min.
4. Collect the precipitate by filtration. Thoroughly wash with water and a mixture of ether and hexane. The yield is 4.3 g (86%).

2.1.6.5. Z-Lys(Fmoc)-Asp(OBut)-Phe-NH_2 (V)

1. Add NMM (0.89 mL, 7.96 mmol), and then IBCF (1.08 mL, 7.96 mmol) to a cold (–20°C) solution of Z-Lys(Fmoc) 4 g (7.96 mmol) in DMF (30 mL).
2. After 5 min stirring, add Asp(OBut)-Phe-NH_2 (2.85 g, 8.5 mmol) and stir the mixture for 1 h at room temperature.
3. The Z-Lys(Fmoc)-Asp(OBut)-Phe-NH_2 precipitates on addition of 1*N* potassium hydrogensulfate (300 mL).
4. Wash the solid with water, saturated $NaHCO_3$, water, and ether, and dry *in vacuo* (6g, 92%):

Boc-Tyr-Lys-Gly-OBz Z-Lys-Asp(OBu^t)-Phe-NH_2
CO-$(CH_2)_2$-CO (VI)

2.1.6.6. Formation of Intermediate (VI)

1. The compound (V) (2.05 g, 2.5 mmol) is partially deprotected in DMF (25 mL) and diethylamine (2.5 mL) for 30 min at room temperature.

2. Concentrate the mixture *in vacuo* to leave a residue that crystallizes on trituration in ether (1.39 g, 93%).
3. This deprotected material (1.39 g, 2.17 mmol) is added to a solution of Boc-Tyr-Lys(CO-CH_2-CH_2-COOH)-Gly-OBz (1.72 g, 2.17 mmol), and BOP (0.96 g, 2.17 mmol) in DMF (20 mL) followed by NMM (0.24 mL, 2.17 mmol). Stir the mixture for 2 h at room temperature.
4. The expected material precipitates on addition of 2% aqueous $NaHCO_3$ solution (300 mL). Collect, wash with water, a 1*N* aqueous potassium hydrogensulfate solution, water, ETOAc, and ether, and dry *in vacuo* (2.13 g, 71%).

Boc-Tyr-Lys-Gly-OBz Z-Trp-Lys-Asp(OBu^t)-Phe-NH_2
└──CO-$(CH_2)_2$-CO─┘ (VII)

2.1.6.7. Formation of Intermediate (VII)

1. Hydrogenate the compound (VI) (2.1 g, 1.53 mmol) overnight at room temperature in DMF (50 mL) containing concentrated HCl (0.14 mL) in the presence of 10% Pd/C catalyst.
2. Filter off the catalyst, and concentrate the filtrate *in vacuo* to leave a residue that solidifies on trituration in ether. Collect, wash with ether, and dry over KOH pellets.
3. To a solution of this partially deprotected peptide in DMF (20 mL), add Z-Trp (0.54 g, 1.6 mmol), BOP (0.71 g, 1.6 mmol), and then NMM (0.35 mL, 3.13 mmol).
4. Stir for 2 h at room temperature, and the expected compound will precipitate on addition of 2% aqueous $NaHCO_3$ solution (300 mL).
5. Collect this compound, wash with water, 1*N* aqueous potassium hydrogensulfate solution, water, ETOAc, and ether, and dry *in vacuo* (2.19 g, 92%).
6. Dissolve this compound in DMF (30 mL) and cool to 0°C. Add K_2CO_3 (0.56 g, 4.05 mmol) in water (15 mL). Stir the mixture for 5 min at 0°C and 90 min at room temperature.
7. Add 1*N* potassium hydrogensulfate solution (10 mL) and water (200 mL) to form a precipitate. Collect by filtration, wash with water, ETOAc, and ether, and purify by silica gel chromatography (chloroform/methanol/HOAc [60:15:5], as the eluent).
8. Pool pure fractions and concentrate *in vacuo* to leave a solid residue that is triturated in ether. Collect and dry *in vacuo* (1.38 g, 77%).

Boc-Tyr-Lys-Gly-Trp-Lys-Asp(OBu^t)-Phe-NH_2
└CO-$(CH_2)_2$-CO┘ (VIII)

2.1.6.8. Cyclization

1. Hydrogenate the compound (VII) overnight at room temperature in DMF (50 mL) containing concentrated HCl (0.07 mL, 0.765 mmol) in the presence of 10% Pd/C catalyst.
2. Remove the catalyst by filtration, and rinse with DMF (25 mL). The total volume of DMF is 75 mL, which is the same as 10 m*M* concentration.
3. Add to this solution BOP (0.37 g, 0.842 mmol) and $NaHCO_3$ (0.32 g, 3.82 mmol).
4. After 5 h stirring at room temperature, concentrate the mixture to a 5-mL volume.
5. The compound precipitates on addition of a 2% aqueous $NaHCO_3$ solution (300 mL). Collect, wash with water, 1*N* aqueous potassium hydrogensulfate solution, water, ETOAc, and ether, and dry *in vacuo* (0.737 g, 82%).

Ac-Tyr-Lys-Gly-Trp-Lys-Asp-Phe-NH_2
CO-$(CH_2)_2$-CO (IX)
(JMV320)

2.1.6.9. Deprotection and Acylation

1. Add the compound (VIII) (700 mg, 0.593 mmol) to a cold (0°C) solution of 2-methyl-indole (780 mg, 5.93 mmol) in TFA (10 mL), and stir under an argon atmosphere for 2 h at room temperature.
2. The expected material precipitates on addition of ether (100 mL). Collect, thoroughly wash with ether, and dry *in vacuo* over KOH pellets. This yield is quantitative.
3. Add to a solution of the above (400 mg, 0.352 mmol) in DMF (4 mL) acetic acid, *N*-hydroxysuccinimide ester (56 mg, 0.352 mmol), and DIEA (62 µL, 0.352 mmol).
4. Stir for 2 h, then add ETOAc (100 mL) to yield a solid. Collect, thoroughly wash with ETOAc, and ether, and dry *in vacuo*. Yield is 81%, 304 mg.
5. Purify the cyclized peptide by HPLC in 20.1 min of A:B (50:50) and lyophilize. A = AcOET/hexanes; 7:3 and B = AcOET.

2.2. Pseudocyclization

Another global conformational constraint is that of pseudocyclization, or the stabilization of secondary structures (α-helix, β-turns, and β-sheets, for example). In addition to the tendencies of the common amino acids to prefer one secondary structure over the other, many uncoded amino acids have been shown to have definite propensities toward stabilizing certain secondary structural elements. One such amino acid is α-amino-isobutyric acid (Aib) (Table 4, **1**). Because of its α,α-disubstituted nature, this amino acid is restricted to a small portion of conformational space, and tends to form α or 3_{10} helices. It also can occupy the i + 1 position of βI- or βIII-type turns, or the i + 2 position of βI or βII turns *(41)*. Aib has been incorporated into several peptides, including ion channels *(42)*, chemotactic peptides *(43,44)*, muramyl dipeptide (MDP) *(45)*, and neuropeptide Y *(46)*. Table 4, entries 1–9 list some other useful α,α-disubstituted amino acids.

In the past few years, many studies have concerned the use of dehydro-amino acid residues in peptides. The conformation of Δ^z-phenylalanine *(16)* in dipeptides and tripeptides has been determined by X-ray *(47)* and NMR *(48)* analyses. Both studies concluded that this residue preferred the i + 1, or the i + 2 position of a β-type turn. Singh et al. *(49)*, and Uma et al. *(50)* observed an analogous situation for small peptides molecules that included Δ^z-leucine. Kaur et al. *(48)*, however, observed NOEs that were consistent with both a turn structure and an extended conformation, and Chauhan *(51)* showed that a hexapeptide with two Δ^z-phenylalanines largely favored an extended conformation. However, in a biological system (Gramicidin S analog {Δ^z-Phe-4.4-Gramicidin S}), Imazu et al. *(52)* showed that this analog retained activity, a characteristic that is dependent on the ability to stabilize a β-sheet structure in this system. Δ^z-Phe has also been analyzed in Aib-containing peptides *(53)*. Δ^z-Phe has been shown to promote other structures, including a βIII' turn and alternating left-right α-helix *(54)*. In contrast to the other dehydro residues, Δ^z-Ala seems to prefer an inverse γ-turn (in $CDCl_3$), whereas in equivalent positions, the normal alanine residue does not *(55)*. In addition to Gramicidin, dehydro-amino acids have been incorporated into other biologically active peptides with interesting results. Examples include bradykinin *(56)*, TRF *(57)*, TRH *(58)*, vasopressins *(59)*, and enkephalins *(60)*. Structures other than amino acids can affect or at least simulate secondary structures. Disulfide linkages in peptides can also simulate α-helix formation, as seen in certain cyclic substance P analogs *(32)*, neuropeptide Y analogs *(61)*, and hirudin analogs *(30)*, and lactam formation tends to simulate α-helix formation in cyclic analogs of GRF *(26)*.

Many non-α,α-disubstituted amino acids stabilize secondary structures. Many of these are listed in Table 3, and will be discussed in Section 2.2.2.

2.2.1. Conformational Constraints Designed to Mimic β and γ Turns

The use of a conformationally restricted δ or γ lactam (*see* Table 5) was proposed as a means of forcing the peptide bond as *trans (41)*. Many constrained lactams have been synthesized to create a covalent mimic of a peptidic turn structure. Unfortunately, these structures can mimic the backbone conformation, but have no ability to display appropriate side-chain functionalities at the corner positions of the "turn". Recent syntheses have concentrated on developing mimics with functional side chains *(62–64)*, such as the lactam (**1**) in Table 5. These structures and variations thereof may be utilized as mimics of some biologically active structures. For example, the (S,R)δ-lactam has been applied to mimic the βII turn at the carboxyl terminus in MIF. The (S,S)δ-lactam has since been shown to prefer the i + 1 position of a βII' turn *(41)*. These techniques of rigid conformational mimicry have been used in several systems, including ACE inhibitors *(65)*, renin inhibitors *(66)*, cyclosporine A analogs (**2**) *(67,68)*, neurokinin antagonists (**3**) *(69)*, and human growth hormone mimics *(70)*. Model peptides are being made to determine the appropriate substitution sites for these units *(71)*, and their conformational preferences in the role of constraining peptides is being studied *(72)*. Other types of mimics include the 3-amino-2-piperidone-6-carboxylic acid *(73)*, the thiazolidine bicyclic system *(74)*, and the 5H-6-oxo-2,3,4,4a,7,7a-hexahydropyrano[2,3,-b]pyrrole *(75)*. For a review of β turn mimics, *see* ref. *76*.

2.2.2. Local Constraints

Thus far, modifications that alter predominantly the peptide backbone configuration have been discussed. However, the proper orientation of the amino acid side-chain group also is critical for biological activity. Thus, it would be useful to be able to constrain the side-chain groups to examine how topological arrangements modulate activity. There are many potential modifications that can impose constraints on the peptide primarily at the side-chain groups.

2.2.2.1. Cyclization

Cyclization can be used as a local constraint as well when it is within a single amino acid residue or between two adjacent residues (for an excellent review on these short-range cyclizations, *see* ref. *77*). Nature uses the amino acid proline towards this end. The properties of proline with respect to peptide structures *(78)* have been examined. Many "pro-

Table 4
Useful Amino Acids in the Development of Conformationally Constrained Peptides

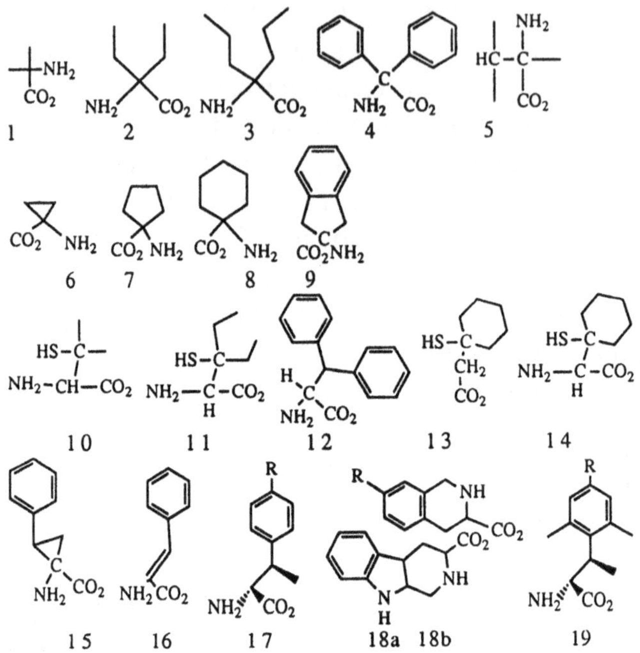

Entry	Biological activities	References
1. Aib (α-aminoisobutyric acid) or MeA (α-methylalanine)	Predominantly α and 3_{10} helix,	*41*
	also i + 1 of βI/III or i + 2 of βI/II turns,	
	in antiamoebin naturally;	*47*
	active in chemotactic peptide;	*43,44*
	active in ion-channel peptides;	*42*
	used in muramyl dipeptide;	*45*
	used in neuropeptide Y	*46*
2. Deg (diethylglycine)	Prefers extended conformation	*41*
3. Dpg (dipropylglycine)	Prefers extended conformation	*41*
4. Diphenylglycine	Prefers fully extended (c-5) region	*114*
5. α-Methyl-Valine	Prefers βI/III and α-helices	*115*
6. Ac_3c	Cyclic α,α disubstituted amino acid	
	prefers bridge region of Φ, Ψ space	*41*
7. Ac_5c (or cycloleucine)	Prefers α/3_{10} helices,	*116*
	less active than parent oxytocin,	*118*
	high chemotatic activity	*117*
8. Ac_6c (aminocyclo-hexanecarboxylic acid)	Gave weakly active Neurokinin A analogs,	*119*
	preferred conformation somewhat like	*41*
	Aib, gave active chemotactic peptides	*43,117*

Table 4 *(continued)*

9. Ind or Ain (aminoindane carboxylic acid)	Helix-forming nature leads to active chemotactic peptides, also leads to enhanced affinities for angiotensin II analogs	*120,121*
10. Pen (penicillamine) (or des-amino Pen)	β Dimethyls constrict disulfide rings, highly active in analogs of Enkephalins, OT, μ opioid antagonists, used in vasopressin and α-HANP	*82,83, 89–92*
11. β,β Diethylcysteine	Used on oxytocin and vasopressin antagonists	*122*
12. Diphenylalanine	Active in Angiotensin II analogs	*121*
13. Des-amino-β,β-pentamethylenepropionic acid	Enhanced oxytocin antagonism	*123*
14. Apmp (α-amino-β,β-pentamethylenepropionic acid)	Used in enkephalin analogs	*124*
15. 2,3 Methanophenylalanine	Fairly strong chymotrypsin inhibition	*125*
16. Dehydrophenylalanine	Prefers i + 1 or i + 2 of β turns,	*47,48*
	can form βIII' turns and α helics,	*54*
	active in gramicidin	*52*
also dehydro-Leu	Prefers i + 2 of β-turns	*49*
dehydro-Ala	Prefers inverse γ turn, incorporated into MDP	*45,55,60*
	Δ 3,4 Pro lead to GnRH antagonists	*7*
	Increase potency in vasopressin analogs	*59*
dehydro-Asp	Active in inhibition of NAALA dipeptidase	*126*
17. β-Methyl-phe	Leads to high δ opioid selectivity	*87*
18a. Tic (tetrahydroisoquinaline carboxylic acid) (Tyr analog, [OH]-Tic also)	Adopts only g– and g+ rotamers, dependent on position in peptide, very highly active in opioids,	*82,83*
	derivitives in ACE inhibitors	*86*
18b. Tca (tetrahydrocarboline carboxylic acid)	Trp analog with biased rotamers	*84*
19. 2'6'-dimethyl-β-Me-Phe (or Tyr)	Methyl groups restrict phenyl ring rotation at x_1 and x_2	—

line-like" amino acids have been synthesized, most of which were designed to limit flexibility or define *cis-trans*-isomerism *(79–81)*. Many other cyclic amino acids that have demonstrated usefulness in constraining peptides are listed in Table 6 (**6–9**, **13–15**, and **18**).

Table 5
Conformational Constraints Designed to Mimic Turn Structures

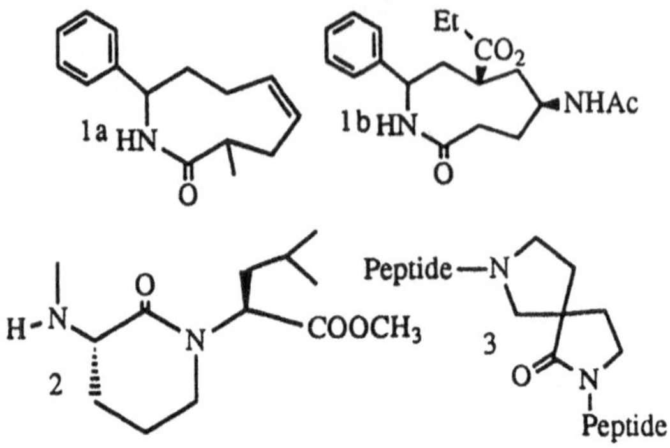

Lactam	Activity	Reference
1a. and 1b.	Lactams designed to have functional side chains in β turn "corner" positions	*63*
2.	A γ lactam used in cyclosporin analogs	*67,68*
3.	A spirolactam exhibited high Neurokinin activity	*69*

2.2.2.2. Rotamer Bias

The cyclic amino acids **18a** {R = H} and **18b** (Table 4) are particularly interesting. These amino acid analogs of Phe and Trp have been cyclized from their free amine to their 2' carbon via a methylene unit. This cyclization will limit the motion of the side-chain, and these residues are constrained by another type of local interaction. From the Newman projections (Fig. 1), it is clear that the Tic residue (1,2,3,4-tetrahydroisoquinoline-3-carboxylic acid) cannot adopt the *trans*-position, and that only the g^- and g^+ rotamers are possible. Moreover, these residues are interesting in that they seem to prefer a single rotamer position dependent on their position in the peptide chain. That is, Tic prefers the g^- when in the amino-terminal position, as shown by NMR *(82,83)*, whereas it prefers the g^+ when in an internal position. Studies have shown that although these cyclizations limit side-chain flexibility, they do not necessarily change the electronics of the residues *(84)*.

The Tic residue has been incorporated in the CTP series as a conformational constraint, as in the peptide TCTAP ([Tic^1,Arg^5]CTP) *(82,83)*. This peptide is the most potent and selective μ opioid antagonist to date. The isoquinolinecarboxylic acid also has been incorporated into deriva-

Table 6
Some Amide Bond Mimics

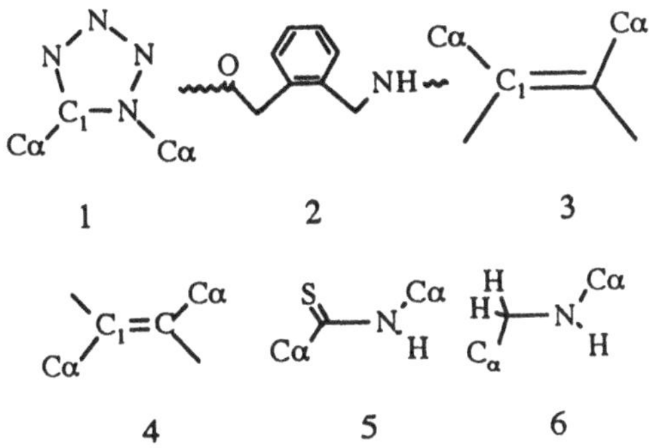

Replacement	Biological activity	Reference
Tetrazole Ψ (CH_4)	Mimics *cis*-peptide bond used in TRH, Bkn, OT, SMS	*41,100* *102–105*
O-AMPA	Designed to mimic *cis*-amide used in somatostatin analogs	*106,107*
cis-olefin	*cis*-amide mimic yet to be applied	NA
trans-olefin	*trans*-amide mimic used in OT and CCK	*100,105,108*
Thioamide	Decreased flexibility enhances receptor selectivity in enkephalins	*41* *112*
Reduced amine	Stabilizes secondary structures used to create antagonists of Bombesin, Substance P, and Secretin	*109–111*

tives of ACE inhibitors *(85)*. The tyrosine version (entry **18a**, R = OH) of Tic has also been made, and is useful in peptide synthesis *(86)*.

Another type of rotamer bias is afforded by β-alkylation. Consider the Newman projections in Fig. 2. It is clear that β methylation of a phenylalanine residue will tend to bias the population of rotamers to a specific rotamer dependent on which isomer is made. The effects of different isomers in binding were evidenced by the differential binding of each of the isomers of β-Me-Phe[4] (BMP, **17**; Table 4) DPDPE. The [(2S,3S) BMP[4]]DPDPE was over 100× more selective than the parent, and is one of the most selective peptides known for the δ opioid receptor *(87)*. The (2S,3R) isomer was also very selective, but only about a third as selec-

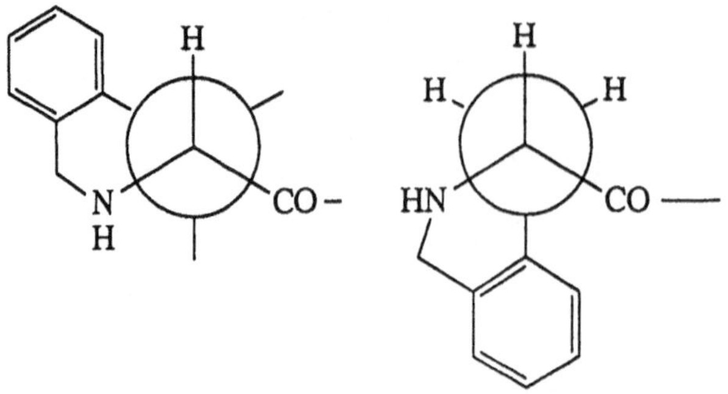

Fig. 1. Tic residue rotamers.

tive as the (2S,3S) isomer *(87)*. All the stereoisomers of β-methyl-Tyr and β-methyl-Trp also have been made in our laboratories.

The dehydro-amino acids (*vide supra*) also can be used as a rotamer bias. For example, Δ^z-phenylalanine has a side chain that is constrained to be between the g^+ and the g^- position, and the Δ^E-phenylalanine has a side chain that is constrained to the *trans* position.

The population of rotamers in a peptide is important to consider, and computational methods can help to predict which rotamers are likely to exist. However, sometimes a rotamer is needed for activity that is not likely to exist in solution. Thus a restriction or bias toward that rotamer is helpful. Studies with hydrophobic amino acids have shown that helix-forming tendency is influenced by side chain rotamer populations because of the destabilization of the helix by β-branching or other sterics in the case of Phe. The altered distribution of rotamer angles presumably correlates to reduction of side-chain conformational entropy in the helix vs the flexible chain *(88)*, and it is clear that rotamer position can determine the relative spacing between side-chain groups that has been determined to be important in many peptides, such as CTP *(82,83)*.

2.2.2.3. Steric Occlusion

One of the most common approaches to conformational constraint is through the addition of bulky groups. Alkylation has been applied at many positions in peptides. Most commonly, these are at the α and β positions of residues and the nitrogen of the backbone. Other positions include the oxygen of serine and threonine, and the ring of phenylalanine, among others.

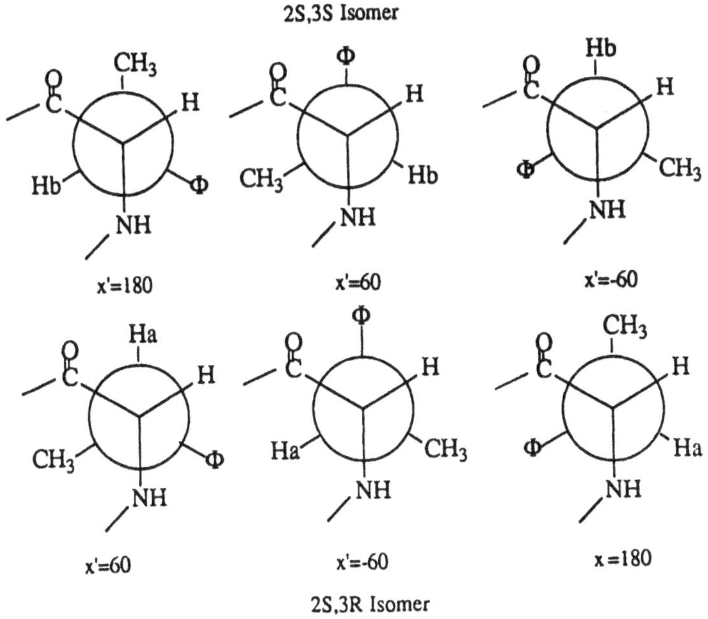

Fig. 2. β-Methyl-Phenylalanine side-chain rotamers.

α-Carbon alkylation-α,α' disubstituted amino acids provide a convenient method to constrain peptides. These disubstituted residues restrict the ϕ and ψ angles to a small region of conformational space and thereby reduce the flexibility of the peptide. Some of these residues (Aib, Ace, and α-methylvaline) have been discussed as they relate to pseudocyclization, and others (Tic, Tca, and so on) as they relate to rotamer bias. Table 4 lists many other α,α'-disubstituted amino acids and their uses.

Steric bulk has also been applied to the β carbon. Beta-methylphenylalanine has already been discussed as it relates to rotamer bias. The Pen residue (β,β dimethylcysteine) (Table 4, **10**) is particularly useful since it can be used in the construction of disulfide linkages (cyclizations). Moreover, the β-methyl groups influence the disulfide bond angles, the helicity of the disulfide bond, and transannular interactions, and therefore can act as both a local and a global constraint. The Pen residue (or des-amino Pen) has been utilized in many peptides, including enkephalins *(89)*, μ opioid antagonists *(82,83)*, vasopressin *(90)*, OT, its original application *(91)*, and the α-human atrial natriuretic peptide *(92)*. The α-human atrial natriuretic peptide containing D-Pen was unusual in that it exhibited full cGMP accumulation, yet was weak in its binding and vasorelaxant activity *(92)*. Amino acids with other β-alkyl groups have also been instructive in OT *(93)* and vasopressin analogs *(94)* (**13**), and in enkephalin analogs (**14**) *(95)*.

The N-alkylation of the backbone in peptides reduces the conformation freedom of the amide bond two- to fourfold, eliminates hydrogen bond-donating ability, and affects the backbone torsional angle. This constraint has been applied to many peptide systems. *N*-Me-norleucine was incorporated into CCK_5 at position 3, and resulted in an increased selectivity for the CCK-B receptor *(96)*. *N*-Me-Phenylalanine replaced Phe^1 in CTOP $[Orn^5]$CTP, and decreased potency *(82,83)*. Analogs of neurokinin N-Me-Val^7 also have been made *(97)*.

The hydroxy oxygen of Ser and Thr residues have provided another site for alkylation. An example of this is the peptide Tyr^1-D-Ser^2-Gly^3-Phe^4-Leu^5-Thr^6. This peptide was *t*-butylated at the Ser^2 and Thr^6 side-chain oxygens yielding the peptide BUBU, which resulted in a highly potent and selective δ-opioid agonist *(98)*.

2.2.3. Peptide Bond Replacements

Peptide bond replacements have been utilized for a variety of reasons *(99)*. Some of these mimics can constrain peptides by forcing the bond to adopt a *cis* or *trans* geometry, by increasing rotational barriers, or by inducing the stabilization of a secondary structure. Also, these replacements can have effects like those of D-amino acid substitution in that they both can stabilize secondary structure, and decrease or eliminate proteolytic degradation. There are many types of amide bond replacements, such as the ketomethylene $\Psi[COCH_2]$, thioester Ψ[COS], depsipeptide Ψ[COO], thioamide Ψ[CSNH], reduced amide $\Psi[CH_2NH]$ (*see* Chapter 12), methylene $\Psi[CH_2CH_2]$, thiomethylene $\Psi[CH_2S]$, olefin Ψ[CH=CH], tetrazole $\Psi[CN_4]$, and retro-inverso Ψ[NHCO], as well as Ψ[NHCONH], $\Psi[CH_2SO]$, Ψ[CONHO], $\Psi CH(CH_3)S]$, and $\Psi[C(=CH_2)CH_2]$, among others. Several of these types have been discovered as naturally occurring. This discussion will pertain only to those that yield conformational constraint on incorporation (Table 6 lists some useful peptide bond surrogates).

2.2.3.1. Tetrazole

Disubstituted 1,5 tetrazoles (**1**, Table 6) have been suggested as a group that may be usable as a *cis*-peptide bond surrogate *(100)*. In such peptides as thyroliberin (TRH) and angiotensin, the *cis* geometry of a peptide bond was implicated in its activity *(101)*. Therefore, a mechanism for the stabilization of this arrangement should yield an active peptide. Tetrazole has been incorporated as a mimic of the *cis* bond in TRH *(102)*, bradykinin *(103)*, somatostatin *(104)*, and oxytocin *(105)*. The CN_4 group was determined to be a good mimic of the *cis* bond as far as geometry and bond lengths are concerned, but the CN_4 group is unlike the peptide bond in that the CN_4 group is more coplanar and has more ψ_1 conformational space available to it, but is restricted to a smaller volume of conformational space near $\phi = -180°$ and $\Psi = 0°$ or 180°.

2.2.3.2. Synthesis of Tetrazole-Containing Analogs of Bradykinin (Scheme 3) *(103)*

The benzyloxycarbonyl-protected Z-L-Ala-L-Ala-OBzl dipeptide was made via the mixed anhydride procedure of Anderson, with isobutyl chloroformate.

2.2.3.3. Z-L-AlaΨ[CN_4]-L-Ala-OBzl

1. Add quinoline (2 mmol) to a stirred solution of PCl_5 (1 mmol) in chloroform (5 mL) at room temperature (a white precipitate forms). Stir the mixture for 20 min before adding the crystalline dipeptide (1 mmol) in portions with stirring such that the temperature never exceeds 20°C.
2. After 30 min at 20°C (the solution is clear), add a benzene solution of hydrazoic acid (3 mL). Stir the reaction mixture at room temperature for 1 h before evaporation.
3. Partition the crude residue between ethyl acetate and water (30 mL each). Wash the organic layer with 1*N* HCl (2 × 15 mL), 1*N* $NaHCO_3$ (2 × 15 mL), water (2 × 15 mL), and saturated NaCl (30 mL).
4. Evaporate the dried (Na_2SO_4) ethyl acetate solution, and purify the residue by flash chromatography. The tetrazole derivative is isolated as a single stereoisomer. The flash chromatography is a DCM/acetone (30:1, v/v) system, and yields white crystals (24.9%).
5. Treat the CN_4 dipeptide (586 mg, 1.43 mmol) in 1 mL of HOAc (with stirring) with 3.5 mL of 30% solution of HBr in HOAc.
6. After 20 min at room temperature, pour the solution into 30 mL of (precooled to –10°C) ether with vigorous stirring.
7. When the oily hydrobromide precipitates, discard the upper phase. Wash the oil with ether (3 × 20 mL) and dry *in vacuo* over KOH to give 490 mg (96.3%) of dipeptide HBr salt as a very hygroscopic glass with the Z-group removed.

Z-Ala-L-Ala-OBzl

↓ 1) PCl_5/quinoline; 2) HN_3, 25%

Z-L-AlaΨ[CN_4]-L-Ala-OBzl

↓ HBr/HOAc, 96%; Boc-Phe-OH, IBCF

Boc-Phe-L-AlaΨ[CN_4]-L-Ala-OBzl

↓ H_2/Pd, 85%

Boc-Phe-L-AlaΨ[CN_4]-L-Ala-OH

↓ H_2N-Phe-Arg(Tos)-polymer, DCC; 4 steps solid phase synthesis

Boc-Arg(Tos)-Pro-Pro-Gly-Phe-L-AlaΨ[CN_4]-L-Ala-Phe-Arg(Tos)-polymer

↓ HF; HPLC

[AlaΨ[CN_4]-Ala]^{6,7}bradykinin

Scheme 3. Synthesis of tetrazole-containing bradykinin analogs *(106)*.

2.2.3.4. Boc-Phe-L-AlaΨ[CN_4]-L-Ala-OBzl

1. Couple Boc-Phe-OH to the HBr salt as follows. Dissolve the Boc-Phe-OH (358 mg, 1.35 mmol) in 5 mL of DCM/DMF (1:1) and cool to –15°C. Treat with 0.21 mL (1.9 mmol) of *N*-methylmorpholine, followed by 0.26 mL (1.9 mmol) of isobutyl chloroformate.
2. Stir the mixture for 10 min, and then add the HBr dipeptide salt (480 mg, 1.35 mmol), followed by 0.21 mL (1.9 mmol) of *N*-methylmorpholine so that the temperature never exceeds –10°C. Stir the reaction for 1 h at –10°C. Warm the reaction slowly to room temperature overnight with stirring.
3. Remove the solvents under reduced pressure, take up the resultant residue in ETOAc (50 mL), and wash with 1*N* $NaHSO_4$ (3 × 20 mL), 1*N* $NaHCO_3$ (3 × 20 mL), water (2 × 20 mL), and saturated NaCl (20 mL). Dry the solution with Na_2SO_4, and evaporate and recrystallize the residue from ETOAc/petroleum ether to yield 560 mg (67%) of white crystals.
4. Hydrogenate a solution of 243 mg (0.46 mmol) of the tripeptide benzyl ester in 20 mL of ETOH, and a few drops of acetic acid overnight in the presence of 100 mg 10% Pd/C.
5. Evaporate the filtered solution, and take up the residue in a small amount of ETOAc and sufficient 1*N* $NaHCO_3$ (1:9).
6. Acidify the aqueous phase with solid sodium bisulfite to pH = 2.5, extract with ETOAc, and dry (Na_2SO_4). Evaporate the ETOAc to yield 208 mg (85%) of crystalline material.
7. To prepare for tetrazole tripeptide incorporation, 1 g of Boc-Arg(Tos)-benzyl ester resin (0.4 mmol/g) is extended by dichloromethane (DCM), 3 × 2 min, 50% TFA/DCM, 5 and 25 min, DCM 3 × 2 min, 10% TEA/DCM, 5 min and 10 min, DCM 3 × 2 min; first coupling: 3 Eq of Boc-amino acid and 3 Eq DCC in DCM; second coupling: 3 Eq Boc-amino acid and 3 Eq of dicylohexylcarbodiimide (DCC) in DMF.
8. Add the tetrazole to the Phe-Arg(Tos)-polymer using DCC/HOBt in DMF.

2.2.3.5. Boc-Arg(Tos)-Pro-Pro-Gly-Phe-L-AlaΨCN_4]-L-Ala-Phe-Arg(Tos)-polymer

The bradykinin sequence was completed by addition of Boc-Gly, Boc-Pro, Boc-Pro, and Boc-Arg(Tos) using solid-phase protocol.

2.2.3.6. [AlaΨ[CN_4]L-Ala]6,7-bradykinin

1. Cleave the peptide from resin with 10 mL HF/anisole (9:1) for 1 h at 0°C (*see* Chapter 4, *PAP*) yielding 150 mg of crude peptide.
2. Purify the peptide via HPLC (A = 0.1% TFA in water; B = 90% acetonitrile/water [0.1% TFA]) with a gradient 15–30% B in 40 min (*see* Chapter 4, *PAP*). Peak 1 is product, and peak 2 (half the size of peak 1) is isomer (D-Ala at

position 6). This conclusion was based on relative abundance and epimerization studies.

Scheme 3 outlines the synthesis of tetrazole analogs of bradykinin *(103)*. This synthetic procedure is superior to previously reported syntheses in that the problem of racemization/epimerization has apparently been overcome in this case. In addition to restriction of the *cis*-conformation, the incorporation of the tetrazole group also may causes changes in other residues in the peptide.

2.2.4. AMPA Group

Van der Elst et al. have also used the AMPA (ortho-[aminomethyl] phenylacetic acid) group (**2**, Table 6) in an effort to mimic a *cis*-amide bond between two Gly residues in a somatostatin analog. This analog was inactive and had a different conformation than the proposed biologically active form. This could have been the result of intramolecular hydrogen bonding of the AMPA spacer *(106)*. The *para* and *meta* versions have also been made *(107)*.

2.2.5. Olefinic Group

The *cis*-double bond (**3**, Table 6) would also be useful as a *cis*-amide mimic, but is synthetically difficult because of a transition from the *cis*-β,γ-carbonyl compound to the more stable trans α,β unsaturated product. The *trans* Ψ[CH=CH] (Table 6, **4**) replacement has demonstrated some utility in synthetic peptides in that it is a good mimic of the *trans*-peptide bond. This is useful when a *trans* orientation about a certain residue is required for activity. Oxytocin analogs have been synthesized with a double bond replacing the Leu-Gly peptide bond. These analogs had lowered, yet prolonged activity *(105)*, and similar effects were seen in cholecystokinin *(100)*. This replacement is topographically similar to the peptide bond *(108)*, but is dissimilar in that it is more lipophilic and more rigid than the normal bond, and has a loss of hydrogen-bonding capacity.

2.2.6. Other Constraining Peptide Replacements

The reduced amide bond has some features unlike that of the normal peptide that include a loss of planarity, altered hydrogen-bonding abilities, and the introduction of a new ionizable moiety. This peptide bond replacement is also less rigid than the normal bond. Interestingly, however, the introduction of the reduced amide bond has been useful in the generation of antagonists, such as in bombesin *(109)*, substance P *(110)*, and secretin *(111)*.

Another important peptide bond mimic for the examination of conformational constraints is the thioamide Ψ[CSNH]. This surrogate displays a higher barrier to rotation than normal amides and thus is useful in restricting the motion of peptides. They also display other unique characteristics, such as altered hydrogen-bonding strengths. These amide bond replacements can also stabilize certain turn structures, and prefer either the i or the i + 2 position of a βII turn, or the i + 2 position of a βIII turn *(41)*. The proportion of peptide bond replacement with thioamides has also been shown to have an effect on receptor selectivity, as was demonstrated in Enkephalin analogs (**112**).

2.2.7. Phenyl Ring Rotational Restriction

It recently has been shown that the relative orientation of the plane of the aromatic ring (Tyr or Phe) may be important for bioactivity in biologically active peptides (Hruby, unpublished). Examples of this include the requirement of a coplanarity between a phenyl ring and an olefin for good platelet activating factor (PAF) antagonism *(113)*. Restriction of rotation may help determine the importance of such orientations. Cyclization is an obvious mechanism, but is not the only one. For example, creation of β-MeTyr and Phe analogs that also are methylated at their 2' (or 2' and 6') positions (*see* Table 4, **19**) enables rotational restriction (*114a*). These Tyr and Phe analogs have been incorporated into the δ opioid agonist DPDPE and deltorphin series (V. J. Hruby and coworkers, unpublished).

3. Summary and Conclusion

Peptides are inherently flexible molecules and, as such, have many conformations available to them. One or possibly more than one of the available conformations will have a specific biological relevance. To determine which conformations are important, it is necessary to confine the peptides to a single region of conformation space, and then determine if that shape is useful. By examining a lot of shapes, it should be possible to determine which are biologically important. Many different methods have been used to achieve constraint, and many successes have been obtained by this approach. However, new synthetic methodologies are needed to create more precise conformations. It should also be pointed out that flexibility should be retained in some cases. That is, the structure, if too rigid, no matter how well designed may not be able to assume another conformation with the desired properties for overall in vivo biological activity, whereas a slightly flexible structure is more likely to be able to make that adjustment. With proper design techniques and synthetic methodologies, one should be able to obtain valuable knowledge about the requirements for peptide receptors, enzyme-active sites, and a wide variety of other biological processes.

Abbreviations

BOP, benzotriazolyloxytris(dimethylamino)-phosphonium hexafluorophosphate; CCK, cholecystokinin; DCC, dicyclohexylcarbodiimide; DCM, dichloromethane; DIEA, *N,N*-diisopropylethylamine; DMF, dimethylformamide; ETOAc, ethyl acetate; ETOH, ethanol; Fmoc, 9-fluorenylmethyloxycarbonyl; HOAc or AcOH, acetic acid; HOBt, hydroxybenzotriazole; HPLC, high performance liquid chromatography; IBCF, isobutylchloroformate; NMM, *N*-Methylmorpholine; OBg, *N*-benzhydrylglycolamide ester; OSu, N-hydroxysuccinamide ester; *p*-MBHA, para-methylbenzhydrylamine; TEA, triethylamine; TFA, trifluoroacetic acid; TLC, thin layer chromatography.

Acknowledgments

This work was supported by grants from the US Public Health Service, The National Institute of Drug Abuse, and the National Science Foundation.

References

1. Hruby, V. J. (1982) Conformational restrictions of biologically active peptides via amino acid side chain groups. *Life Sci.* **31,** 189–199.
2. Hruby, V. J., Al-Obeidi, F., and Kazmierski, W. (1990) Emerging approaches in the molecular design of receptor-selective peptide ligands: conformational, topographical and dynamic considerations. *Biochem.* **268,** 249–262.
3. Wood, S. P., Tickle, I. J., Treharne, A. M., Pitts, J. E., Mascarenhas, Y., Li, J. Y., Husain, J., Cooper, S., Blundell, T., Hruby, V. J., Buku, A., Fischman, A. J., and Wyssbrod, H. R. (1986) Crystal structure analysis of deamino-oxytocin: conformational flexibility and receptor binding. *Science* **232,** 633–636.
4. Brewster, A. I. R. and Hruby, V. J. (1973) 300-MHz nuclear magnetic resonance study of oxytocin in aqueous solution: conformational implications. *Proc. Natl. Acad. Sci. USA* **70,** 3806–3809.
5. Brewster, A. I. R., Hruby, V. J., Glasel, J. A., and Tonelli, A. E. (1973) Proposed conformations of oxytocin and selected analogs in dimethyl sulfoxide as deduced from proton magnetic resonance studies. *Biochemistry* **12,** 5294–5304.
6. Hill, P. S., Smith, D. D., Slaninova, J., and Hruby, V. J. (1990) Bicyclization of a weak Oxytocin agonist produces a highly potent Oxytocin antagonist. *J. Amer. Chem. Soc.* **112,** 3110–3113.
7. Struthers, R. S., Tanaka, G., Koerber, S. C., Solmajer, T., Baniak, E. L., Gierasch, L. M., Vale, W., Rivier, J., and Hagler, A. T. (1990) Design of biologically active,

conformationally constrained GnRH antagonists. *Proteins: Structure, Function, and Genetics* **8(4)**, 295–304.
8. Dutta, A. S., Gormley, J. J., Mclachlan, P. F., and Woodburn, J. R. (1989) Conformationally restrained cyclic peptides as antagonists of LHRH. *Biochem. Biophys. Res. Commun.* **159(3)**, 1114–1120.
9. Coy, D. H., Jiang, N. Y., Kim, S. H., Moreau, J. P., Lin, J. T., Frucht, H., Qian, J. M., Wang, L. W., and Jensen, R. T. (1991) Covalently cyclized agonists and antagonist analogs of Bombesin and related peptides. *J. Biol. Chem.* **266(25)**, 16,441–16,447.
10. Al-Obeidi, F., Hruby, V. J., Hadley, M. E., Sawyer, T. K., and de la Castrucci, A. M. (1990) Design, synthesis, and biological activities of a potent and selective α-melanotropin antagonist. *Inter. J. Pep. Prot. Res.* **35,** 228–234.
11. Pelton, J. T., Gulya, K., Hruby, V. J., Duckles, S. P., and Yamamura, H. I. (1985) Conformationally restricted analogs of somatostatin with high μ-opiate receptor specificity. *Proc. Natl. Acad. Sci. USA* **82,** 236–239.
12. Spear, K. L., Brown, M. S., Reinhard, E. J., McMahon, E. G., Olins, G. M., Paloma, M. A., and Patton, D. R. (1990) Conformational restriction of angiotensin II: cyclic analogues having high potency. *Med. Chem.* **33(7)**, 1935–1940.
13. Barbato, G., D'auria, G., Paolillo, L., and Zanotti, G. (1991) Heterodetic bicyclic decapeptide c(Glu-Leu-Pro-Gly-Ser-Ile-Pro-Ala)-cyclo-(1γ,5β)-Phe-Gly. *Int. J. Peptide Protein Res.* **37(5)**, 388–398.
14. Skala, G., Smith, C. W., Taylor, C. J., and Ludens, J. H. (1984) A conformationally constrained Vasopressin analog with antidiuretic antagonist activity. *Science* **226,** 443–445.
15. Spur, K. L., Reinho, E. J., McMahon, E. G., Palomo, M. A., and Patton, D. R. (1989) Conformationally restricted analogues of atriopeptin (103-125) amide. *J. Med. Chem.* **32,** 67–72.
16. Veber, D. F., Holly, F. W., Paleveda, W. J., Nutt, R. F., Bertstrand, S. J., Torchiana, M., Glitzer, M. S., Saperstein, R., and Hirschmann, R. (1979) Conformationally restricted bicyclic analogs of somatostatin. *Proc. Natl. Acad. Sci. USA* **75,** 2636–2640.
17. Schiller, P. W. and DiMaio J. (1983) Aspects of conformational restriction in biologically active peptides, in *Peptides: Structure and Function. Proceedings of the Eighth American Peptide Symposium* (Hruby, V. J. and Rich, D. H., eds.), Pierce Chemical Co., Rockford, IL, pp. 269–278.
18. Hruby, V. J., Kao, L.-F., Pettitt, M., and Karplus, M. (1988) The conformational properties of the delta opioid peptide [D-Pen^2,D-Pen^5]-enkephalin in aqueous solution determined by NMR and energy minimization calculations. *J. Am. Chem. Soc.* **110,** 3351–3359.
19. Chew, C., Villar, H. O., and Loew, G. H. (1991) Theoretical study of the flexibility and solution conformation of the cyclic opioid peptides (D-Pen^2,D-Pen^5)-enkephalin. *Mol. Pharmacol.* **39(4)**, 502–510.
20. Smith, P. E., Dang, L. X., and Pettitt, B. M. (1991) Simulation of the structure and dynamics of the bis(penicillamine) enkephalin zwitterion. *Amer. Chem. Soc.* **13(1)**, 67–73.

21. Sawyer, T. K., Sanfilippo, P. J., Hruby, V. J., Engel, M. H., Heward, C. B., Burnett, J. B., and Hadley, M. E. (1980) [Nle^4, D-Phe^7]-α-melanocyte stimulating hormone: a highly potent α-melanotropin with ultralong biological activity. *Proc. Natl. Acad. Sci. USA* **77,** 5754–5758.
22. Al-Obeidi, F., Hadley, M. E., Pettitt, B. M., and Hruby, V. J. (1989) Design of a new class of superpotent cyclic alpha-Melanotropins based on quenched dynamics simulations. *J. Amer. Chem. Soc.* **111,** 3413–3416.
23. Trivedi, D., Grummel, M. J., Gysin, B., Sanderson, D., Brendel, K., and Hruby, V. J. (1990) Conformational restriction of glucagon leads to an analog with two distinct binding modes: further evidence that glycogenolysis follows two signal transduction pathways, in *Peptides: Chemistry, Structure, and Biology* (Rivier, J. E. and Marshall, G. R., eds.), Escom, Leiden, pp. 201–202.
24. Charpentier, B., Dor, A., Roy, P., England, P., Pham, H., Durieux, C., and Roques, B. P. (1989) Synthesis and binding affinities of cyclic and related linear analogs of CCK-8 selective for central receptors. *J. Med. Chem.* **32(6),** 1184–1190.
25. Rodriguez, M., Lignon, M. F., Galas, M. C., Amblard, M., and Martinez, J. (1990) Cyclic cholecystokinin analogs that are highly selective for rat and guinea pig central CCK receptors. *Molec. Pharmacol.* **38(3),** 333–341.
26. Felix, A. M., Heimer, E. P., Wang, C.-T., Lambros, T. J., Fournier, A., Mowles, T. F., Maines, S., Campbell, R. M., Wegrzynski, B. B., Toome, V., Fry, D., and Madison, V. (1988) Synthesis, biological activity, and conformational analysis of cyclic GRF analogues. *Int. J. Peptide Protein Res.* **32(6),** 441–454.
27. Mierke, D. F., Schiller, P. W., and Goodman, M. (1990) A conformational comparison of two stereoisomeric cyclic dermorphin analogs employing NMR and computer simulations. *Biopolymers* **29(6–7),** 943–952.
28. Chorev, M., Roubini, E., McKee, R. L., Gibbons, S. W., Goldman, M. E., Caulfield, M. P., and Rosenblatt, M. (1991) Cyclic parathyroid hormone related protein antagonists: lysine-13 to aspartic acid-17 (i + 4) side chain to side chain lactamization. *Biochemistry* **30(24),** 5968–5974.
29. Holzemann, G., Pachler, K. G. R., Eberhart, B., Holzel, H., Kraft, M., and Barnickel, G. (1991) Synthesis, conformational studies, and molecular dynamics calculations of two cyclic tetrapeptides with 17-membered and 18-membered rings. *Int. J. Peptide Protein Res.* **37(4),** 283–292.
30. Krstenansky, J. L., Owen, T. J., Yates, M. T., and Simon, M. (1988) Design, synthesis, and antithrombin activity for conformationally restricted analogs of peptide anticoagulants based on the carboxyl-terminal region of the leech peptide, hirudin. *Biochim. Biophys. Acta* **957(1),** 53–59.
31. Heavner, G. A., Audhya, T., Doyle, D., Tjoeng, F. S., and Goldstein, G. (1991) Biologically active conformations of thymopentin: studies with conformationally restricted analogs. *Int. J. Peptide Protein Res.* **37(3),** 198–209.
32. Ploux, O., Lavielle, S., Chassaing, G., Julien, S., Marquet, A., D'Orleans-Juste, P., Dion, S., Regoli, D., Beaujouan, J.-C., Bergstrom, L., Torrens, Y., and Glowinski, J. (1987) Interaction of tachykinins with their receptors studied with cyclic analogues of substance P and Neurokinin B. *Proc. Natl. Acad. Sci. USA* **84(22),** 8095–8099.

33. Veber, D. F., Freidinger, R., Perlow, D. S., Paeveda, W. J., Jr., Holly, F. W., Strachan, R. G., Nutt, R. F., Arison, B. H., Homnick, C., Randall, W. C., Glitzer, M. S., Saperstein, R., and Hirschmann, R. (1981) A potent cyclic analog of somatostatin. *Nature* **292,** 55–57.
34. Almquist, R. G., Kadambi, S. R., Yasuda, D. M., Weitl, F. L., Polgar, W. E., and Toll, L. R. (1989) Paralytic activity of (des-Glu-1) conotoxin GI analogs in the mouse diaphragm. *Int. J. Peptide Protein Res.* **34(6),** 455–462.
35. Spinella, M. J., Malik, A. B., Everitt, J., and Andersen, T. T. (1991) Design and synthesis of a specific Endothelin 1 antagonist: effects on pulmonary vasoconstriction. *Proc. Natl. Acad. Sci. USA* **88(16),** 7443–7446.
36. Lebl, M., Fric, I., Sugg, E. E., Cody, W. L., and Hruby, V. J. (1987) Conformational studies of carba-analogs of oxytocin, in *Peptides 1986* (Theodoropoulos, D., ed.), Proceedings of the 19th European Peptide Symposium, Walter de Gruyte, Berlin, pp. 341–344.
37. Mosberg, H. I. and Omnaas, J. R. (1985) Dithioether-containing cyclic peptides. *J. Amer. Chem. Soc.* **107(10),** 2986–2987.
38. Siemion, I. Z., Szewczuk, Z., Herman, Z. S., and Stachura, Z. (1991) To the problem of biologically active conformation of enkephalin. *Mol. Cell. Biochem.* **34,** 23–29.
39. Thaisrivongs, S., Blinn, J. R., Pals, D. T., and Turner, S. R. (1991) Conformationally constrained renin inhibitory peptides: cyclic (3-1)-1-(carboxymethyl)-L-propyl-L-phenylalanyl-L-histidinamide as a conformational restriction at the P-2-P-4 tripeptide portion of the angiotensin template. *J. Med. Chem.* **34(4),** 1276–1282.
40. Ozeki, E., Kimura, S., and Imanishi, Y. (1989) Conformation and complexation with metal ions of cyclic hexapeptides: cyclo-(L-Leu-L-Phe–L-Pro)$_2$ and cyclo-(L-Cys(Acm)-L-Phe-L-Pro)$_2$. *Int. J. Peptide Protein Res.* **34(2),** 111–117.
41. Toniolo, C. (1989) Structure of conformationally constrained peptides: from model compounds to bioactive peptides. *Biopolymers* **28,** 247–257.
42. DeGrado, W. F. and Lear, J. D. (1990) Conformationally constrained alpha-helical peptide models for protein ion channels. *Biopolymers* **29(1),** 205–214.
43. Semus, S. F., Becker, E. L., Toniolo, C., and Freer, R. J. (1988) A computer graphic determination of the chemotactic peptide preferred conformation. *Biochem. Biophys. Res. Commun.* **157(2),** 569–574.
44. Iqbal, M., Balarm, P., Showell, H. J., Freer, R. J., and Becker, E. L. (1984) Conformationally constrained chemotactic peptide analogs of high biological activity. *FEBS Lett.* **165(2),** 171–174.
45. Kundu, B., Rizvi, S. Y., Srivastava, V. P., Haq, W., and Mathur, K. B. (1990) Synthesis of some conformationally constrained analogues of muramyl dipeptide (MDP). *Indian Chem., Section B.* **29(5),** 409–413.
46. Beck-Sickinger, A. G., Jung, G., Gaida, W., Koeppen, H., Schorrenberg, G., and Lang, R. (1990) Structure/activity relationships of carboxyl-terminal neuropeptide Y peptide segments and analogues composed of sequence 1-4 linked to 25-36. *Eur. J. Biochem.* **194(2),** 449–456.
47. Singh, T. P., Narula, P., Chauhan, V. S., and Kaur, P. (1989) Crystal structure and molecular conformation of the peptide N-Boc-Gly-dehydro-Phe-methylamide. *Biopolymers* **28(7),** 1287–1294.

48. Kaur, P., Uma, K., Balarm, P., and Chauhan, V. S. (1989) Synthetic and conformational studies on dehydrophenylalanine containing model peptides. *Int. J. Peptide Protein Res.* **33(2),** 103–109.
49. Singh, T. P., Narula, P., Chauhan, V. S., Sharma, A. K., and Hinrichs, W. (1989) Structure of N-Boc-L-Pro-dehydro-Leu-methylamine. *Int. J. Peptide Res.* **33(2),** 167–172.
50. Uma, K., Balarm, P., Kaur, P., Sharma, A. K., and Chauhan, V. S. (1989) Conformation of peptides containing Z-alpha,beta-dehydroleucine: a comparison of Boc-Pro-delta-Z-Leu-Gly-NHEt and Boc-Pro-delta-Z-Phe-Gly-NHEt. *Int. J. Biol. Macromol.* **11(3),** 169–171.
51. Chauhan, V. S., Kaur, P., and Balarm, P. (1989) Conformations of dehydrophenylalanine containing model peptides: NMR studies of an acyclichexapeptide with two Δ-Z-Phenylalanine residues. *Biopolymers* **28(3),** 763–772.
52. Imazu, S., Shimohigashi, Y., Kodama, H., Sakaguchi, K., Waki, M., Kato, J., and Izumiya, M. (1988) Conformationally stabilized gramicidin S analog containing dehydrophenylalanine in place of phenylalanine: synthesis and antimicrobial activity. *Int. J. Peptide Protein Res.* **32(4),** 298–306.
53. Uma, K. and Balarm, P. (1989) Peptide design: an analysis of studies using alpha-amino-isobutyric acid and Z-alpha,beta-dehydro-phenylalanine (Delta-Z-Phe). *Indian J. Chem. B.* **28(9),** 705–710.
54. Ciajolo, M. R., Tuzi, A., Prates, C. R., Fissi, A., and Pieroni, O. (1990) Crystal and molecular structure of Boc-D-Ala-Δ-Phe-Gly-Δ-Phe-D-Ala-OMe: a 3_{10} helical dehydropeptide. *Biopolymers* **30(9–10),** 911–920.
55. Gupta, A. and Chauhan, V. S. (1990) Synthetic and conformational studies on dehydro-alanine containing model peptides. *Biopolymers* **30(3–4),** 395–404.
56. Fisher, G. H., Berryer, P., Ryan, J. W., Chauhan, V., and Stammer, C. H. (1981) Dehydrophenylalanyl analogs of bradykinin: synthesis and biological activities. *Arch. Biochem. Biophys.* **211(1),** 269–275.
57. Grim, M. D., Chauhan, V., Shimohigashi, Y., Kolar, A. J., and Stammer, C. H. (1981) Synthesis of dehydroliberin, ΔzPhe2-TRF. *J. Org. Chem.* **46,** 2671–2673.
58. Felix, A. M., Wan, C. T., Liebman, A. A., Delaney, C. M., Mowles, T., Berrghardt, B. A., Charnecki, A. M., and Meienhofer, J. (1977) Synthesis, biological activity, and tritiation of L-3,4-dehydroproline-containing peptides. *Int. J. Peptide Protein Res.* **10,** 299–310.
59. Wyssbrod, H. R., Fischman, A. J., Willbold, W. M., Jr., Smith, C. W., Walter, R., and Schwartz, I. L. (1981) Comparative study of the neurohypophyseal hormone [8-arginine]vasopressin and its analog [7-(3,4-dehydroproline),8-arginine]vasopressin in aqueous solution. *Int. J. Peptide Protein Res.* **17,** 48–55.
60. Shimohigashi, Y. and Stammer, C. H. (1982) Dehydro-enkephalins VI. Dehydroalanine^3-enkephalin: a potent enkephalin analog for the δ opiate receptor. *Int. J. Peptide Protein Res.* **20,** 199–206.
61. McLean, L. R., Buck, S. H., and Krstenansky, J. L. (1990) Examination of the role of the amphipathic alpha-helix in the interaction of Neuropeptide Y and cyclic analogues with cell membrane receptors and dimyristlyolphosphatidylcholine. *Biochemistry* **29(8),** 2016–2022.

62. Garvey, D. S., May, P. D., and Nadzan, A. M. (1990) 3,4-Disubstituted gamma-lactam rings as conformationally constrained mimics of peptide derivatives containing aspartic acid or norleucine. *J. Org. Chem.* **55(3),** 936–940.
63. Olsen, G. L., Voss, M. E., Hill, D. E., Kahn, M., Madison, V. S., and Cook, M. (1990) Design and synthesis of a protein beta-turn mimetic. *J. Amer. Chem. Soc.* **112(1),** 323–333.
64. Wolf, J. P. and Rapoport, H. (1989) Conformationally constrained peptides: chirospecific synthesis of 4-alkyl-substituted gamma-lactam-bridged dipeptides from L-aspartic acid. *J. Org. Chem.* **54(13),** 3164–3173.
65. Hayashi, K., Nunami, K., Kato, J., Yoneda, N., Kubo, M., Ochiai, T., and Ishida, R. (1989) Studies on angiotensin converting enzyme inhibitors, 4: synthesis and ACE inhibitory activities of 3-acyl-1-alkyl-2-oxoimidazole-4-carboxylic acid derivatives. *J. Med. Chem.* **32,** 289–297.
66. Thaisrivongs, S., Pals, D. T., Turner, S. T., and Kroll, L. T. (1988) Conformationally constrained renin inhibitory peptides: gamma lactam bridged dipeptide isostere as conformational restriction. *J. Med. Chem.* **31,** 1369–1376.
67. Lee, J. P., Dunlap, B., and Rich, D. H. (1990) Synthesis and immunosuppressive activities of conformationally restricted cyclosporin lactam analogs. *Int. J. Pept. Prot. Res.* **35(5),** 481–494.
68. Aebi, J. D., Guillaume, D., Dunlap, B. E., and Rich, D. H. (1988) Synthesis, conformation, and immunosuppressive activity of a conformationally restricted Cyclosporine lactam analogue. *J. Med. Chem.* **31,** 1805–1815.
69. Ward, P., Ewan, G. B., Jordan, C. C., Ireland, S. J., Hagan, R. M., and Brown, J. R. (1990) Potent and highly selective neurokinin antagonists. *J. Med. Chem.* **33,** 1848–1851.
70. Ede, N. J., Lim, N., Rae, I. D., Ng, F. M., and Hearn, M. T. W. (1991) Synthesis and evaluation of constrained peptide analogues related to the amino-terminal region of human growth hormone. *Peptide Res.* **4(3),** 171–176.
71. Valle, G., Crisma, M., Toniolo, C., Yu, K.-L., and Johnson, R. L. (1989) Crystal state structural analysis of two gamma-lactam restricted analogues of Pro-Leu-Gly-NH_2. *Int. J. Peptide Protein Res.* **33(3),** 181–190.
72. Paul, P. K. C., Burney, P. A., Campbell, M. M., and Osguthorpe, D. J. (1990) The conformational preferences of gamma-lactam and its role in constraining peptide structure. *J. Computer-Aided Mole. Design.* **4(3),** 239–254.
73. Kemp, D. S. and Sun, E. T. (1982) Amino acid derivatives that stabilize secondary structure of polypeptides, I: synthesis of LL-3-amino-piperidone-6-carboxylic acid (LL-Acp), A novel beta turn forming amino acid. *Tet. Lett.* **23,** 3759–3761.
74. Nagai, U. and Seto, K. (1985) Synthesis of a bicyclic dipeptide with the shape of a beta-turn central part. *Tet. Lett.* **26,** 647–650.
75. Krstenansky, J. L., Baranowski, R. L., and Currie, B. L. (1982) A new approach to conformationally restricted peptide analogs: rigid beta-bends, 1: enkephalin as an example. *Biochem. Biophys. Res. Comm.* **109,** 1368.
76. Ball, J. B. and Alewood, P. F. (1990) Conformational constraints: nonpeptide beta-turn mimics. *J. Mol. Recog.* **3(2),** 55–64.

77. Toniolo, C. (1990) Conformationally restricted peptides through short-range cyclization. *Int. J. Peptide Protein Res.* **35,** 287–300.
78. Sankararamakrishnan, R. and Viveshwara, S. (1990) Conformational studies on peptides with proline in the right-handed helical region. *Biopolymers* **30(3–4),** 287–298.
79. Yamazaki, T., Probstl, A., Schiller, P., and Goodman, M. (1991) Biological and conformational studies of (valine-4) morphiceptin and (D-valine-4) morphiceptin analogs. Incorporating cis-2-aminocyclopentanecarboxylic acid as a peptidomimetic for proline. *Int. J. Peptide Protein Res.* **37(5),** 364–381.
80. Koskinen, A. M. P. and Rapoport, H. (1989) Synthesis of 4-substituted prolines as conformationally constrained amino acid analogues. *J. Org. Chem.* **54(8),** 1859–1866.
81. Zagari, A., Nemethy, G., and Scheraga, H. A. (1990) The effect of the L-azetidine-2-carboxylic acid residue on protein conformation: I. Conformations of the residue and of dipeptides. *Biopolymers* **30(9–10),** 951–960.
82. Kazmierski, W. M., Yamamura, H. I., and Hruby, V. J. (1991) Topographic design of peptide neurotransmitters and hormones on stable backbone templates: relation of conformation and dynamics to bioactivity. *J. Amer. Chem. Soc.* **113,** 2275–2283.
83. Kazmierski, W., Wire, W. S., Lui, G. K., Knapp, R. J., Shook, J. E., Burks, T., Yamamura, H. I., and Hruby, V. J. (1988) Design and synthesis of somatostatin analogues with topographical properties that lead to highly potent and specific μ opioid receptor antagonists with greatly reduced binding at somatostatin receptors. *J. Med. Chem.* **31,** 2170–2177.
84. Tilstra, L., Sattler, M. C., Cherry, W. R., and Barkly, M. D. (1990) Fluorescence of a rotationally constrained tryptophan derivitive 3-carboxy-1,2,3,4-tetrahydro-2-carboline. *J. Amer. Chem. Soc.* **112(25),** 9176–9182.
85. Hausin, R. J., and Codding, P. W. (1991) Molecular and crystal structures of MDL27,467a hydrochloride and quinapril hydrochloride, two ester derivitives of potent angiotensin converting enzyme inhibitors. *J. Med. Chem.* **34(2),** 511–517.
86. Tourwe, D., Toth, G., Lebl, M., Verscheuren, K., Knapp, R., Davis, P., Van Binst, G., Yamamura, H. I., Burks, T. K., Kramer, T., and Hruby, V. J. (1992) A high yield simple synthesis of 1,2,3,4-tetrahydro-7-hydroxyisoquinoline-3-carboxylic acid (OH-Tic) a conformationally constrained tyrosine analog and its incorporation into opioid peptides, in *Peptides: Chemistry and Biology* (Smith, J. A. and Rivier, J. E., eds.), Escom, Leiden, pp. 307–308.
87. Hruby, V. J., Toth, G., Gehrig, C. A., Kao, L.-F., Knapp, R., Lui, G. K., Yamamura, H. I., Kramer, T. H., Davis, P., and Burks, T. F. (1991) Topographically designed analogues of [D-Pen, D-Pen^5]enkephalin. *J. Med. Chem.* **34,** 1823–1830.
88. Padmanabhan, S., Marqusee, S., Ridgeway, T., Laue, T. M., and Baldwin, R. L. (1990) Relative helix-forming tendencies of nonpolar amino acids. *Nature* **344,** 268–270.
89. Mosberg, H. I., Hurst, R., Hruby, V. J., Gee, K., Yamamura, H. I., Galligan, J. J., and Burks, T. F. (1983) *Bis*-penicillamine enkephalins possess highly improved specificity toward δ opioid receptors. *Proc. Natl. Acad. Sci. USA* **80,** 5871–5873.

90. Manning, M., Lammek, B., Kruszynski, M., Seto, J., and Sawyer, W. H. (1982) Design of potent and selective antagonists of the vasopressor responses to arginine vasopressin. *J. Med. Chem.* **25(4),** 408–414.
91. Schulz, H. and du Vigneaud, V. (1966) Synthesis of 1-L-penicillamine-oxytocin, 1-D-penicillamine-oxytocin, and 1- deaminopenicillamine-oxytocin, potent inhbitors of the oxytocic response of oxytocin. *J. Med. Chem.* **9(5),** 647–650.
92. Minamitake, Y., Furuya, M., Kitajima, Y., Takehisa, M., and Tanaka, S. (1990) Synthesis and biological properties of alpha-human atrial natriuretic peptide analogs with a constrained or stereochemically modified cyclic moiety. *Chem. Pharm. Bull. (Tokyo)* **38(7),** 1920–1926.
93. Nestor, J. J., Ferger, M. F., and du Vigneaud, V. (1975) [1-β-Mercapto-β,β-pentamethylenepropionic acid]oxytocin, a potent inhibitor of oxytocin. *J. Med. Chem.* **18(3),** 284–287.
94. Kruszynski, M., Lammek, B., and Manning, M. (1980) [1-β-Mercapto-β,β-cyclopentamethylenepropionic acid), 2-(*O*-methyl)tyrosine)arginine vasopressin and [1-(β-mercapto)-β,β,(cyclopentamethylenepropionic acid)]arginine vasopressin, two highly potent antagonists of the vasopressor response to arginine-vasopressin. *J. Med. Chem.* **23,** 364–368.
95. Bryan, W. M., Callahan, J. F., Codd, E. E., Lemieux, C., Moore, M. L., Schiller, P. W., Walker, R. F., and Huffman, W. F. (1989) Cyclic enkephalin analogues containing α-amino, β-mercapto-β,β- pentamethylenepropionic acid at positions 2 or 5. *J. Med. Chem.* **32,** 302–304.
96. Hruby, V. J., Fang, S., Knapp, R., Kazmierski, W., Lui, G. K., and Yamamura, H. I. (1990) Cholecystokinin analogues with high affinity and selectivity for brain vs. peripheral membrane receptors, in *Peptides: Chemistry, Structure, and Biology* (Rivier, J. and Marshall, G. R., eds.), Escom, Leiden, pp. 53–55.
97. Rovero, P., Pestellini, V., Patacchini, R., Santicioli, P., Maggi, C. A., and Meli, A. (1987) Synthesis and biological activity of N-methylated analogues of neurokinin A. *Neuropeptides* **10(4),** 355–359.
98. Belleney, J., Gacel, G., Fournie-Zaluski, M. C., Maigret, B., and Roques, B. P. (1989) Delta opioid receptor selectivity induced by conformational constraints in linear enkephalin-related peptides: proton 400 MHz NMR study and theoretical calculations. *Biochemistry* **28(18),** 7392–7400.
99. Spatola, A. F. (1983) Peptide backbone modifications: a structure-activity analysis of peptides containing amide activity analysis of peptides containing amide bond surrogates and conformational constraints, in *Chemistry and Biochemistry of Amino Acids, Peptides and Proteins,* vol. 7 (Weinstein, B., ed.), Marcel Dekker, New York, pp. 267–357.
100. Marshall, G. R., Humblet, C., van Oplenbosch, N., and Zabrocki, J. (1981) Peptide bond modification and its effect on conformational mimicry, in *Peptides: Synthesis, Structure, Function* (Rich, D. H. and Gross, E., eds.), Pierce Chemical Co., Rockford, IL, pp. 669–672.
101. Liakopoulou-Kyriakides, M. and Galardy, R. E. (1979) s-Cis and s-Trans isomerism of the His-Pro peptide bond in angiotensin and thyroliberin analogues. *Biochemistry* **18,** 1952–1957.

102. Smith, G. D., Zabrocki, J., Flak, T. A., and Marshall, G. R. (1991) Conformational mimicry: II. An obligatory cis amide bond in a small linear peptide. *Int. J. Peptide Protein Res.* **37(3),** 191–197.
103. Zabrocki, J., Dunbar, J. B., Jr., Marshall, K. W., Toth, M. V., and Marshall, G. R. (1992) Conformational mimicry. 3. Synthesis and incorporation of 1,5-disubstituted tetrazole dipeptide analogues into peptides with preservation of chiral integrity: bradykinin. *J. Org. Chem.* **57,** 202–209.
104. Zabrocki, J., Slomczynska, U., and Marshall, G. R. (1990) Synthesis of a somatostatin analog containing a tetrazole cis-amide bond surrogate, in *Peptides: Chemistry, Structure, and Biology* (Rivier, J. E. and Marshall, G. R., eds.), Escom, Leiden, pp. 195–197.
105. Lebl, M., Slaninova, J., and Johnson, R. L. (1989) Analogues of oxytocin containing a pseudopeptide Leu-Gly bond of *cis* and *trans* configuration. *Int. J. Peptide Protein Res.* **33(1),** 16–21.
106. Vander Elst, P., Van den Berg, E., Pepermans, H., Vander Auwera, L., Zeeuws, R., Tourwe, D., and Van Binst, G. (1987) Synthesis and conformational study of a cyclic hexapeptide analogue of Somatostatin: cyclo(Phe-D-Trp-Lys-Thr-O-AMPA). *Int. J. Peptide Protein Res.* **29(3),** 318–330.
107. Van der Elst, P., Gondol, D., Wynants, C., Tourwe, D., and Van Binst, G. (1987) Synthesis and conformational study of two cyclic analogues of somatostatin containing an AMPA-spacer unit. *Int. J. Peptide Protein Res.* **29,** 331–346.
108. Shue, Y.-K., Correra, G. M., Jr., and Nadzen, A. M. (1987) Amide bond surrogates: a general synthetic route to trans carbon-carbon double bond isosteres. *Tet. Lett.* **28,** 3225–3228.
109. Coy, D., Wang, L. H., Jiang, N. Y., and Jensen, R. (1990) Short chain bombesin pseudopeptides with potent bombesin receptor antagonist activity in rat and guinea pig pancreatic acinar cells. *Eur. J. Pharm.* **190(1–2),** 21–38.
110. Qian, J. M., Coy, D. H., Jiang, N. Y., Gardner, J. D., and Jensen, R. T. (1989) Reduced peptide bond pseudopeptide analogues of Substance P: a new class of Substance P receptor antagonists with enhanced specificity. *J. Biol. Chem.* **64(28),** 16,667–16,671.
111. Haffar, B. M., Hocart, S. J., Coy, D. H., Mantey, S., Chiang, H. C. V., and Jensen, R. T. (1991) Reduced peptide bond pseudopeptide analogs of Secretin: a new class of Secretin receptor antagonists. *J. Biol. Chem.* **266(1),** 316–322.
112. Sherman, D. B., Spatola, A. F., Wire, W. S., Burks, T. F., Nguyen, T. M. D., and Schiller, P. W. (1989) Biological activities of cyclic Enkephalin pseudopeptides containing thioamides as amide bond replacemements. *Biochem. Biophys. Res. Commun.* **162(3),** 1126–1132.
113. Guthrie, R. W., Kaplan, G. L., Mennona, F. A., Tilley, J. W., Kierstead, R. W., O'Donnell, M., Crowley, H., Yaremko, B., and Welton, A. F. (1990) Propenyl carboxamide derivitives as antagonists of platelet activating factor. *J. Med. Chem.* **33(10),** 2856–2864.
114. Crisma, M., Valle, G., Bonora, G. M., DeMenego, E., Toniolo, C., Lelj, F., Barone, V., and Fraternali, F. (1990) Structural versatility of peptides from C-

alpha,alpha-disubstituted glycines: preferred conformation of the C-alpha,alpha-diphenylglycine residue. *Biopolymers* **30(1–2),** 1–12.

114a. Jiao, D., Russell, K. C., and Hruby, V. J. (1993) Locally constrained tyrosine analogues with restricted sidechain dynamics. *Tetrahedron* **49,** 3511–3520.

115. Valle, G., Crisma, M., Toniolo, C., Polinelli, S., Boesten, W. H. J., Schoemaker, H. E., Meijer, E. M., and Kamphuis, J. (1991) Peptides from chiral C-alpha,alpha-disubstituted glycines: crystallographic characterization of conformation of C-alpha-methyl, C-alpha-isopropylglycine (alpha-methyl Valine) in simple derivatives and model peptides. *Int. J. Peptide Protein Res.* **37(6),** 521–527.

116. Santini, A., Barone, V., Bavaso, A., Benedetti, E., DiBlasio, B., Fraternali, F., Lelj, F., Pavone, V., Pedone, C., Crisma, M., Bonora, G. M., and Toniolo, C. (1988) Structural versatility of peptides from C-alpha,alpha-dialkylated glycines: a conformational energy calculation and X-ray diffraction study of homopeptides from 1-aminocyclopentane-1-carboxylic acid. *Int. J. Biol. Macro.* **10(5),** 292–299.

117. Sukumar, M., Raj, P. A., Balarm, P., and Becker, E. L. (1985) A highly active Chemotactic peptide analog incorporating the unusual residue 1-aminocyclohexanecarboxylic acid at position 2. *Biochem. Biophys. Res. Commun.* **128(1),** 339–344.

118. Hlavacek, J., Pospisek, J., Slaninova, J., Chan, W. Y., and Hruby, V. J. (1987) Oxytocin analogues with non-coded amino acid residues in position 8: [8-neopentylglycine]oxytocin and [8-cycloleucine]oxytocin. *Coll. Czech. Chem. Comm.* **52,** 2317–2325.

119. Rovero, P., Pestellini, V., Rhaleb, N. E., Dion, S., Roussi, N., Tousignant, C., Telemaque, S., Drapeau, G., and Regoli, D. (1989) Structure-activity studies of neurokinin A. *Neuropeptides* **13(4),** 263–270.

120. Gavuzzo, E., Lucente, G., Mazza, F., Zechini, G. P., Paradisi, M. P., Pochetti, G., and Torrini, I. (1991) Synthesis and properties of chemotactic peptide analogs: I. Crystal structure and molecular conformation of HCO-Met-Leu-Ain-OMe. *Int. J. Peptide Protein Res.* **37(4),** 268–276.

121. Hsieh, K., Lattan, T. R., and Speth, R. C. (1989) Topographic probes of angiotensin and receptor: potent angiotensin II agonist containing diphenylalanine and long-acting antagonists containing biphenylalanine and 2-Indane amino acid in position 8. *J. Med. Chem.* **32,** 898–903.

122. Dyckes, D. F., Nestor, J. J., Jr., Ferger, M. F., and du Vigneaud, V. (1974) [1-β-Mercapto-β,β-diethylpropionic acid]-8-lysine-vasopressin, a potent inhibitor of 8-lysine-vasopressin and of oxytocin. *J. Med. Chem.* **17,** 250–252.

123. Kruszynski, M., Lammek, B., Manning, M., Seto, J., Haldar, J., and Sawyer, W. H. (1980) [1-β-Mercapto-β,β-cyclopentamethylenepropionic acid), 2-(*O*-methyl)tyrosine] arginine-vasopressin and [1-β-mercapto-β,β-cyclopentamethylenepropionic acid)]arginine-vasopressine, two highly potent antagonists of the vasopressor response. *J. Med. Chem.* **23,** 364–368.

124. Bryan, W. M., Callahan, J. F., Codd, E. E., Lemieux, C., Moore, M. L., Schiller, P. W., Walker, R. F., and Huffman, W. F. (1989) Cyclic enkephalin analogues

containing α-amino β-mercapto-β,β-pentamethylenepropionic acid at position 2 or 5. *J. Med. Chem.* **32,** 302–304.

125. Osawa, T., Kodama, H., Yoshioka, K., and Shimohigashi, Y. (1990) Enzyme-inhibitory conformation of dipeptides containing sterically constrained amino acid 2,3 methanophenylalanine. *Peptide Res.* **3(1),** 35–41.
126. Subasinghe, N., Schulte, M., Chan, M. Y. M., Roon, R. J., Koerner, J. F., and Johnson, R. L. (1990) Synthesis of acyclic and dehydroaspartic acid analogues of Ac-Asp-Glu-OH and their inhibition of rat brain *N*-acetylated alpha-linked acidic dipeptidase (NAALA dipeptidase). *J. Med. Chem.* **33(10),** 2734–274.

CHAPTER 12

Solid-Phase Synthesis of Peptides Containing the CH_2NH Reduced Bond Surrogate

Michael W. Pennington

1. Introduction

There has been considerable interest in the past decade in the design and development of competitive peptide agonists and antagonists for numerous peptide-receptor systems. Systematic side-chain replacement is often the first step in the design process of higher-affinity ligands. Modification of the peptide backbone is another step in the design process, but requires more information about the stability and structure of the peptide. The most common types of backbone alterations are the incorporation of N-methylated amino acids and/or D amino acids into the peptide. These changes can help stabilize the peptide against proteases, disrupt or induce secondary structural elements, and eliminate hydrogen bonds. In addition to these approaches, replacement of peptide bonds by peptide bond surrogates, such as a CH_2NH reduced peptide bond, can also be performed. The initial approach to these type of compounds was exceedingly more difficult to accomplish and required considerable solution synthesis *(1)*. In this chapter, the protected dipeptide containing the reduced bond was prepared in solution by reductive alkylation with the protected α-amino aldehyde and subsequently coupled to the peptide chain. Low coupling yields were often described when this method was attempted probably because of isosteric components present during the reaction *(2)*.

From: *Methods in Molecular Biology, Vol. 35: Peptide Synthesis Protocols*
Edited by: M. W. Pennington and B. M. Dunn

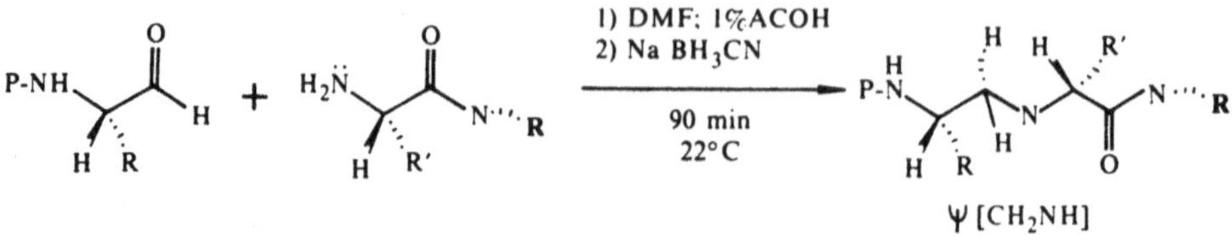

Fig. 1. Reductive alkylation scheme for the synthesis of CH_2NH isostere.

Since this early report, a procedure was developed by Sasaki and Coy *(2)* that greatly simplified the entire process by exploiting the solid-phase resin to eliminate the laborious postreaction work-up steps. This approach relies on the same overall scheme of reductive alkylation using N-protected α-amino aldehydes in the presence of $NaBH_3CN$ (Fig. 1). However, the reduced bond is formed *in situ* on the resin with no need for subsequent work-up.

This chapter is aimed at introducing this technology to peptide chemists who may have little or no experience working with or preparing noncommercially available amino acid derivatives. This chapter employs the same overall strategy as developed by Sasaki and Coy *(2)* to prepare the reduced bond isostere CH_2NH on a solid-phase support. In addition, a description of the preparation of the starting material, N-protected α amino aldehyde is presented. This chapter assumes some knowledge of basic organic and peptide chemistry techniques.

2. Materials

1. Boc-amino acid derivatives may be obtained from Bachem Bioscience Incorporated (King of Prussia, PA). Boc-α-amino alcohols can also be purchased from the same vendors as custom synthesis products.
2. Dimethyl sulfoxide (DMSO), pyridine-sulfur trioxide complex (pyr.SO_3), isobutyl chloroformate, sodium cyanoborohydride, sodium borohydride, 1,2 dimethoxyethane (DME), *N*-methyl morpholine (NMM),*O,N*-Dimethyl-hydroxylamine HCl, triethyl amine (TEA), dicyclohexylcarbodiimide (DCC), lithium aluminum hydride (LAH), and potassium hydrogen sulfate benzo-triazole-1-yloxytris-[dimethylamino]-phosphonium hexafluorophosphate (BOP) may be obtained from Aldrich (Milwaukee, WI).
3. All solvents, such as diethyl ether, tetrahydrofuran, dimethyl formamide, dichloromethane, and methanol, must be anhydrous. These can be purchased from Fisher (Fair Lawn, NJ), Burdick Jackson (McGaw Park, IL), J. T. Baker (Phillipsburg, NJ), or Aldrich. These solvents should be dried over Zeolite (molecular sieves) or redistilled prior to use.

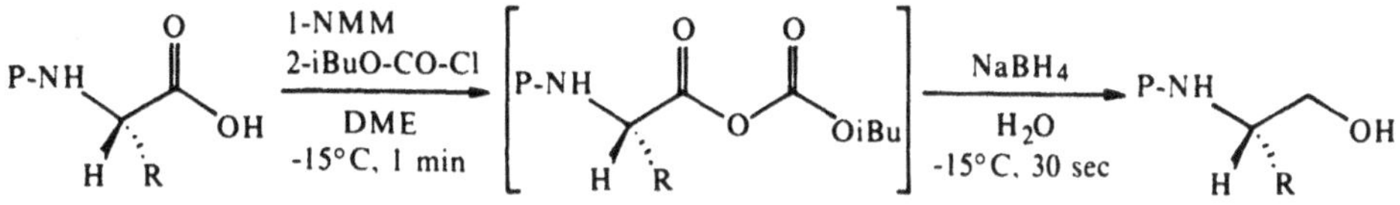

Fig. 2. Synthesis of N-protected α-amino alcohol from α-amino acids.

3. Methods

To successfully introduce a CH_2NH reduced peptide bond into a solid-phase synthetic scheme, it is necessary to prepare the N-protected α-amino aldehyde of the desired residue. The aldehyde derivative must be freshly prepared and promptly used because of its limited stability. The aldehyde derivative may be conveniently prepared from either the Boc-α amino alcohol *(3–5)* or from the Boc amino acid directly *(6)*. Each of these procedures has been used successfully, and a detailed description of preparation from the protected α-amino alcohol as well as from the N-protected α-amino acid is presented below because of their relative simplicity.

3.1. Preparation of N-Protected α-Amino Alcohol

Using a mixed anhydride obtained from the Boc-α-amino acid by reaction with isobutyl chloroformate in 1,2 dimethoxyethane, reacted with 1.5 Eq of aqueous sodium borohydride results in the Boc-α-amino alcohol (Fig. 2) *(7)*.

1. Cool 20 mL of DME to –15°C using a salt-ice bath.
2. Add 20 mmol of N-protected α-amino acid to the cold DME.
3. Successively add 2.2 mL (20 mmol) of NMM and 2.72 mL (20 mmol) of isobutyl chloroformate.
4. After 1–2 min, the precipitated NMM HCl salt is removed by filtration and washed 5 times in 2 mL of cold DME. The filtrate and the washings are combined in a 1-L flask and placed back in the salt-ice slurry.
5. Dissolve 1.140 g (30 mmol) of sodium borohydride in 10 mL of water, and add this in one batch to the cold DME solution. This reaction will quickly evolve gas.
6. Quickly add 500 mL of water. The resulting protected α-amino alcohol will usually precipitate, and can be collected over a Buchner funnel and washed with water and hexane. Occasionally, the compound must be extracted with ethyl acetate or *n*-butanol.

3.2. Preparation of the Protected α-Amino Aldehyde

The protected α-amino alcohol must now be oxidized to the aldehyde before it can be coupled to the resin-bound peptide chain. Other proce-

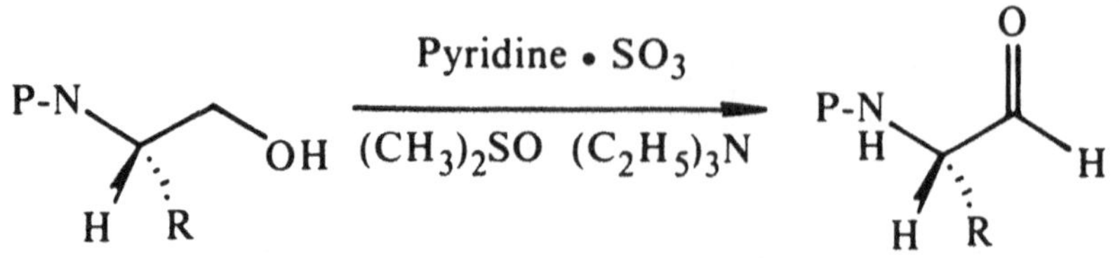

Fig. 3. Synthesis of N-protected amino aldehydes from α-amino alcohol.

dures exist for accomplishing this oxidation *(4,5)*, but low yields resulting from a purification step over silica gel, which resulted in racemization of the amino aldehyde, hamper these methods. The method described by Hamada and Shiori *(3)* utilizes a combination of sulfur trioxide–pyridine complex and DMSO in the presence of TEA (also known as the Parikh-Doering oxidation) to accomplish this oxidation (Fig. 3).

1. Dissolve the protected α-amino alcohol (10 mmol) in 30 mL of anhydrous DMSO containing TEA (3.035 g, 30 mmol). Stir this solution at room temperature, purge the vessel with nitrogen, and maintain an inert atmosphere.
2. Initiate the oxidation by adding a solution of sulfur trioxide–pyridine complex dissolved in 30 mL of anhydrous DMSO over 3–5 min.
3. The reaction progress can be conveniently monitored by TLC using chloroform-ethyl acetate (9:1) as the mobile phase.
4. Stir the reaction for 30–90 min under inert atmosphere, and follow reaction by TLC.
5. Pour entire reaction mixture into 300 mL of ice-water.
6. Extract this mixture with diethyl ether (4 × 200 mL) saving each of the organic layers.
7. The combined organic layers are then washed successively with 10% aqueous citric acid (2 × 200 mL), water (2 × 200 mL), and saturated sodium bicarbonate (2 × 200 mL), and subsequently dried over $MgSO_4$.
8. The solvent is removed by rotary evaporation under reduced pressure resulting in the protected α-amino aldehyde. Certain amino aldehydes can be recrystallized, such as Boc-Phe-al, and Boc-Ala-al, Boc-Tyr(Bzl)-al, whereas several are oils, such as Boc-Leu-al, Boc-Val-al, and Boc-Pro-al.

3.3. Synthesis of the Boc-Amino Aldehyde from the Boc-Amino Acid by Reduction of the N'*-Methoxy-*N*-Methyl-α-(Boc-Amino)-Carboxamide*

As an alternative to preparation of the Boc-amino alcohol, one can also prepare the aldehyde by reduction of the protected intermediate *N'*-

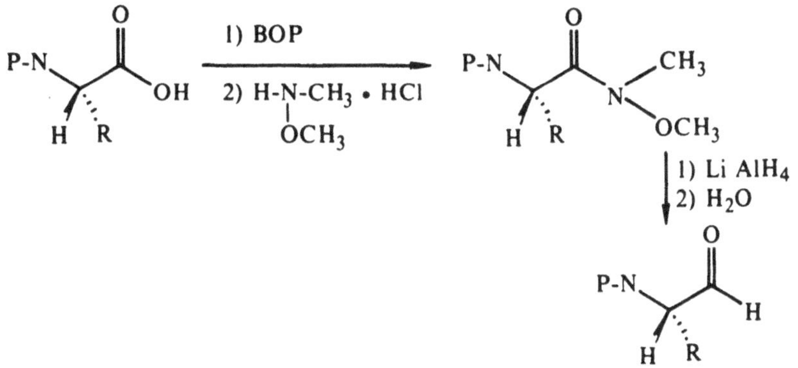

Fig. 4. Synthesis of Boc-amino aldehyde by reduction of the *N*′-methoxy-*N*-methyl-α-(Boc-amino)-carboxamide.

methoxy-*N*-methyl-α-(Boc-amino)-carboxamide with lithium aluminum hydride *(6)*. These products are obtained in a relatively good yield with a high degree of optical purity (Fig. 4).

1. Dissolve 10 mmol of Boc-amino acid in 50–100 mL of DCM.
2. Add 10 mmol (1.012 g) of triethylamine to the stirred amino acid solution.
3. Dissolve 10 mmol (4.42 g) of the BOP reagent to this solution, and allow to mix for 10 min.
4. Dissolve 11 mmol (1.113 g) of *O,N*-dimethylhydroxylamine HCl to this solution, and allow to mix for 2 h at room temperature. Monitor the pH of the solution and maintain a pH of 7.0 by adding drops of TEA.
5. The reaction mixture is diluted to 250 mL with DCM and subsequently extracted (3 × 75 mL) (3*N* HCl, followed by (3 × 75 mL) of saturated $NaHCO_3$ solution, followed by (3 × 75 mL) of saturated NaCl solution.
6. The organic layer is retained and dried over $MgSO_4$.
7. The solvent is removed *in vacuo*.
8. A portion of the resulting product the *N*-methyl-α-(Boc-amino)-carboxamide (2.5 mmol) is dissolved in either 25 mL of diethyl ether or tetrahydrofuran depending on the relative solubility.
9. Reduction of this carboxamide is initiated by adding 2.5 mmol (95 mg; 5 Eq) of lithium aluminum hydride to the stirred solution. The reduction is complete within 0.5 h.
10. The mixture is then hydrolyzed with 10 mL 0.35*M* $KHSO_4$ in water.
11. This solution is subsequently diluted with ether and extracted as described previously in step 5.
12. The organic layer is retained and dried over $MgSO_4$.

13. The solvent is subsequently removed *in vacuo* resulting in the Boc-amino aldehyde.

3.4. Synthesis of the Reduced Bond CH_2NH on the Resin-Bound Peptide

At this point, the reduced peptide bond may be introduced to the resin-bound peptide. The N-terminal-protecting group of the resin-bound peptide must be removed prior to performing the reductive alkylation step. The CH_2NH peptide bond is subsequently introduced by the reductive alkylation reaction between the protected α-amino aldehyde and the amine on the resin-bound peptide using sodium cyanoborohydride in an acidified DMF solution by the procedure developed by Sasaki and Coy *(2)*.

1. Dissolve the protected α-amino aldehyde (4 Eq) in DMF containing 1% acetic acid by volume.
2. Check the resin-bound peptide by the Kaiser test *(8)* to ensure that the N-terminal residue has been deblocked.
3. Add the above solution to the resin-bound peptide and mix for 5 min.
4. Add the sodium cyanoborohydride (4 Eq) portionwise, and continue to mix the solution for 1.5 h.
5. Complete the synthetic cycle by washing peptidyl-resin with DMF (2 × 20 mL), DCM (2 × 20 mL), ETOH (2 × 20 mL), and DCM (2 × 20 mL). Check for completeness of reaction by performing a Kaiser test *(8)*. A negative test indicates a successful coupling. A positive test requires a recoupling step by repeating steps 1, 3, 4, and 5.
6. Following successful coupling, solid-phase assembly of the remainder of the peptide can be resumed by conventional Merrifield strategy *(9)* (*see* Notes 1–5).

4. Notes

1. Capping protocols should not be employed from this point to prevent capping the reactive secondary amine that has been created at the CH_2NH bond. The secondary amine generated has poor reactivity because of steric factors. However, a small reactive acylating species, such as acetic anhydride, will react much more readily than a Boc-amino acid derivative.
2. The unmasked secondary amine has been shown not to be a serious problem in the synthesis of several peptides using a DCC-mediated acylation *(10–12)*. However, a less hindered amino acid, such as Boc-Gly, was reactive at this site. Sasaki and Coy *(2)* found that Boc-Gly coupling at the secondary amine was reduced by using the HOBT ester instead of the DCC-generated symmetric anhydride.

3. Amino acid analysis results will reflect a loss of the dipeptide generated by incorporating the CH_2NH bond. This a nonhydrolyzable bond that is stable to acid as well as to proteases.
4. The CH_2NH bond is stable to HF cleavage conditions (*see* Chapter 4) and requires no special scavengers.
5. Protection of the reduced bond secondary amine can be accomplished if desired by reacting the peptidyl resin with benzyl chloroformate (carbobenzoxy-chloride) prior to deprotection of the Boc group.

References

1. Szelke, M., Leckie, B., Hallet, A., Jones, D. M., Sueiras, J., Atrash, B., and Lever, A. F. (1982) Potent new inhibitors of human renin. *Nature* **299,** 555–557.
2. Sasaki, Y. and Coy, D. H. (1987) Solid phase synthesis of peptides containing the CH_2NH peptide bond isostere. *Peptides* **8,** 119–121.
3. Hamada, Y. and Shiori, T. (1982) New methods in organic synthesis: 29: a practical method for the preparation of optically active N-protected α-amino aldehydes and peptide aldehydes. *Chem. Pharm. Bull.* **30,** 1921–1924.
4. Pfitzner, K. and Moffatt, J. (1970) Sulfoxide-carbodiimide reactions. I. A facile oxidation of alcohols. *J. Am. Chem. Soc.* **87,** 5661–5670.
5. Stanfield, C. F., Parker, J. E., and Kanellis, P. (1981) Preparation of protected amino aldehydes. *J. Org. Chem.* **46,** 4797–4798.
6. Fehrentz, J. and Castro, B. (1983) An efficient synthesis of optically active α-(*t*-butyloxycarbonylamino)-aldehydes from α-amino acids. *Synthesis* 676–678.
7. Rodriguez, M., Llinares, M., Doulut, S., Heitz, A., and Martinez, J. (1991) A facile synthesis of chiral N-protected α-amino alcohols. *Tetrahedron Lett.* **32,** 923–926.
8. Kaiser, E., Colescott, R., Bossinger, C., and Cook, P. (1970) Color test detection of free terminal amine groups in the solid-phase synthesis of peptides. *Anal. Biochem.* **34,** 595–598.
9. Merrifield, R. B. (1963) Solid phase peptide synthesis: the synthesis of a tetrapeptide. *J. Am. Chem. Soc.* **85,** 2149–2153.
10. Martinez, J., Bali, J., Rodriguez, M., Castro, B., Magous, R., Laur, J., and Ligon, M. (1985) Synthesis and biological activities of some pseudo-peptide analogues of tetragastrin: the importance of the peptide backbone. *J. Med. Chem.* **28,** 1874–1879.
11. Coy, D. H., Heinz-Erian, P., Jiang, N., Sasaki, Y., Taylor, J., Moreau, J., Wolfrey, W. T., Gardner, J., and Jensen, R. (1988) Probing peptide backbone function in bombesin. *J. Biol. Chem.* **263,** 5056–5060.
12. Haffar, B., Hocart, S., Coy, D., Mantey, S., Chiang, V., and Jansen, R. (1991) Reduced peptide bond pseudopeptide analogues of secretin. *J. Biol. Chem.* **266,** 316–322.

CHAPTER 13

Approaches to the Asymmetric Synthesis of Unusual Amino Acids

Victor J. Hruby and Xinhua Qian

1. Introduction

This chapter will focus on critically evaluating some selected new methods to achieve the asymmetric synthesis of α-amino acids. It is not meant to be comprehensive, but rather to reflect some of the possibilities opening up in this area.

The study of α-amino acids is one of fundamental importance to many areas of chemistry and its relation to molecular biology. Nature only utilizes 20 amino acids in the production of polypeptides on genes, yet the combinations of these amino acids have provided a wonderful diversity of chemical structures and functions. The number of naturally occurring or synthetically derived nonproteogenic α-amino acids is rapidly increasing and, depending on definitions, might already have exceeded 1000. In addition, because of the significant advances in the technology of synthesizing polypeptides (e.g., the solid-phase Merrifield method), the possibilities in the design and synthesis of new enzymes, hormones, synthetic immunostimulants, drugs, and countless other important biopolymers have been dramatically increased. Certain nonproteogenic amino acids have already proven of considerable experimental value in probing amino acid chemistry and function, as a tool for enhanced understanding of the roles and functions of proteins, in understanding the chemistry and biochemistry of interactions in living systems, and as analogs of naturally occurring hormones, neurotransmitters, growth factors, enzyme inhibitors, neuromodulation, immunomodulators, and many other biologically significant compounds.

From: *Methods in Molecular Biology, Vol. 35: Peptide Synthesis Protocols*
Edited by: M. W. Pennington and B. M. Dunn

In the past several years, the asymmetric synthesis of α-amino acids and their derivatives has become a highly active area of research, and several reviews have appeared on the chemistry and biochemistry of α-amino acids and their uses. In 1988 and 1989, two important reviews and discussions on asymmetric synthesis of α-amino acids appeared *(1,2)* Both were extensive and well organized. Hence, this chapter will cover primarily studies since that time with special emphasis on α-amino acids that can be used in the design of conformationally and topographically constrained peptides.

2. Synthetic Methods

Since so many methodologies have been established for the asymmetric α-amino acids synthesis, it often is difficult to select the most appropriate methodology for constructing the amino acid of immediate interest. Thus far, despite all the methodologies available, there is no one single best method that a laboratory may use to solve every amino acid problem that may be encountered. In a recent report *(3)* Schmidt et al. summarized four general methodologies used for the preparation of optically active, nonribosomal α-amino acids and α-alkylamino acids:

1. Alkylation or amination of optically active enolates;
2. Alkylations with optically active, electrophilic glycine compounds;
3. Diastereoselective Strecker and Ugi reactions with optically active amines; and
4. Diastereoselective hydrogenation of α, β-dehydroamino acid derivatives.

Except for the last method, nearly all of the above processes give rise predominantly or exclusively to compounds in either the *S*- or the *R*-series. In these cases, the enantiomer often is considerably more difficult to obtain because usually only one of the two enantiomers of the optically active auxiliary reagent employed (amino acids, amino alcohols) is "cheaply" available. Meanwhile, numerous miscellaneous methods have been developed, some of which are very efficient and show excellent control of both stereochemistry and regiochemistry. Biologically related methods also are very attractive these days. Enzymatic, chemoenzymatic, and cell-free biosynthesis of nonproteinogenic α-amino acids has shown good results *(4–9)*. Since the α-amino acids generally are not the final goal of the research, it often is important that the method is amenable to rapid implementation.

3. Syntheses of α-Amino Acids

3.1. Asymmetric Derivatization of Glycine and Other Proteogenic α-Amino Acids

Derivatization of glycine is one of the most frequently used approaches for the preparation of α-amino acids. Since glycine is the simplest amino acid, derivatization of glycine (including "double derivatization" to make α, α'-disubstituted α-amino acids) can, in principle, provide an infinite variety of α-amino acids. Numerous approaches have been devised and previously reviewed *(1–3)*. Here, we will focus on the work that has been done during the past 3 yr.

Williams' group has been a leader in this area with the development of several specific methodologies *(10)*. Most recently, Williams and associates reported an asymmetric synthesis of 2,6-diamino-6-(hydroxymethyl) pimelic acid using this approach *(11)*. (5R, 6S)-4- (benzyloxycarbonyl)-5,6-diphenyl-2,3,5,6-tetrahydro-1,4-oxazin-2-one (**1**) underwent electrophilic alkylation to give the allyoxazinone **3** (Scheme 1), which then underwent another electrophilic alkylation at the same position to give the (methoxymethyl)homoallyoxazinone **4** in high yield. The adduct **4** was ozonolized to aldehyde **5**, which was coupled with enolate **6** prepared from the enatiomer of starting material **1** (5S, 6R)-4-(benzyloxycarbonyl)-5,6-diphenyl-2,3,5,6-tetrahydro-1,4-oxazin-2-one (**2**). The enantioselectivity is good (Scheme 1). After separation, the major product of coupling reaction, **7**, was treated with phenyl chlorothionoformate along with bis (trimethylsilyl) amide to give the thionoformate **9**. Reduction of **9** provided the product **10** (and **11**; some racemization occurred in this step). After separation, the dilactone **10** was hydrogenated to give the amino acid **12**, which was then directly converted into (2S, 6R)-2,6-diamino-6-(hydroxymethyl)pimelic acid **13**. (2S, 6S)-**13** also was synthesized by the same protocol, starting with **2**; the diastereoselectivities was better in the latter case.

Based on their experiences *(12)*, Belokon and associates recently reported an asymmetric synthesis of a series of α-amino acids via alkylation of the chiral nickel(II) Schiff base complex of glycine and alanine *(13)*. The overall yield of the reaction is good, but the diastereoselectivity is not high. At about the same time, they also reported another asymmetric synthesis of 4-substituted proline derivative via condensation of "Glycine" with olefins *(14)*. The "Glycine" here is a chiral Ni(II) complex of

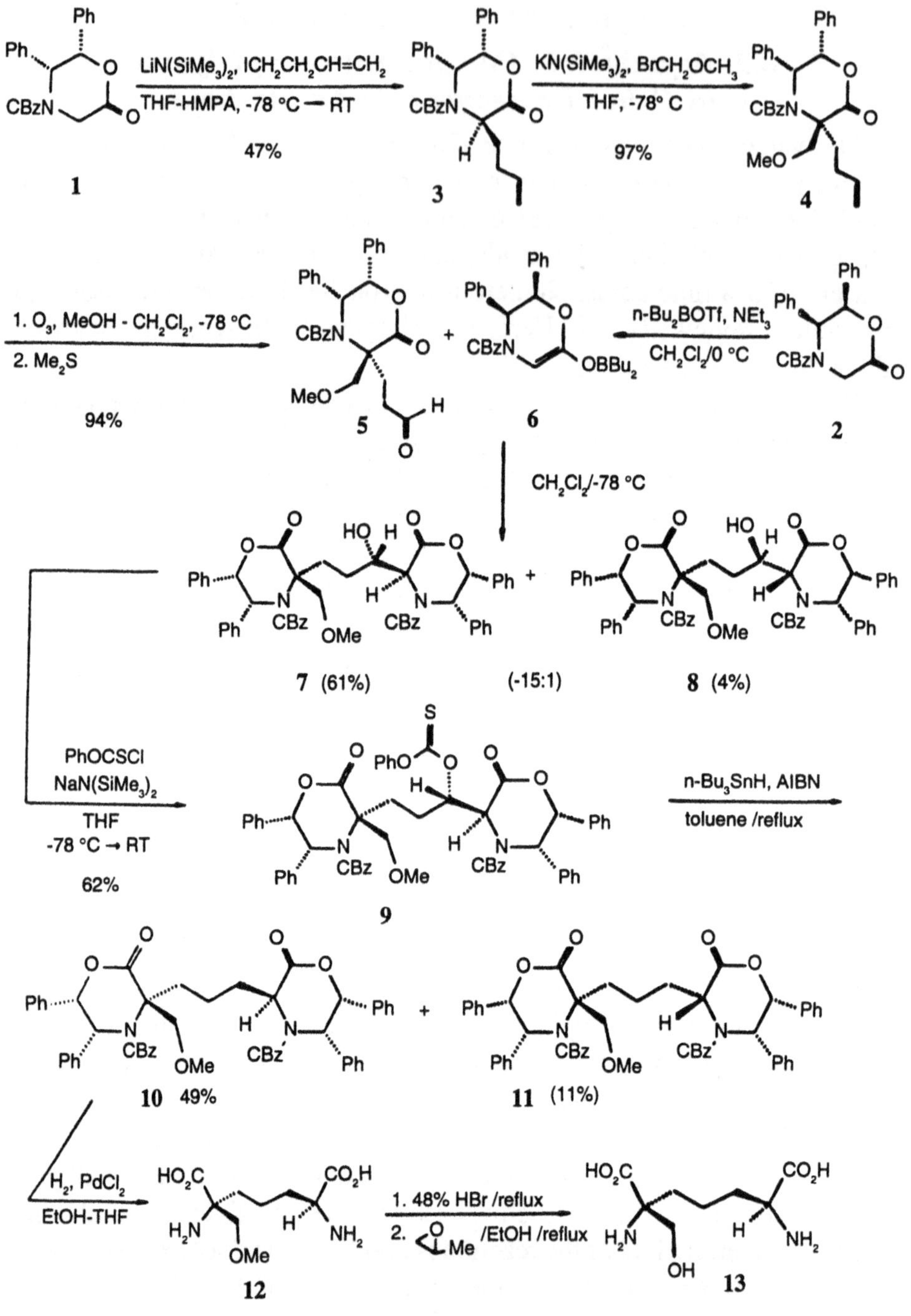
LiN(SiMe3)2, ICH2CH2CH=CH2
THF-HMPA, -78 °C → RT
47%
KN(SiMe3)2, BrCH2OCH3
THF, -78° C
97%
1
3
4
1. O3, MeOH - CH2Cl2, -78 °C
2. Me2S
94%
5
6
2
n-Bu2BOTf, NEt3
CH2Cl2/0 °C
OBBu2
CH2Cl2/-78 °C
7 (61%)
(-15:1)
8 (4%)
PhOCSCl
NaN(SiMe3)2
THF
-78 °C → RT
62%
9
n-Bu3SnH, AIBN
toluene /reflux
10 49%
11 (11%)
H2, PdCl2
EtOH-THF
12
1. 48% HBr /reflux
2. /EtOH /reflux
13

Scheme 1.

glycine. The diastereoselectivity at C_α (90%) and C_β is high. Nebel and Mutter reported a stereoselective synthesis of isovaline (IVA) and IVA-containing dipeptides *(15)*. The starting material, which acts as the chiral inducer, is a methylated glycine chiral template (oxazolidinone derivative). The key step is the alkylation of the α-position of an α-methylated glycine template under basic condition. The synthetic route is only three steps and provided isovaline with high yield along with high diastereoselectivity (>99%). Guanti and associates synthesized β-hydroxy-α-amino acids by utilizing dibenzylaminoacetates as synthetic equivalents of glycine *(16)*. The starting dibenzylaminoacetates were treated with LDA to afford lithium enolates and, following acidic aldol condensation with silyl ketene acetals, yielded predominant *syn* adducts with selectivity from 5:1 to 32:1. The best results (in terms of yields and stereoselectivity) were from an acylation–reduction process (aldol reaction with an acid chloride). The selectivity of the *syn* isomer was >13:1. Dellaria and Santarsiero *(17)* reported the enantioselective synthesis of α-amino acids derivatives via stereoselective alkylation of a homochiral glycine enolate. A simple one-pot, three-step deprotection provided the final α-alkylated glycine.

Schollkopf's group has been one of the pioneers in this area *(18,19)*. Recently, they reported a new asymmetric synthesis of (2R, 3S)-*threo*-3-arylserine derivatives *(20)*, using a titanium derivative of the bislactim ether of *cyclo*(-L-Val-Gly) as a chiral template. Simple hydrolysis afforded (2R, 3S)-*threo*-arylserine methyl ester. Mittendorf and Hartwig *(21)* reported a synthesis of 2,3-diamino acids using bislactim ethers (Schollkopf-type glycine enolates) as chiral auxiliary to achieve asymmetric α-alkylation (from 84 to >95%). The electrophile they used for alkylation was dibromomethane. Another key step in their synthesis was the azide displacement to the product (the bromide) of asymmetric alkylation in order to obtain the three-amino functionality. Very recently, Hamon and associates reported a very interesting way to derivatize glycine *(22)*. The chiral auxiliary they used was (–)-8-phenylmenthol. The key reaction is the asymmetric alkylation of the 8-phenylmenthyl ester of the N^α-Boc derivative of 2-bromoglycine by treatment with allyl-*tri-n*-butylstannanes. The diasteroselectivity is excellent.

Seebach's group is also one of the leaders in this area *(23,24)*. Recently, they reported a stereoselective synthesis of MeBmt by employing a new chiral glycine enolate derivative that is a chiral oxazolidineones derivative *(25)*. The electrophile they used for asymmetric alkylation was (2R, 4E)-2-methyl-4-hexenal. This aldol addition reaction was complicated and gave an unexpected product. The diastereoselectivity is good at the reaction center, but the yield of the major product was not excellent, even though MeBmt was obtained with a reasonable overall yield of 30–39% in a four-step process.

3.2. Asymmetric Hydrogenation Method

Asymmetric hydrogenation is one of the major methodologies for the synthesis of amino acids. The substrates of this type of hydrogenation usually are Schiff's bases *(9)* or α,β-unsaturated derivatives (dehydro-α-amino acids). The catalysts of this type of hydrogenation usually are chiral rhodium complexes. Because this methodology can provide a short and efficient route to various amino acids, it is very attractive to the industry, and a lot of this work has been patented. Several recent notable applications of this technology are worthy of note.

Takahashi and Achiwa reported the synthesis of a series of α-amino acid derivatives via asymmetric hydrogenation of (*Z*)-2-acetamidoarylic acid derivatives *(26)*. The catalyst they used was (2S, 4S)-MOD-BPPM rhodium complex. The reaction and proposed mechanism are shown in Scheme 2. In this case, the α-amino acids have the *R* configuration, and reasonably high e.e. values (58.6–98%, varies from substrate) were obtained. Shioiri and associates reported a synthesis of derivatives of β-hydroxy-α-amino acids *(27)*. They used different starting materials to synthesize 4-alkoxycarbonyloxazole derivatives, which are synthons for β-hydroxy-α-amino acids. The 4-alkoxycarbonyloxazole can undergo rearrangement under acidic conditions to give 5-substituted 3-amino-tetronic acid. This α,β-unsaturated acid was treated with 5% rhodium-alumina in ethyl acetate at 120 atm and ambient temperature for 24 h to afford the optically active 5-substituted lyxo-1,4-lactone. Both the diastereoselectivity and the yield of the reaction are high. In the same paper, they also reported the asymmetric transformation of α-amino acids to related β-amino acids. Gladiali and Pinna reported a synthesis of (–)-(R)-α-methylserine via a regioselective hydroformation *(28)*. The starting material they used was methyl *N*-acetamidoarylate (MAA). It was treated with CO/H_2 (1:1) at 80°C, 100 atm for 70 h with the catalyst of $HRd(CO)(PPh_3)_3$/chelating diphosphine to afford an α-amino alde-

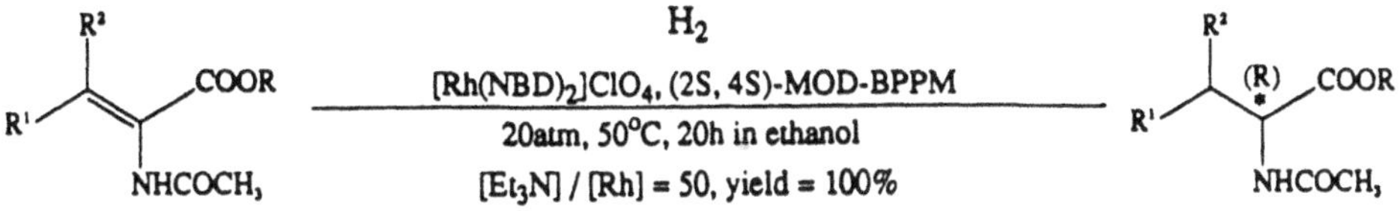

(2S, 4S)-MOD-BPPM: (2S, 4S)-N-(ᵗButoxy-carbonyl)-4-{[bis(4'-methoxy-3',5'-dimethylphenyl)]phosphino}-2-{[[bis(4'-methoxy-3',5'-dimethylphenyl)]phosphino]methyl}pyrrolidine.

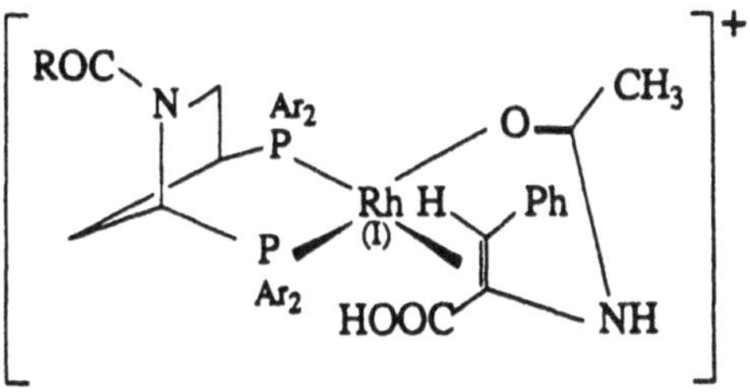

Transition State of the Asymmetric Hydrogenation

Scheme 2.

hyde. The diastereoselectivity of the reaction is >9:1 whereas the enantiomeric excess is <60%. The aldehyde obtained underwent reduction to provide (–)-(R)-α-methylserine. Very recently, Genet and associates reported a "practical production" of D- and L-threonine *(29)*. They started with a derivative of 3-oxobutyrate. After electrophilic amination, they obtained an oxime (Schiff's base type) that underwent normal catalytic hydrogenation to afford 2-racemic α-amino-3-oxobutyrate. This 2-acylamino-3-oxobutyrate was reduced to β-hydroxy α-amino acid derivatives through a dynamic kinetic resolution in rhodium- and ruthenium-catalyzed hydrogenation. The results and the reaction mechanisms of the hydrogenation by utilizing these two catalysts have been compared. It appears that the hydrogenation with the catalyst of chiral biphosphine ruthenium complex provides better results. Schmidt et al. have reported the synthesis of the derivatives of trihydroxynorleucines [(2S,4S,5S)- and (2R,4S,5S)-2-amino-4,5,6-trihydroxyhexanoic acid] *(30)*. Starting with the corresponding α,β-didehydro compounds, they used the optically active homogeneous catalyst $[Rh(COD)(DIPAMP)]^+BF^-$

to realize the asymmetric hydrogenation and obtained the protected α-amino acids.

3.3. Nucleophilic Amination of α-Substituted Acids and Electrophilic Amination of Enolates

These two approaches have been attractive to chemists, because they can provide highly selective asymmetric synthesis of α-amino acids and can be used to obtain important unusual amino acids. The general ideas of these strategies are quite straightforward. Nucleophilic amination involves an S_N2 reaction between an α-substituted acid or its precursor and an amino nucleophile. Usually the substituents at the optically active α position of the acids (or precursors) are halogens or hydroxyl groups. The nucleophile can vary, as will be seen in the following examples. The idea of electrophilic amination of enolates involves the reaction of an enolate with a nitrogen-centered electrophile. The stereochemistry is usually controlled by the neighboring chiral auxiliary or the neighboring chiral centers. Because of the nature of nitrogen atoms, it is hard to make an electrophilic nitrogen. Although considerable effort has been made already, much more is needed. Indeed thus far, the only highly successful electrophilic amination appears in nature. Oppolzer's group has been one of the leading groups working on both electrophilic amination of enolates and nucleophilic amination of α-substituted acid derivatives *(31)*. Recently, Oppolzer and Tamura reported the asymmetric synthesis of α-amino acids via electrophilic amination *(32)*. They used a new chiral sultam **14** (Scheme 3) as a chiral auxiliary. They coupled **14** with an acid chloride to give **15**, in which R corresponds to the carbon skeleton of the desired amino acids. The hydroxylamine **16** was obtained via electrophilic amination by treating **15** with an electrophilic nitrosochloride (the nitrosochloride also acts as an indicator of the reaction). The aminations were shown to give ca. 100% selectivity within the limits of ^{1}H-NMR analysis. The hydroxyl amine **16** was reduced to the amine **17**, followed by removal of the chiral auxiliary and basic hydrolysis to give (R)-α-amino acids **18**. (S)-α-amino acids **(18)** are equally accessible by using the available antipode of auxiliary **14** (entries b and i in Table 1). The synthesis is highly stereoselective with high yields, and can provide a variety of pure (R)- and (S)-α-amino acids. In the same paper, they also reported an interesting route for the synthesis of β-chiral α-amino acids as shown in Scheme 4. The two chiral centers in hydroxylamine **20** were

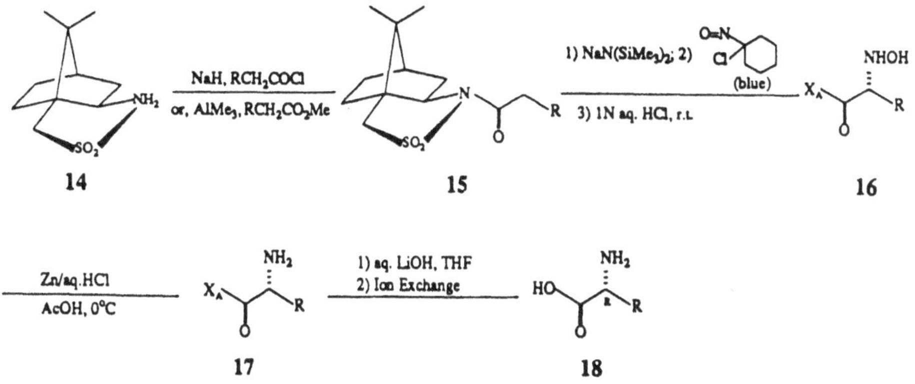

Scheme 3.

Table 1
Transformation of *N*-Acylbornylsultams
into Enantiomerically Pure α-Amino Acids **18**

	R	Yield% 16	Yield% 17	Amino acids	Yield%	e.e.%	Config.
a	CH_3	80	83	**18**	>99	>99	R
b	CH_3	82	78	(S) - **18**	99	>99	S
c	CH_2=CH—CH_2	69	78	**18**	>99	>99	R
d	Me_2CH	70	85	**18**	>99	>99	R
e	Me_2CH—CH_2	87	97	**18**	>99	99	R
f	$PhCH_2$	78	93	**18**	94	>99	R
g	Ph	77	95	**18**	97	>99	R
h	*p*-MeOph	72	84	**18**	97	>99	R
i	*p*-MeOPh	73	90	(S) - **18**	99	>99	S

generated via addition of *N*-crotonoylsultam **19** by a nucleophile (ethylmagnesium bromide) or by an electrophile (the nitrosochloride). This step was achieved with high stereoselectivity (99% e.e. at C[α], 90% e.e. at C[β]). The conditions of converting the hydroxylamine **20** to an α-amino acid are the same as in Scheme 3. They ended up with (S,S)-isoleucine **21**.

Takano and associates reported a concise route for synthesis of (S)-phenylalanine *(33)*. They used (R)-epichlorohydrin as starting material. Optically active epoxides often have been used as building blocks in the synthesis of β- or γ-hydroxy α-amino acids. After lengthening the car-

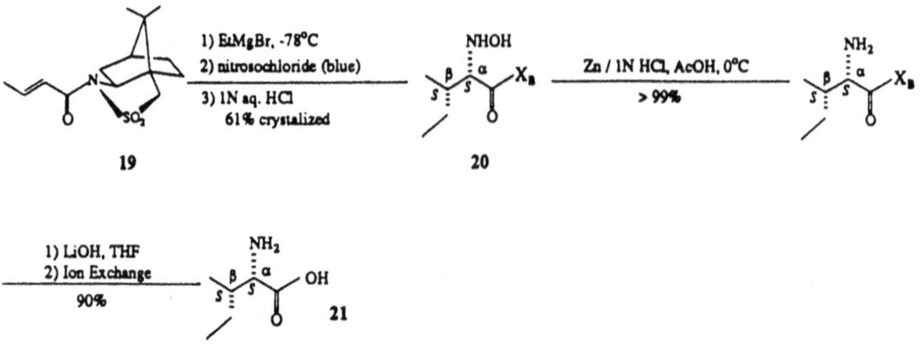

Scheme 4.

bon backbone, they obtained an internal acetyl that has an allylic hydroxyl group. The hydroxyl group was replaced by the treatment with phthalimide under basic conditions. The inversion of chirality at the reaction center is expected. The adduct then underwent reduction, oxidation (to acid), and finally deprotection to afford (S)-phenylalanine. Jung and Jung reported a "rapid" synthesis of β-hydroxy-α-amino acids *(34).* They started with an allylic alcohol that underwent Sharpless asymmetric resolution to provide optically active alcohols as key intermediates. The nucleophilic amination was achieved by treating the epoxide with benzoyl isocyanate. The adduct was treated with NaH to obtain optically active oxazolidinones. Following oxidation of the hydroxy group, opening of the oxazolidinone ring provided L-threonine, β-hydroxyphenylalanine, or β-hydroxyleucine depending on the reagents used. Schmidt and associates prepared α-amino-β-hydroxy acids by using nucleophilic amination *(3).* They started with an allylic alcohol that underwent Sharpless oxidation. The epoxide alcohol obtained was converted to an imide ester in excellent yield by treatment with tricholoroacetonitrile/DBU. The imide nitrogen then underwent regioselective intramolecular nucleophilic attack to open the epoxide ring. The oxazolines were converted to oxazolidinones that underwent Jones' oxidation, hydrolysis, and deprotection to afford *N*-Boc-α-amino-β-hydroxy acid esters. The stereochemistry was well controlled by Sharpless asymmetric epoxidation. Wagner and associates also reported a synthesis of a β-hydroxy α-amino acid *(35).* They started with (R,R)-(+)-tartaric acid. After regioselective protection, they obtained a triflate (C-2) with all other functionalities protected by benzyl groups. The triflate was exposed to tetramethylguandinium azide to afford an azide that underwent reduction. A series of regioselective deprotection and reprotection steps provided N^{α}-Boc-(2S,3R)-3-hydroxy aspartic acid. Evans' group is one of the leading groups in this area. They have done outstanding work on nucleophilic amination and electrophilic amination of enolates using the same chiral auxiliary *(36)*, and in a recent paper have summarized this work *(37)*. They did the nucleophilic amination through an azide displacement with a chiral bromide (S_N2 mechanism). The diastereoselectivity of the former bromination depends on the reagent used to make the enolates. They also have examined direct azidation to the enolates. The selection of quenching reagent for this reaction is critical. They achieved both high yield and protection from racemization by using glacial acetic acid at low temperature for quenching. Recently, in our own laboratory, we have synthesized several β-methyl α-amino acid analogs. Using in part Evans' methodology *(38–41)* asymmetric synthesis of all four isomers of β-methylphenylalanine *(42)* (Scheme 5) was achieved. We used *S*-(+)-3-phenylbutyric acid **22** as starting material, which was attached to the chiral auxiliary **23** derived from D-phenylalanine to afford *N*-acyl oxazolidinone **24**. Oxazolidinone **24** was converted to a boron enolate **25** by use of dibutylborontriflate in dichloromethane. Stereoselective bromination was accomplished using NBS, and S_N2 displacement of the resulting crude bromide by tetramethylguanidium azide gave the diastereoisomeric azide **26** with high stereoselectivity (Table 2). Removal of the chiral auxiliary was effected by hydrolysis using LiOH in the presence of hydrogen peroxide, followed by reduction (10% Pd-C, 1:1 AcOH:H_2O) of the resulting azido acid **27** which gave *threo*-L-β-methylphenylalanine. We provide here the details of the asymmetric synthesis of a β-methylphenylalanine.

3.3.1. General Procedure for the Preparation of N*-Acyloxazolidinone: Illustrated by the Preparation of (4R)-3-(3'S)-3'-(Phenylbutanoyl)-4-(Phenylmethyl)-2-Oxazolidinone,* **24** (42)

1. To a stirred solution of 19.8 g (0.11 mol) of *S*-(+)-3-phenylbutyric acid in 450 mL of freshly distilled THF, add 15.3 mL (0.11 mol) of triethylamine under an atmosphere of argon.
2. Cool the mixture to –78°C, and add 14.2 mL (0.115 mol) of trimethylacetylchloride using a cannula. Stir the resulting white suspension for 10 min at –78°C, 1 h at 0°C, and recool to –78°C.

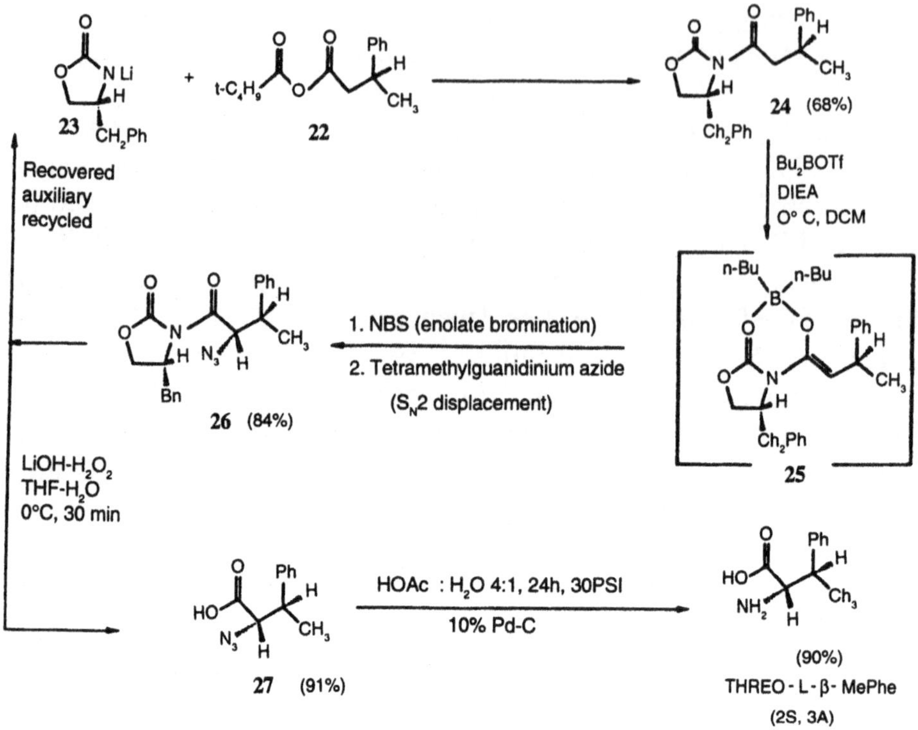

Scheme 5.

3. Meanwhile, in a different flask, prepare a solution of metallated D-oxazolidinone (**23,** Scheme 5) by the dropwise addition of 69 mL of *n*-butyllithium (1.6*M* in hexane) to a –78°C solution of 19.4 g of the D-auxiliary *(43)* in 450 mL of dry THF. Stir the mixture for 20 min at –78°C.
4. Transfer the lithiated chiral auxiliary via a cannula into the reaction flask containing the preformed mixed anhydride at –78°C. Stir the mixture at 0°C for 1 h and allow to warm to 23°C in 16 h.
5. Quench the mixture with 300 mL of saturated ammonium chloride solution. Evaporate THF *in vacuo*. Extract the product with (3 × 300 mL) of dichloromethane.
6. Wash the organic layer with 1*N* sodium hydroxide (2 × 100 mL) and 1*N* sodium bisulfate (1 × 100 mL), dry (anhd. magnesium sulfate), filter, and evaporate to give 30 g of colorless solid.
7. Purify by silica gel chromatography (elution with 15–30% ethyl acetate in hexane) to give 24.2 g (yield, 68%) of the desired compound *(34)* as a colorless solid, mp 82–84°. $[\alpha]^{23}D = -38.4°$ (c 0.5, $CHCl_3$). ^1H-NMR ($CDCl_3$, 250 MHz) δ 1.35 (d, J = 6.8 Hz, 3H), 2.59 (dd, J = 14.8, 9.4 Hz,

Table 2
Diastereoselectivities of All Four Individual Isomers of β-Methylphenylalanine

Reactions	Diastereoselectivities of β-methylphenylalanine
L-auxiliary + (S)-(+)-phenylbutyric acid	(2R, 3R):(2S, 3S) = 95:5
L-auxiliary + (R)-(–)-phenylbutyric acid	(2R, 3R):(2S, 3S) = 99:1
D-auxiliary + (S)-(+)-phenylbutyric acid	(2R, 3R):(2S, 3S) = 99:1
D-auxiliary + (R)-(–)-phenylbutyric acid	(2R, 3R):(2S, 3S) = 95:5

1H), 3.1–3.2 (m, 2H), 3.3–3.5 (m, 2H), 4.1–4.2 (m, 2H), 4.61–4.67 (m, 1H), 7–7.3 (m, 10H).

3.3.2. General Procedure for Asymmetric Bromination of N*-Acyloxazolidinone and Subsequent Displacement by Azide: Illustrated by the Preparation of (4R)-3-(2'S,3'S)-2'-Azido-3'-(Phenylbutanoyl)-4-(Phenylmethyl)-2-Oxazolidinone,* **26**

1. Cool a solution of 26 g (0.08 mol) of *N*-acyloxazolidinone **24** in 180 mL of dichloromethane to –78°C.
2. Transfer a solution of 19.7 mL (0.112 mol) of freshly distilled diiso-propylethylamine, followed by 111 mL of di-*n*-butylborontriflate (1*M* solution in DCM), via a cannula. Stir the mixture for 1 h at 0°C and then cool to –78°C.
3. Meanwhile in another flask, cool a suspension of 18.5 g of *N*-bromo-succinimide (0.10 mol) in 250 mL of dichloromethane to –78°C.
4. Transfer the boron enolate solution at –78°C via a cannula.
5. Stir the mixture at –78°C for 2 h.
6. Quench the mixture with 260 mL of aq. sodium bisulfate solution, and wash with 250 mL of water. Dry the organic layer (over sodium sulfate), filter, and evaporate to give the crude bromide as a brown oil that is used in the next step without purification.
7. From ^{1}H-NMR of this crude material, the ratio of major and minor isomers of the two diastereoisomeric bromides is found to be 94:6 (by integration of the two doublets corresponding to the diastereomeric bromides at δ 6.2).
8. Purify a small amount of this bromide (Scheme 5) by silica gel chromatog-raphy (elution with 90% hexane and 10% ethyl acetate). From the eluant, analytically pure bromide crystallizes on standing. The bromide has the fol-lowing physical characteristics: mp 94–95°. $[\alpha]^{23}_D = -38°$ (c 0.5, $CHCl_3$).
9. Dissolve the crude bromide from the above reaction in 100 mL of acetonitrile, and add 51 g (0.32 mol, 5.5 Eq) of tetramethylguanidium azide

in one portion at 0°C. Warm the mixture to ambient temperature, and stir for 16 h.

10. Monitor the reaction (by ^{1}H-NMR) by the disappearance of signal (doublet) for the proton α to Br at δ 6.2 and appearance of signals for the proton α to the azide at δ 5.36.
11. Quench the reaction by the addition of 200 mL of saturated aq. sodium bicarbonate.
12. Extract the resulting mixture three times with dichloromethane (3 × 100 mL). Wash with water (3 × 100 mL), 6*N* HCl (1 × 100 mL), water (1 × 100 mL), 0.1*N* sodium bicarbonate (1 × 100 mL), and brine (1 × 100 mL).
13. Dry the organic extracts (anhd. sodium sulfate), filter, and evaporate *in vacuo*. Purify the resulting α-azido carboximide by silica gel chromatography (elution with 90% hexane and 10% ethylacetate) to give 17.9 g (84%) of azide **26** as a colorless solid. mp 84–86°. $[\alpha]^{23}D = +80.8°$ (c 1.1, $CHCl_3$). CIMS (isobutane), m/z (relative intensity) $M^+ + 1 = 365$ (2%), $M^+ + 1 - N_2 = 337$ (8%), $M^+ + 1 - N_3H = 322$ (15%); IR ($CHCl_3$): 2103, 1771, 1689 cm^{-1}. ^{1}H-NMR ($CDCl_3$, 250 MHz): δ 1.32 (d, J = 7 Hz, 3H, β-CH_3); 2.58 (dd, J = 14.9, 9.4 Hz, 1H); 3.10–3.2 (m, 2H); 4.11 (m, 2H); 4.60 (m, 1H); 5.23 (d, J = 9.2 Hz, α-H, 1H); 7.1–7.3 (m, 10H).

*3.3.3. General Procedure for the Removal of Chiral Auxiliary: Illustrated by the Preparation of (2S)-Azido-(3R)-Phenylbutanoic Acid, **27***

1. Cool a solution of 12 g (0.032 mol) of acylazide **26** in 450 mL of THF and 175 mL of water to 0°C, and treat with 12.2 mL (0.13 mol) of 31% hydrogen peroxide, followed by 2.8 g of lithium hydroxide monohydrate (0.064 mol).
2. Stir the mixture for a total of 30 min.
3. At this time, thin-layer chromatography (hexane: ethyl acetate:acetic acid = 8:1.9:0.1) indicates complete disappearance of the starting material.
4. Quench the reaction with a solution of 0.5*N* sodium bicarbonate. Remove tetrahydrofuran *in vacuo*.
5. Extract with dichloromethane (5 × 100 mL) to give the recovered chiral auxiliary. Cool the aqueous layer to 0°C and acidify with 6*N* hydrochloric acid.
6. Extract with ethyl acetate (5 × 200 mL), dry (anhd. sodium sulfate), filter, and remove solvent to leave the azido acid as an oil.
7. Purify by silica gel chromatography (elution with 7:2.9:0.1 = hexane:ethyl acetate:acetic acid) to give 6 g (91%) of pure azido acid **27** as a light-yellow oil. $[\alpha]^{23}_D = -11°$ (c 1.0, $CHCl_3$). TLC, $R_f = 0.57$ (elution with 7:2.9:0.1 = hexane:ethyl acetate:acetic acid). CIMS (isobutane), m/z (rela-

tive intensity) $M^+ + 1 = 206$ (38%). ^{1}H-NMR ($CDCl_3$, 250 MHz): δ 1.37 (d, J = 7.2 Hz, 3H, β-CH_3); 4.06 (d, J = 7 Hz, 1H, α-H); 7.26–7.33 (m, 5H, aryl-H); 9.1 (s, 1H, —COOH). IR (film): 2600–3400 cm^{-1} (br, —OH); 2113 cm^{-1} (s, N_3); 1712 cm^{-1} (s, C=O).

3.3.4. *Threo-L-(2S, 3R)-β-Methylphenylalanine*

1. To a solution of 2.7 g of azido acid **27** from the above reaction, in 110 mL of glacial acetic acid, add 30 mL of water in a Parr hydrogenation vessel.
2. Bubble a stream of argon through this solution for 5 min, and add 1 g of 10% Pd/C.
3. Hydrogenate the mixture at 30 psi for 24 h, then add 100 mL of water, and filter off the catalyst.
4. Add to the filtrate 20 mL of hydrochloric acid, and remove the solvents *in vacuo*.
5. Add 300 mL of anhd. ether to the residue. Filter the precipitated solid by suction filtration, and dry to give 2.2 g (80%) of the amino acid as its hydrochloride salt.
6. Purify a small amount of this amino acid by ion-exchange chromatography (*see* Chapter 2, *PAP*) (Amberlite, IR 120, H^+).
7. Elute with 10% ammonium hydroxide. The analytical data of the purified *threo*-L-(2S, 3R)-β-methylphenylalanine are listed below: mp 190–192°. $[\alpha]^{23}_D = -5.3°$ (c 0.75, H_2O), Lit (–5.8°, c 1.0, H_2O). CIMS (isobutane), m/z (relative intensity) $M^+ + 1 = 180$ (100%). ^{1}H-NMR (250 MHz, D_2O, dioxane as std at δ 3.55): δ 1.18, (d, J = 7.3 Hz, 3H, β-CH_3); 3.33 (m, 1H, β-H); 3.73 (d, J = 4.9 Hz, 1H, α-H); 7.15–7.25 (m, 5H, aryl hydrogens).
8. Thin layer chromatography of this compound on a chiral TLC plate shows only one enantiomer $R_f = 0.65$ (4:1:1 = acetonitrile:methanol:water). HPLC analysis of the *N*-acetyl derivative of this amino acid shows >99:1 ratio of *threo* to *erythro* isomers.

Recently, Font and associate reported an enantioselective synthesis of both (–)-*erythro*- and (–)-*threo*-γ-hydroxynorvaline *(44)*. They started with D-ribonolactone to prepare 5-deoxy-D-ribonolactone by the methods of Papageorgiou and Benezra *(45)*. The obtained 5-deoxy-D-ribonolactone was converted to a tosylate, which was readily displaced by azide. The azide subsequently underwent reduction, and ring opening of the lactone to afford (–)-erythro-γ-hydroxynorvaline; (–)-threo-γ-hydroxynorvaline was prepared using a similar process. Very recently, Font and associates synthesized (–)-4,5-dihydroxy-D-*threo*-L-norvaline using the same starting material *(46)*. The stereochemistry of the azide displacement is very unusual in that retention of configuration was obtained. Similar processing of azide afforded the final γ,δ-dihydroxy-D-threo-L-norvaline. We would like to propose a mechanism for this unexpected retention of configuration. We think it may be the result of solvent involvement in the reaction. Since the authors did not mention the experimental details, we suppose that they did not isolate the triflate intermediate; pyridine is a good nucleophile, and since it is the solvent in the reaction, it may participate in the overall reaction to cause the double inversion of the configuration at the reaction center.

Frejd and associates reported a nice synthesis of γ-hydroxyisoleucine *(47)* using an asymmetric epoxide, benzyl 2,3-anhydro-4-*O*-(ᵗbutyldimethylsilyl)-β-L-ribopyranoside, as starting material. They opened the epoxide ring by regioselective methylation using trimethyl aluminum. The hydroxy compound obtained was converted to another epoxide asymmetrically, and it in turn underwent amination by treatment with $Ti(O^iPr)_2(N_3)_2$ via an S_N2 mechanism. The azide obtained was oxidized to the α-azido acid, which was reduced to the final (2R,3R,4R)-γ-hydroxyisoleucine. The C-2 diastereomeric γ-hydroxyisoleucine also was prepared by a slightly different process in which the amination reagent they used was HN_3, DEAD, and Ph_3P. In this regard, it is worth mentioning that recently Fleming and Sharpless reported selective transformations of *threo*-2,3-dihydroxy esters *(48)*. This transformation can be used for the asymmetric synthesis of β-hydroxy α-amino acids starting from α,β-unsaturated esters following by oxidation to *threo*-2,3-dihydroxy esters. The obtained esters were transformed to α-hydroxy sulfonate esters with high regioselectivity (at C-2). These esters are ready for azide displacement to give α-azido-3-hydroxy esters, which are the precursors of β-hydroxy α-amino acids. Corey and Chai *(49)* also reported an asymmetric synthesis of β-hydroxy α-amino acids precursors using the reagent shown below:

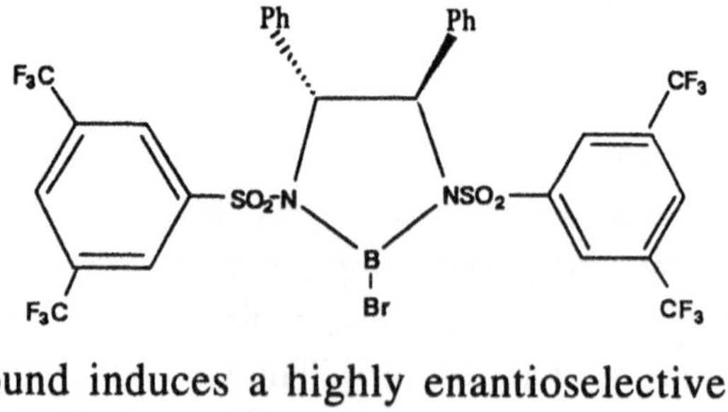

This compound induces a highly enantioselective aldol reaction between achiral aldehydes and ᵗbutylbromoacetate. The α-bromo-β-hydroxy

^{t}butyl esters obtained can undergo azide displacement and subsequent conversion to β-hydroxy α-amino acids.

3.4. Asymmetric Strecker Syntheses

The Strecker method is one of the traditional methods to prepare α-amino acids *(50)*. Although it is relatively convenient compared to most other methods, surprisingly, not much work has focused on this methodology. The basic idea for asymmetric synthesis via this method was the formation of a chiral Schiff base by condensation between an optically active amine and an aldehyde, or by condensation between an optically active aldehyde and an amine. Subsequent addition of HCN followed by hydrolysis should afford optically active α-amino acids. Harada and Okawara contributed based on the modifications of this methodology in the 1960s and 1970s *(51)*, and more recently, Kunz et al. have provided new insights *(52–54)*, including the use of carbohydrates as chiral templates. They found that the asymmetric induction was dependent on the solvent when using carbohydrate templates. For example, when using pivaloyl-D-galactosylamine **28** (Scheme 6) as a chiral template, (R)-diastereomeric amino nitriles were obtained in excess if the reactions were carried out in isopropanol in the presence of zinc chloride *(52)*, but (S)-diastereoisomeric amino nitriles were preferred if the reactions were carried out in chloroform. Based on these results, they used 2,3,4-tri-*O*-pivaloyl-2-D-arabinopyranosylamine **29** *(54)* (Scheme 6), which is a pseudo mirror-image of **28** as a chiral template (Scheme 6). **29** was condensed with aldehyde **30** in the presence of zinc chloride in THF to afford (α-D, 2S)-**31** with high selectivity (*see* Table 3). Treatment of (S)-**31** with hydrogen chloride in methanol followed by addition of water led to removal of the *N*-formyl group and the cleavage of the *N*-glycoside, which on acid hydrolysis gave the pure (S)-α-amino acid **33** (Table 4). Very recently, Kunz and associates reported another application of this methodology *(55)* starting with 2,3,4-tri-*O*-acetyl-α-arabinopyranosyl azide, a compound first prepared by Paulsen et al. *(56)*. This azide was then converted to the amine, the acetyl-protecting groups were changed to pivaloyl-protecting groups, and the chiral amine was condensed with aldehydes to give *N*-arabinosylimines, which were coverted to α-amino nitriles (L/D = 7–10:1). The nitriles were purified by recrystallization to obtain the pure L-amino nitriles (83–84% yield), which underwent acidic hydrolysis to provide L-phenylglycine and other L-amino acids. The

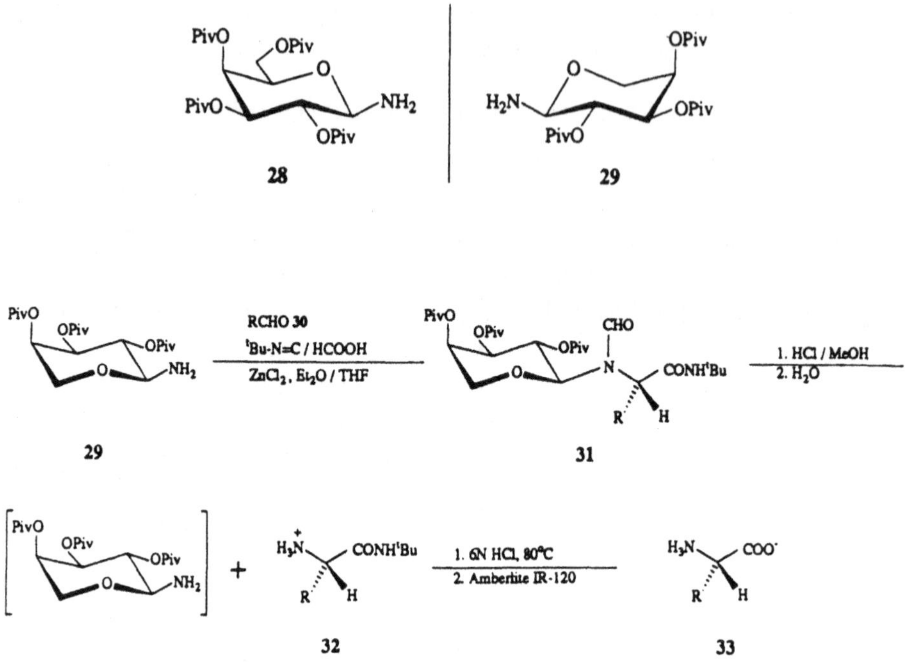

Scheme 6.

above chiral amine also can be converted to *N*-formyl-*N*-arabinosyl amino acid amides with better diastereoselectivity of (L/D = 20–30:1). These L-amino acid amides could be purified by either crystallization or flash chromatography in high yield (85–91%). The free enantiomerically pure L-amino acids were easily released from the carbohydrate templates by a two-step acidic hydrolysis. We provide here a specific example for the synthesis of L-2-(4-chlorophenyl)-α-amino acid **40** (Scheme 7).

3.4.1. α-D-Arabinopyranosyl Azide

1. Add 1*N* NaOMe in MeOH (1 mL) to a solution of 2,3,4-tri-*O*-acetyl-α-D-arabinopyranosyl azide **34** (0.1 mol) in MeOH (200 mL).
2. After 2 h, neutralize the solution using ion-exchange resin IR 200 (H^+ form 3 g), filter, and evaporate the solvent *in vacuo*. α-D-arabinopyranosyl azide: yield 100%; mp 93°; $[\alpha]^{22}_D = -21.2°$ (c = 1, H_2O). ^{13}C-NMR ($CDCl_3$/DMSO-d_6/TMS): δ = 67.66, 67.85, 70.20, 72.50 (C-2–C-5), 90.65 (C-1).

3.4.2. 2,3,4-Tri-O-Pivaloyl-Glycosyl Azide, ***35***

1. Add pivaloyl chloride (40 mL) dropwise to a solution of the α-D-arabino-pyranosyl azide (0.1 mol) in pyridine (150 mL) at 0°C.

Table 3
Diastereoselective Ugi Synthesis
of *N*-Arabinopyranosyl Amino Acid Amides **31(a–e)**

Product	R	Reaction temp., °C/time, h	Kinetic ratio (C-2)S:(C-2)R	Yield% of pure (C-2)S-**31**
31a	$(CH_3)_3C$—	–25/72	97:3	85
31b	$PhCH_2$—	–78/24	97:3	87
31c	*p*-$ClC_6H_4CH_2$—	–25/24	98:2	91
31d	2-furyl	–25/24	96:4	85
31e	2-thienyl	–25/24	4:96	85 (C-2)R-**31**

Table 4
(S)-Amino Acids **33** via Hydrolysis of
N-Arabinopyranosyl (S)-Amino Acid Amides **31(a–e)**

Product	R	Overall yield, %	$[\alpha]_D^{22}$
33a	$(CH_3)_3C$—	70	+8.5 (c 2, 1.5*N* HCl)
33b	$PhCH_2$—	82	–33.5 (c 0.5, H_2O)
33c	*p*-$ClC_6H_4CH_2$—	85	+139.5 (c 1, 1*N* HCl)

2. After 24 h at room temperature, evaporate pyridine and pivaloyl chloride *in vacuo*, dissolve the residue in $CHCl_2$ (200 mL), wash with 2*N* HCl (100 mL), sat. aq. $NaHCO_3$ (5 × 50 mL) and H_2O (100 mL), dry ($MgSO_4$), and concentrate *in vacuo*.
3. Recrystallize from MeOH to deliver pure compound: 2,3,4-tri-*O*-pivaloyl-α-D-arabinopyranosyl azide **35:** yield 89%; mp 90°; $[\alpha]^{22}_D$ = + 0.93° (c = 1, $CHCl_3$); $C_{20}H_{33}N_3O_7$ calc (%). C 56.19, H 7.78, N 9.83; found C 56.15, H 7.77, N 9.92. ^{1}H-NMR ($CDCl_3$/TMS): δ = 4.54 (d, 1H, $J_{1,2}$ = 9.7 Hz, 1-H), 5.08 (dd, 1H, $J_{3,4}$ = 3.3 Hz, 3-H), 5.19 (dd, 1-H, 2-H), 5.23 (m, 1H, 4-H).

3.4.3. Tri-O-Pivaloyl-Glycosylamine, 36

1. Hydrogenate a solution of the *O*-pivaloylated glycosyl azide **35** (0.1 mol) in MeOH (250 mL, containing 1–5% of CH_2Cl_2) under atmospheric pressure in the presence of Raney Ni (10 g).
2. After 3 h (TLC control), remove the catalyst by centrifugation, evaporate the solvent *in vacuo*, and recrystallize the remaining residue from MeOH. 2,3,4-tri-*O*-pivaloyl-α-D-arabinopyranosylamine **36**: yield 88%; mp 106°C; $[\alpha]^{22}_D$ = –46.7° (c = 1, $CHCl_3$). ^{1}H-NMR ($CDCl_3$/TMS): δ = 4.02 (d, 1H, $J_{1,2}$ = 8.4 Hz, 1-H), 5.07 (dd, 1H, $J_{3,2}$ = 10.2 Hz, $J_{3,4}$ = 3.3 Hz; 3-H), 5.01 (dd, 1H, 2-H), 5.18 (m, 1H, 4-H).

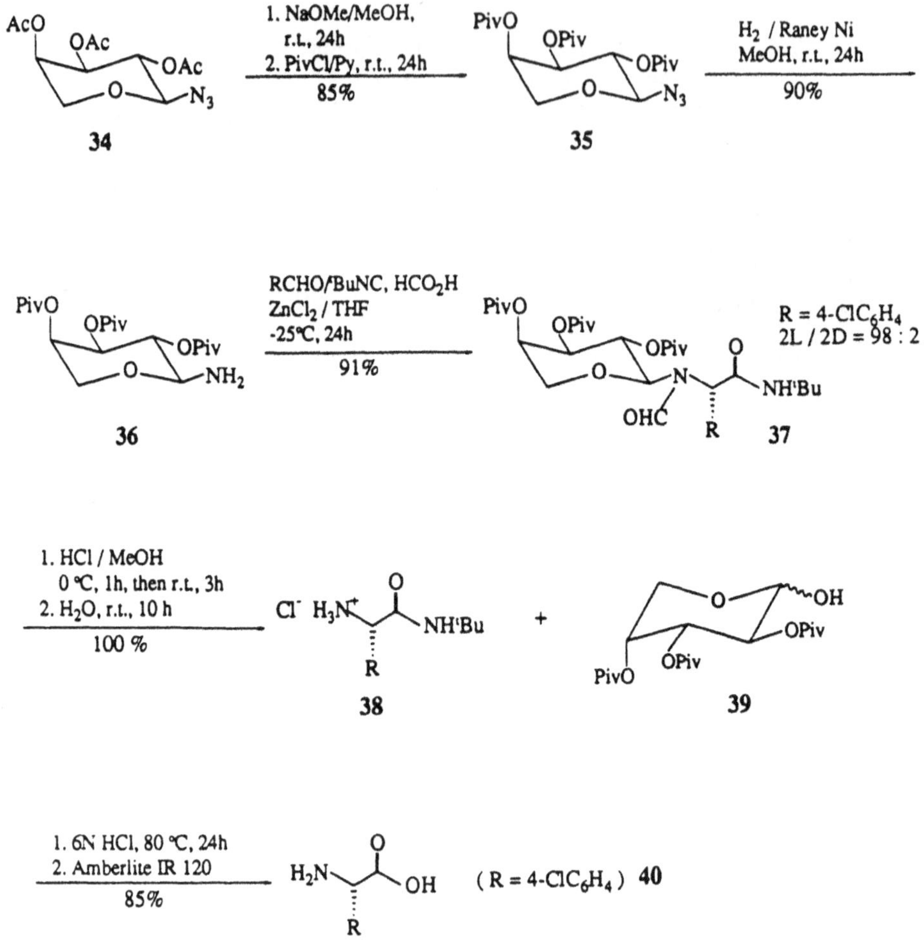

Scheme 7.

3.4.4. N-*Formyl*-N-*Glycosyl Amino Acid* N'-tert-*Butylamide*, **37**

1. Add $ZnCl_2$ (4 mmol, as 2.2 molar solution of the ET_2O complex in CH_2Cl_2) to a solution of the glycosylamine **36** (4 mmol), the *p*-chloro-benzaldehyde (4.1 mmol), formic acid (4.4 mmol), and *t*-BuNC (4.2 mmol) in THF (30 mL), cooled to –25°C.
2. Monitor the reaction by TLC (light petroleum ether/ETOAc).
3. After complete disappearance of **36**, evaporate the solvent *in vacuo*, dissolve the residue in CH_2Cl_2 (50 mL) extracted with sat. aq. $NaHCO_3$ (2 × 100 mL) and with H_2O and dry ($MgSO_4$).
4. Evaporate the solvent *in vacuo*. The crude mixture of diastereomers obtained almost quantitatively (2L:2D = 98:2) is investigated by HPLC (on

120-5m C18 [reverse phase] in MeOH/20% H_2O). Recrystallize or purify by flash chromatography to deliver the pure *N*-formyl-*N*-(2,3,4-tri-*O*-pivaloyl-α-D-arabinopyranosyl)-L-amino acid *N'*-*tert*-butylamides **37** in high yield. *N*-formyl-*N*-(2,3,4-tri-*O*-pivaloyl-α-D-arabinopyranosyl)-L-amino acid *N'*-*tert*-butylamides **37**: yield 91%; mp 202°C; $[\alpha]^{20}_D = -36.8°$ (c = 1, MeOH); $C_{33}H_{49}ClN_2O_9$ (653.2): satisfactory elemental analysis obtained, C m 0.1, H m 0.15, N m 0.1; ^{1}H-NMR ($CDCl_3$/TMS): δ = 5.17 (d, $J_{1,2}$ = 9.4–9.6 Hz, 1-H), 5.01 (s, a-CH).

3.4.5. Hydrolysis of **37** *to Give L-2-(4-Chlorophenyl)-2-Amino Acid,* **40**

1. Add a saturated solution of HCl in MeOH (3 mL) to the *N*-glycosyl-L-amino acid amide **37** (2 mmol) dissolved in dry MeOH (10 mL). Stir the mixture 1 h at 0°C and 3 h at room temperature. Add H_2O (2 mL), and stir the mixture for 10 h.
2. Evaporate the solvent, and dissolve the residue in H_2O (25 mL).
3. Extract the solution with pentane (2 × 20 mL). From the dried pentane solution, tri-*O*-pivaloyl-D-arabinopyranose **39** is recovered almost quantatively (>96%).
4. Evaporate the aqueous solution to dryness to give the amino amide **38** quantitatively.
5. Heat in 6*N* HCl at 80°C for 24 h.
6. Evaporate the solution to dryness, and distill off toluene (2 × 10 mL) from the residue. Then dissolve in water, and load on an ion-exchange column (Amberlite IR 120).
7. Wash the resin to neutral reaction of the eluent, and then elute the amino acid with aq. NH_4OH (3%).
8. Evaporate the ammonium salt solution *in vacuo* to give the L-amino acid **40** in crystalline form. Yield 85%; $[\alpha]^{20}_D = +139.5°$ (c = 1, 1N HCl). Data for the mp, elemental analysis, and NMR were not reported.

Chakraborty and associates synthesized optically pure L- and D-α-amino acids via diastereoselective Strecker synthesis *(57)* (Scheme 8) using α-phenylglycinol as chiral auxiliary. Imine **43** was generated from the condensation of respective aldehyde **41** and R-(–)-2-phenylglycinol **42**; **43** was then treated with trimethyl silyl cyanide to afford (1S, 1'R)-**44a** as the major product. The diastereoselectivities were good (*see* Table 5). After separation of the diastereoisomers **44a** and **44b**, the major isomer (1S, 1'R)-**44a** was converted to its *N*-substituted α-amino ester **45** by the treatment of saturated methanolic hydrochloric acid. Finally, the chiral auxiliary was easily removed from **45** by oxidative cleavage with

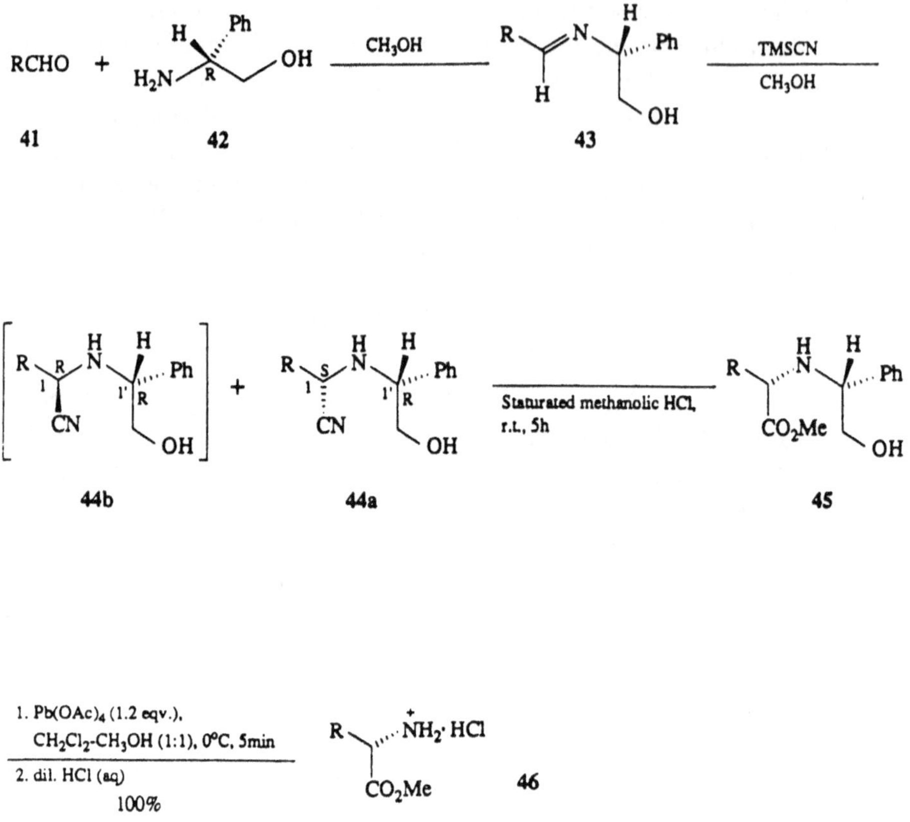

Scheme 8.

lead tetraacetate to give the amino acid **46**. Also recently, Cainelli and Panunzio synthesized a series of cyclic α-amino acids by a modified Strecker synthesis *(58)*.

3.5. Homologation of the β-Carbon

Homologation on the β-carbon of α-amino acids is a useful approach to many α-amino acids. L-Aspartic acid and proline derivatives are frequently used starting compounds in this approach, since their β or γ positions can be activated by electron-withdrawing groups. Since the strategy of this methodology is actually based on modifications of already available optically active α-amino acids, the idea is straightforward. Although the chemistry is usually not as sophisticated as seen in total syntheses of α-amino acids (amination methods, Stecker synthesis, and so forth), it is very attractive. Homologation of the γ-carbon also is cataloged in this

Table 5
Diastereoselective Strecker Synthesis
with R-(–)-2-Phenylglycinol **42** and Various Aldehydes

Entry	Aldehyde	Diastereoselectivity (1S,1'R)-**44a**:(1R,1'R)-**44b**	Total yield, %
1	Benzaldehyde (**41a**)	82:18	92
2	*p*-Tolualdehyde (**41b**)	85:15	90
3	*p*-Methoxybenzaldehyde (**41c**)	90:10	95
4	Isobutyraldehyde (**41d**)	84:16	95
5	Pivalaldehyde (**41e**)	88:12	92

section because it utilizes a similar strategy. Rapoport's group is one of the leading groups developing this methodology *(53)*, and recently they synthesized 2,3-diamino acids using this strategy (Scheme 9) *(60)* starting with aspartic acids. After protecting the carboxylic acid groups and amine group, the aspartic ester **47** was treated with KHMDS and BnX (X = Cl or Br) to provide adducts **48(a,b)** via electrophilic addition. Several aspartic acid derivatives were made with differences at the β-ester group. The diastereoselectivity of adduct **48(a,b)** can reach 25:1 (Fig. 1). Selective cleavage of the β-ester in **48(a,b)** followed by Curtius degradation using diphenylphosphorylazide (DPPA) yielded the cyclic 2,3-diamino derivatives **49(a,b)**. The *N*-protection was removed by treatment with THF, followed by acidic hydrolysis to afford the final β-substituted 2,3-diamino acids **50** and **51**.

Sasaki and associates reported a synthesis of β,γ-unsaturated α-amino acids *(61)*. They started with (2R)-2-$^{\alpha}$Boc-amino-3-phenylsulfonyl-1-(2-tetrahydropyranyloxy)propane or its (2S)-antipode. The C-1 hydroxy group and the protected amine group at C-2 position are the precursors of the acid group and amino group in the final amino acids, respectively. The β-homologation was achieved by electrophilic addition of aldehydes to the β-(C-3) position. The adducts underwent elimination of water and then oxidation to provide exclusively L-Z- or D-Z-propenylglycine. Baldwin et al. also have published extensively on "β-homologation" and have synthesized β,γ-unsaturated α-amino acids using this strategy *(62)*. They started with a diester of aspartic acid, α-^{t}Butyl-β-methyl-*N*-Z(S)-aspartate, which has an activated β-carbon. β-Homologation was achieved via electrophilic addition of ketones under basic conditions. Following elimination of water, decarboxylation, and hydrolysis, *E*- β,γ-

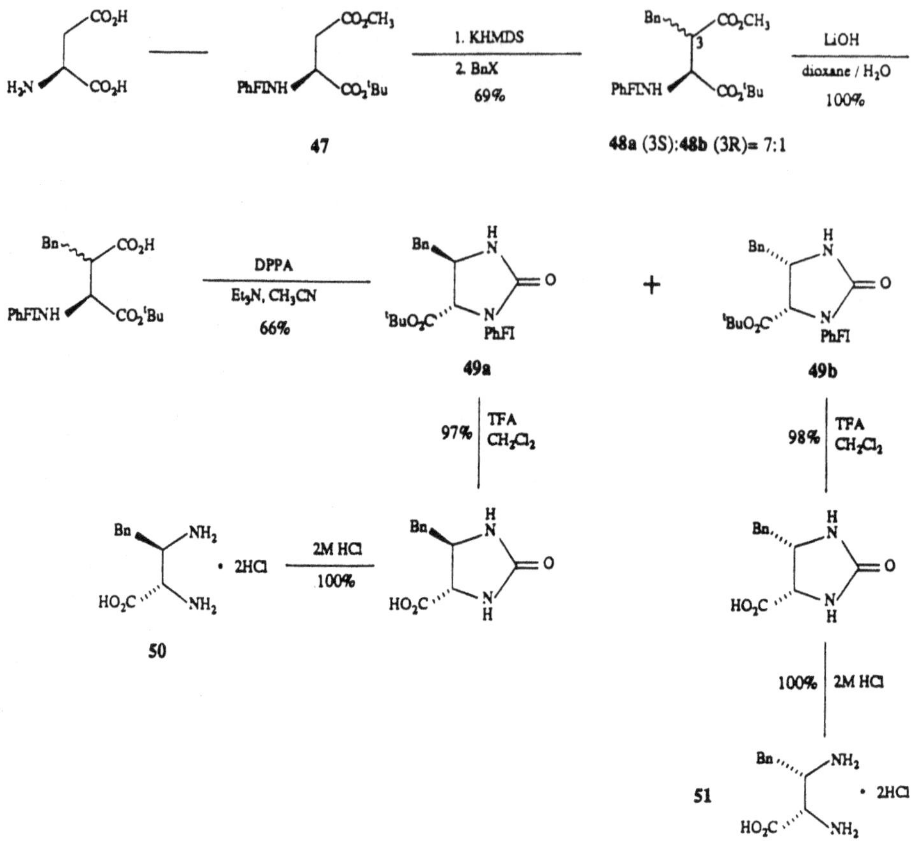

Scheme 9.

unsaturated α-amino acids (E:Z = 9:2) were obtained in good yields. In another recent paper, they reported a similar synthesis of β-alkylated aspartic acids using the same starting material, but different electrophiles (including alkyl halides) *(63)*. Parry and Lii also used the β-homologation strategy for the synthesis of *trans*-(+)-1-propenyl-L-cysteine sulfoxide *(64)*. The key step in this synthesis was the nucleophilic ring opening of 2-amino-β-propiolactone by an optically active thiol under basic condition. Finally, there are also some examples of γ-homologations. Baldwin and associates used dibenzyl *N*-trityl-(S)-glutamate as an γ-anion synthon to synthesize γ-carboxyglutamic acid and other γ-alkylated α-amino acids, via electrophilic substitution *(65)*. The electrophiles they used were carbonyl compounds, but diastereoselectivity was not high at the γ-posi-

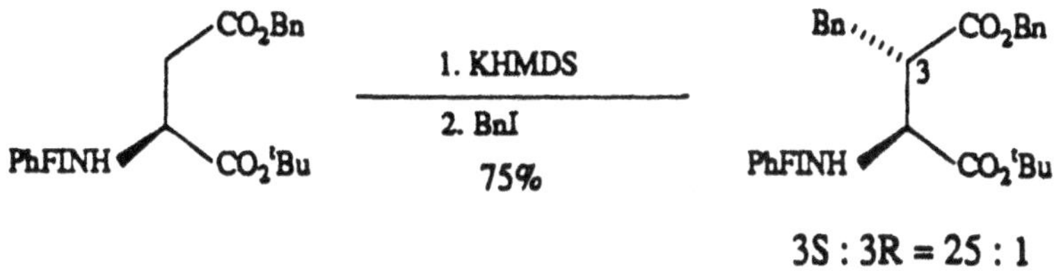

Fig. 1. Asymmetric alkylation at the b-carbon.

tion. Further work was done by the same group, in which they asymmetrically synthesized γ,δ-unsaturated α-amino acid by using the same synthon, α-tButyl γ-methyl *N*-trityl-(S)-glutamate. Baldwin and associates have tried to use (L)-pyroglutamic acid as a chiral starting material, to synthesize α-amino acids *(66)*. They first generated lactam enolate, followed by addition of electrophiles at the C-4 position in the ring system, although the diastereoselectivity is not high at the two new chiral centers (γ-C and δ-C). Elimination of water, ring opening, and deprotection provided the γ,δ-unsaturated α-amino acids. Hanson and asssociate used a six-membered ring γ,δ-unsaturated α-amino acid *(baikian)* as starting material to stereoselectively synthesized Δ^4-pipecolic acid *(67)*. Actually the γ-homologation product in this synthesis is not the goal, but only a key intermediate that can introduce an asymmetric alkylation on the ring system and recover the γ,δ-unsaturated functionality. This synthesis is an elegant application of γ-homologation. Hudlicky and Merola reported a synthesis of (–)-D-*erythro*- and (+)-L-*threo*-4-fluoroglutamic acid by the γ-isomerization of protected L-Hyp *(68)*. The stereocontrol step here is an oxidation by utilizing RuO_4 as oxidation reagent.

Very recently, El Hadrami and Lavergne synthesized (2S,4S,6S)-2-amino-6-hydroxy-4-methyl-8-oxodecanoic acid (AHMOD) *(69)*, a constituent of Leucinostatines that might be considered a pseudopeptide. They started with vinyl magnesium bromide and (R)-3-bromo-2-methyl propanol (Scheme 10) to obtain the vinyl alcohol **52**, which was converted to iodide **53**. Iodide **53** can alkylate the optically active Schiff base **54** under the basic condition through a Michael reaction. The chiral part of the Schiff base **54** works as a chiral auxiliary to achieve the diastereoselective alkylation. The adduct **55** underwent cleavage of the chiral auxiliary via acidic hydrolysis. The obtained terminal-vinyl-α-

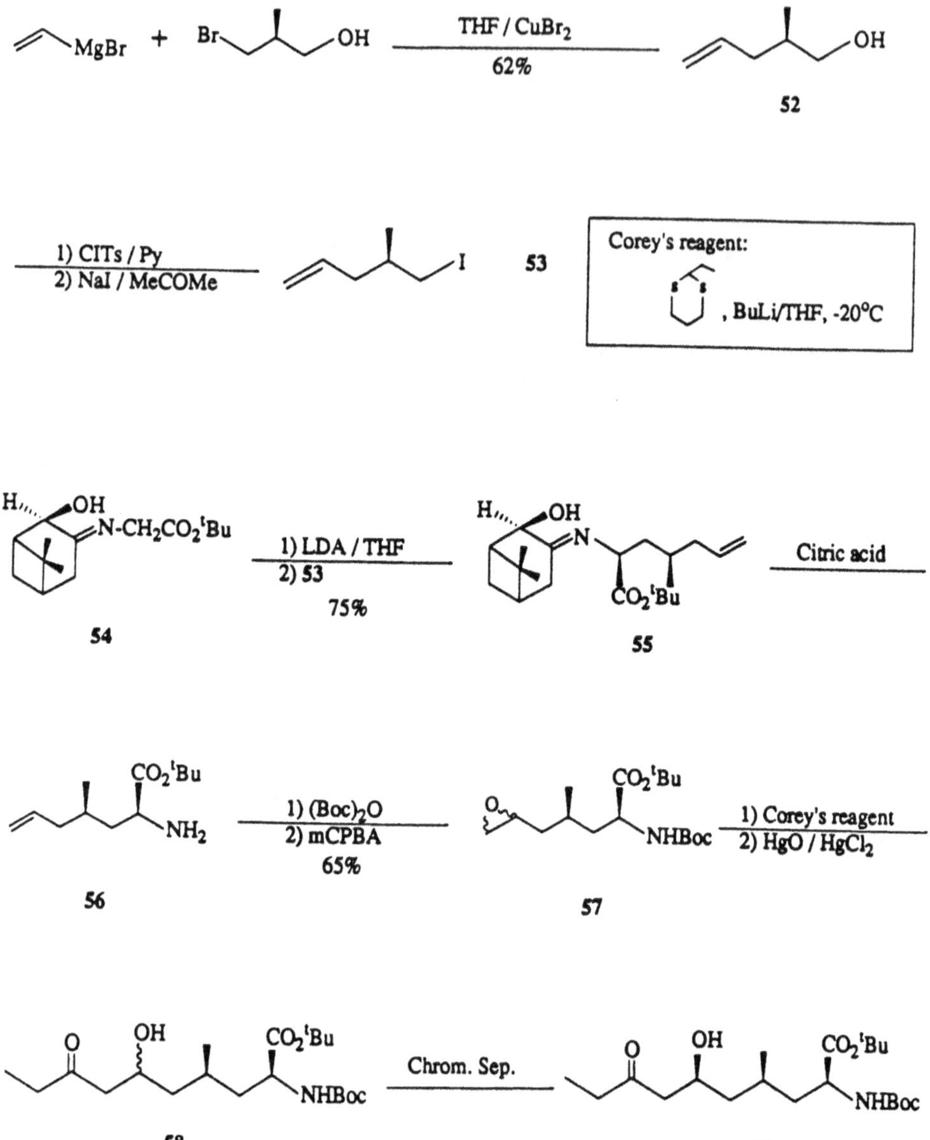

Scheme 10.

amino ester **56** was oxidized, followed by protection of the amino group to afford epoxide **57**. The epoxide ring was opened by Corey's reagent (prepared from 2-ethyl-1,3-dithiane) via nucleophilic attack. The adduct underwent deprotection of the carbonyl function by the treatment of mercuric ions *(70)* to provide (6-*dl*)-AHMOD **58**. Final column chromatography separation afforded the (2S,4S,6S) diastereomer **59**.

3.6. Total Synthesis of α-Amino Acids

Urbach and Henning synthesized (1S,3S,5S)- and (1R,3S,5R)-azabicyclo-[3.3.0]octane-3-carboxylic acid from L-serine *(71)*. One five-membered ring product was directly obtained from the starting material, 3-bromocyclopentene. The other fused *N*-contained five-membered ring was achieved by an intramolecular radical cyclization of the appropriate optically active olefinic α-amino acid derivative. Unfortunately, the diastereoselectivity of the latter example was not as good as the former example; the former, with the ratio of (S,S,S) isomer to (R,S,R) isomer, is 1.25:1. Mulzer and associates did interesting work on the synthesis of the nonproteogenic amino acid (2S,2S,4S)-3-hydroxy-4-methylproline (HMP) and its enantiomer *(72)*. They started with the optically active tetrol derivatives that are readily available from D-mannitol. The key intermediates are optically active azido epoxides and their cyclic successor, diastereomerically pure 1-aza-bicyclo[3.1.0]hexanes, which underwent Staudinger aminocyclization to afford proline derivatives that could be converted to the final α-amino acids. The aminocyclization step was both stereo- and regiocontrolled. Hamada and coworkers synthesized γ-azetidinyl-β-hydroxy-α-amino acid as a precursor of mugineic acid *(73)*. The starting material they used was *O,O'*-isopropylidene-(R)-glyceric acid. There were two key steps in this total synthesis. One was the selective catalytic hydrogenation of a α,β-unsaturated five-membered lactone derivative; the other step was the catalytic hydrogenation of an aldehyde with the *p*-toluenesulfonate salt of benzyl (S)-2-azetidinecarboxylate by use of sodium cyanoborohydride to give the γ-azetidinyl-hydroxy-α-amino acid to mugneic acid in another earlier paper *(74)*. Schmidt's group accomplished quite a few total syntheses on complicated amino acids, which actually are cyclized small peptides *(75)*. Very recently, Schmidt and associate synthesized three different protected (2S,4R)-4-hydroxyornithines *(76)*. The key intermediates in this total synthesis are oxazolidine aldehyde, which is derived from (S)-malic acid in three steps, and a didehydroamino acid derivative that can undergo stereoselective catalytic hydrogenation to afford the final product: hydroxyornithine ester derivative. The catalyst for the hydrogenation was (R,R)-$[Rh(1,5\text{-}COD)(DIPAMP)]^+BF_4^-$; the obtained diastereoselectivity varies from 75:25 to 100:1.

Ⓟ — 10% cross-linking polyacrylic resin with a loading of 1 meg of aldehyde function per gram

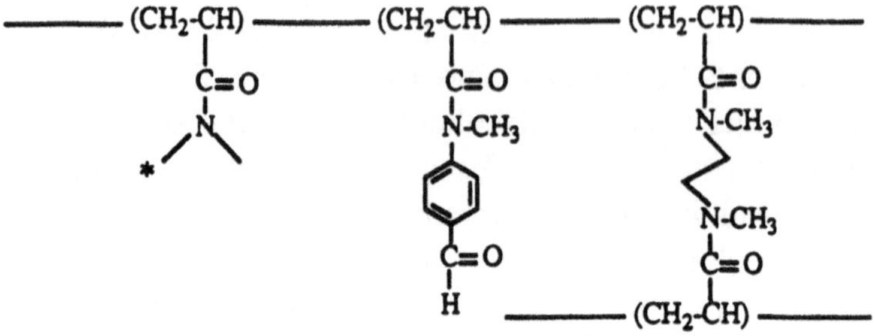

Fig. 2. Schematic structure of the polymer used in supermolecular asymmetric induction. (*indicates the location of chiral pedants).

3.7. Other Methods

α-Amino acids can be synthesized by many other different methodologies because these methodologies cannot be catalogued as a single separate class. Miscellaneous methods are very important not only because they involve many different kinds of chemistry, but also because of this recent exposure during the past few years *(77–93)*. We would like to introduce one interesting example in detail here. Daunis and coworkers synthesized α-amino acids via supramolecular asymmetric induction *(86)* Previously, similar ideas had been brought up by Saito and Harada *(81)*. The chiral inducer consists of crosslinked (10%) polyacrylic resin (with a loading of 1 mEq of aldehyde function/gram) and chiral pendant (Fig. 2). The chiral pendants they used were *N*-methyl α-phenylethylamine, prolinol, and prolinol methyl ester. Acid-catalyzed condensation of *t*-butyl glycinate with the above polymer **60** gave Schiff base **61** (Scheme 11). Compound **61** treated with LDA in THF afforded enolate **62**, which subsequently reacted with alkyl halide, followed by hydrolysis to give the crude amino acid hydrochloride **63**. In their paper, they estimated that there are about three to four chiral pendants surrounding the alkylation reaction center to give the chiral induction. Amino acid **63** was treated with hexamethyldisilazane to give the bistrimethylsilyl derivative **64**, which was then converted to pure α-amino acid **65** by the treatment of excess of methanol. Some of these results are shown in Table 6.

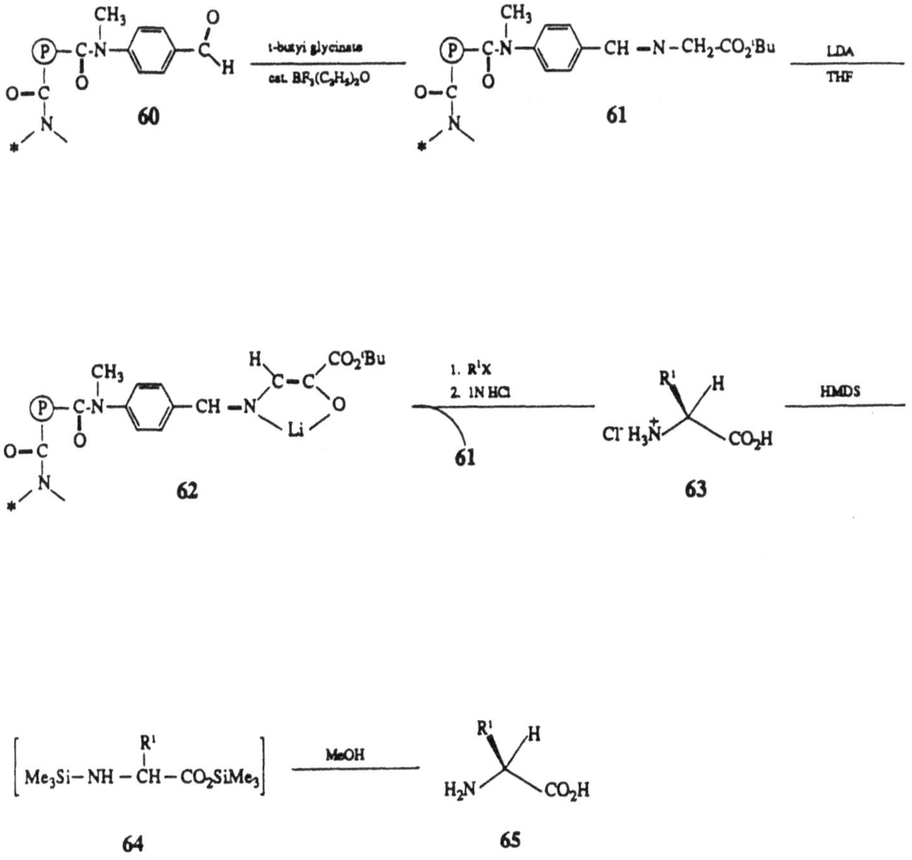

Scheme 11.

4. Use of α-Amino Acids

There are many potential uses for proteogenic α-amino acids, and here we will mention the two most common applications.

4.1. Utilization in Organic Synthesis

As already implied in the previous sections, there are many examples of using α-amino acids as starting materials as well as key intermediates in the asymmetric synthesis of other α-amino acids. We will give just a few such examples of employing α-amino acids in organic synthesis in this section. Easton and associates reported a regioselective formation of amidocarboxy-substituted free radicals *(94)*. The α-amino acids they used were *N*-benzolylvaline methyl ester and *N*-benzoylsarosine methyl ester. Rapoport et al. have used cysteine as a side chain to modify

Table 6
Diastereoselectivities of the Synthesis Using Prolinol as Chiral Pendant

Chiral pendant	Entry	Alkylating agent	Temp., °C	Yields, % **63**	**65** e.e.% (S)
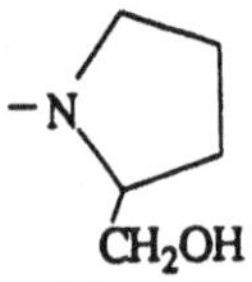	1	CH_3I	–78	75	88
	2	CH_3I	20	85	82
	3	*i*-C_3H_7I	–78	77	89
	4	*i*-C_3H_7I	20	84	84

Phycocyanobilin *(95)*. Jefford and associates synthesized (–)-Indolizidine 167B and (+)-Monomorine by using D-norvaline and L-alanine as starting materials *(96)*.

4.2. Utilization in Biological and Pharmacological Studies

The most important and widespread use of α-amino acids is in biological and pharmacological research. The biggest application in this area is the study of the chemistries, functions, and biological properties of peptides and proteins. As seen in the rest of this book, these studies have been so extensive that it is not possible to write a review about this application of α-amino acids in this short chapter. However, we want to give a few new examples here. Ueda and associates synthesized phenoxyacetyl-*N*-(hydroxydioxocyclobutenyl)cycloserine *(97)*. The L-isomer of this compound is thought to be a rational analog of lactivin, which, although not a β-lactam, has been shown to bind to penicillin-binding proteins, thus indicating a similar mode of action to the β-lactam antibiotics. In a recent report, Baldwin and associates gave further evidence for the involvement of a monocyclic β-lactam in the enzymatic conversion of β-L-α-amino-dipoyl-L-cysteinyl-D-valine into isopenicilin N *(98)*. Ouazzani and coworkers synthesized the enantiomeric α-amino phosphonic acids, phosphonic analogs of homoserine derivatives *(99)*. Such amino acids are believed to be the most important substitutes of the corresponding α-amino acids in the biological systems. Finally, in our laboratories *(100–103)*, we have successfully incorporated unusual α-amino acids (β-methyl phenylalanine, Tic, and D-Tic, and so on) for the conformational and topographical design of cyclic peptides in order to improve their biological profiles, and to obtain a more rational approach to conformation–activity relationships. The experimental results are optimistic.

5. Conclusion

In this discussion, we have reviewed several established methodologies for asymmetric synthesis of α-amino acids. In addition, there are many miscellaneous methods that are being developed, and some of them can provide pure optically active α-amino acids in high yields. We expect to see increased activity in the field.

Abbreviations

AcOH, acetic acid; AHMOD, 2-amino-6-hydroxy-4-methyl-8-oxodecanoic; AIBN, acidazo-(bis)-isobutyronitrile; BnX, benzyl halides (X = Cl, Br, I); Boc, *tert*-butyloxycarbonyl; $(Boc)_2O$, di-*tert*-butyldicarbonate; ^tBu, ^nBu (n-Bu), *tert*-butyl, butyl, respectively; Cbz, benzyloxycarbonyl; DBU, 1,8-diazabicyclo[5.4.0]undec-7-ene; de, diastereomeric excess; DEAD, *N,N*-α-diethylazodicarboxylate; DPPA, diphenylphosphonylazide; e.e., enatiomeric excess; ET, ethyl; HMP, 3-hydroxy-4-methylproline; HMPA, hexamethylphosphoramide; L-Hyp, L-4-hydroxyproline; LDA, lithium diisopropylamide; MAA, methyl *N*-acetamidoarylate; MeBmt, (4R)-4-[(E)-2-butenyl]-4, N-dimethyl-L-threonine; *m*CPBA, *meta*-chloroperoxybenzoic acid; Me, methyl; MIC, minimum inhibitorial concentration; (2S, 4S)-MOD-BPPM, *see* Scheme 2; Ph, phenyl; PIV, pivaloyl; ^iPr,Pr, *iso*-propyl, propyl, respectively; [Rh-(1,5-COD) (DIPAMP)]$^+$BF$^-$, rhodium-(1,5-cyclooctadiene)-{1,2-ethanediylbis-[(o-methoxyphenyl)phosphine]}; TFA, trifluoro acetic acid; TfO (OTf), triflate; THF, tetrahydrofuran; Tic (or D-Tic), 1,2,3,4-tetrahydroisoquinoline-carboxylic acid.

References

1. O'Donnell, M. J. (ed.) (1988) α-Amino acid synthesis (Tetrahedron Symposium-in-Print). *Tetrahedron* **44,** 5253–5614.
2. Williams, R. M. (1989) *Synthesis of Optically Active α-Amino Acids.* Pergamon, Oxford.
3. Schmidt, U., Respondek, M., Lieberknecht, A., Werner, J., and Fisher, P. (1989) Amino acids and peptides; 70. Optically active α-amino acids, *N*-Boc-amino-aldehydes and α-amino-β-hydroxy acids from 2,3-epoxy alcohols. *Synthesis,* 256–261.
4. Trigalo, F., Buisson, D., and Azerad, R. (1988) Chemoenzymatic synthesis of conformationally rigid glutamic acid analogs. *Tetrahedron Lett.* **29(47),** 6109–6112.
5. Lalonde, J. J., Bergbreiter, D. E., and Wong, C.-H. (1988) Enzymatic kinetic resolution of α-nitro α-methyl carboxylic acids. *J. Org. Chem.* **53,** 2323.
6. Schumacher, D. P., Clark, J. E., Murphy, B., and Fischer, P. A. (1990) An efficient synthesis of florfenicol. *J. Org. Chem.* **55,** 5291–5294.

7. Clark, J. E., Fischer, P. A., and Schumacher, D. P. (1991) An enzymatic route to florfenicol. *Synthesis* **10,** 891–894.
8. Parry, R. J. and Ju, S. (1991) The biosynthesis of sinefungin: investigation using a cell-free system. *Tetrahedron* **47(31),** 6069.
9. Vekemans, J. A. J. M., Versleijen, J. P. G., and Buck, H. M. (1991) NADH model reduction of C=N substrates: enantioselective synthesis of D- and L-phenylglycinates. *Tetrahedron: Asym.* **2(10),** 949–952.
10. Williams, R. M. and Zhai, W. (1988) Versatile, stereocontrolled, asymmetric synthesis of *E*-vinyl glycine derivatives. *Tetrahedron* **44,** 5425–5430.
11. Williams, R. M., Im, M.-N., and Cao, J. (1991) Asymmetric synthesis of 2,6-diamino-6-(hydroxymethyl)pimelic acid: assignment of stereochemistry. *J. Am. Chem. Soc.* **113,** 6967–6981.
12. Belokon, Y. N., Zel'ter, I. E., Bakhmutov, V. I., Saporovskaya, M. B., Ryzhov, M. G., Yanovsky, A. I., Struchkov, Y. T., and Belikov, V. M. (1983) Asymmetric synthesis of threonine and partial resolution and retroracemization of α-amino acids via copper (II) complexes of their Schiff base with (*S*)-2-*N*-(*N'*-benzylprolyl) aminobenzaldehyde and (*S*)-2-*N*-(*N'*-benzylprolyl)aminoacetophenone. Crystal and molecular structure of copper (II) complexes of glycine Schiff base with (*S*)-2-*N*-(*N'*-benzylprolyl)aminoacetophenone. *J. Am. Chem. Soc.* **105,** 2010–2017.
13. Belokon, Y. N., Bakhmutov, V. I., Chernoglazova, N. I., Kochetkov, K. A., Vitt, S. V., Garbalinskaya, N. S., and Belikov, V. M. (1988) General method for the asymmetric synthesis of α-amino acids via alkylation of the chiral nickel(II) Schiff base complexes of glycine and alanine. *J. Chem. Soc.* **2,** *Perkin Trans.* **I,** 305–312.
14. Belokon, Y. N., Bulychev, A. G., Pavlov, V. A., Feforova, E. B., Tsyryapkin, V. A., Bakhmutov, V. I., and Belikov, V. M. (1988) Synthesis of enatio- and diastereomerically pure substituted prolines via condensation of glycine with olefins activated by a carbonyl group. *J. Chem. Soc.* **8,** *Perkin Trans.* **I,** 2075–2083.
15. Nebel, K. and Mutter, M. (1988) Stereoselective synthesis of isovaline(IVA) and IVA-containing dipeptides for use in peptide synthesis. *Tetrahedron* **44(15),** 4793–4796.
16. Guanti, G., Banfi, L., Narisano, E., and Scolastico, C. (1988) Dibenzylaminoacetates as useful synthetic equivalents of glycine in the synthesis of α-amino-β-hydroxyacids. *Tetrahedron* **44(12),** 3671–3684.
17. Dellaria, J. F. and Santarsiero, B. D. (1989) Enantioselective synthesis of α-amino acid derivatives via the stereoselective alkylation of a homochiral glycine enolate synthon. *J. Org. Chem.* **54,** 3916–3926.
18. Schollkopf, U., Hausberg, H. H., Hoppe, I., Segal, M., and Reiter, U. (1978) Asymmetric synthesis of α-alkyl-α-aminocarboxylic acids by alkylation of 1-*chiral*-substituted 2-imidazolin-5-ones. *Angew Chem. Int. Ed. Engl.* **17,** 117–119.
19. Schollkopf, U., Tiller, T., and Bardenhagen, J. (1988) Asymmetric synthesis via heterocyclic intermediates—XXXIX. Asymmetric synthesis of (enantiomerically and diastereomerically virtually pure) methyl 2-amino-4,5-epoxy-3-hydroxy-alkanoates and methyl 2-amino-3-hydroxy-4,5-methylene-alkanoates by the *bis*lactimether method. *Tetrahedron* **44,** 5293–5305.

20. Schollkopf, U. and Beulshausen, T. (1989) Asymmetric syntheses of diastereomerically and enantiomerically pure (2R,3S)-*threo*-3-arylserine methyl esters by the *bis*lactim ether method asymmetric synthesis of chloramphenicol. *Liebigs Ann. Chem.* **3,** 223–225.
21. Mittendorf, J. and Hartwig, W. (1991) Enantioselective synthesis of 2,3-diamino acids by the bislactim ether method. *Synthesis* 939–941.
22. Hamon, D. P. G., Massy-Westropp, R. A., and Razzino, P. (1991) Enantioselective synthesis of α-amino acids via carbon-carbon bond forming radical reactions. *J. Chem. Soc., Chem. Comm.* **10,** 722–724.
23. Seebach, D. and Naef, R. (1981) Enantioselective generation and diastereoselective reactions of chiral enolates derived from α-heterosubstituted carboxylic acids. *Helv. Chim. Acta* **64,** 2704–2708.
24. Gander-Coquoz, M. and Seebach, D. (1988) Hertelling enantiomerenreiner, α-alkylierter lysin-, ornithin- und tryptophan-derivate. *Helv. Chim. Acta* **71,** 224–236.
25. Blaser, D., Ko, S. Y., and Seebach, D. (1991) A stereoselective synthesis of MeBmt employing a new chiral glycine enolate derivative. *J. Org. Chem.* **56,** 6230–6233.
26. Takahashi, H. and Achiwa, K. (1989) Efficient asymmetric hydrogenations of (*Z*)-2-acetamidoacrylic acid derivatives with the cationic rhodium complex of (2S,4S)-MOD-BPPM. *Chem. Lett.* **2,** 305–308.
27. Hamada, Y., Kawai, A., Matsui, T., Hara, O., and Shiori, T. (1990) 4-Alkoxycarbonyloxazoles as β-hydroxy-α-amino acid synthons: efficient, stereoselective synthesis of 3-amino-2,3,6-trideooxyhexoses and a hydroxy amino acid moiety of AI-77-B. *Tetrahedron* **46(13/14),** 4823–4846.
28. Gladiali, S. and Pinna, L. (1990) Completely regioselective hydroformylation of methyl *N*-acetamidoacrylate by chiral rhodium phosphine catalysts. *Tetrhedron: Asymm.* **1(10),** 693–696.
29. Genet, J. P., Pinel, C., Mallart, S., Juge, S., Thorimbert, S. and Laffitte, J. A. (1991) Asymmetric synthesis. Practical production of D & L threonine. Dynamic kinetic resolution in rhodium and ruthenium catalyzed hydrogenation of 2-acylamino-3-oxobutyrates. *Tetrahedron: Asymm.* **2(7),** 555–567.
30. Schmidt, U., Lieberknecht, A., Kazmaier, U., Griesser, H., Jung, G., and Metzger, J. (1991) Amino acids and peptides; 75. Synthesis of di- and trihydroxyamino acids—construction of lipophilic tripalmitoyldihydroxy-α-amino acids. *Synthesis* **1,** 49–55.
31. Oppolzer, W. and Moretti, R. (1988) Enantioselective synthesis of α-amino acids from α-halogenated 10-sulfonamido-isobornyl esters. *Tetrahedron* **44,** 5541–5552.
32. Oppolzer, W. and Tamura, O. (1990) Asymmetric synthesis of α-amino acids and α-*N*-hydroxyamino acids via electrophilic amination of bornanesultam-derived enolates with 1-chloro-1-nitrosocyclohexane. *Tetrahedron Lett.* **31(7),** 991–994.
33. Takano, S., Yanase, M., and Ogasawara, K. (1989) A concise route to (*S*)-phenylalanine from (*R*)-epichlorohydrin. *Heterocycles* **29(9),** 1825–1828.
34. Jung, M. E. and Jung, Y. H. (1989) Rapid synthesis of β-hydroxy-α-amino acids, such as L-threonine, β-hydroxyphenylalanine, and β-hydroxyleucine, via an

application of the sharpless asymmetric epoxidation. *Tetrahedron Lett.* **30(48),** 6637–6640.
35. Wagner, R., Tilley, J. W., and Lovey, K. (1990) An improved synthesis of a protected (2S,3R)-3-hydroxyaspartic acid suitable for solid-phase peptide synthesis. *Synthesis* **9,** 785–786.
36. Evans, D. A., Britton, T. C., Dorow, R. L., and Dellaria, J. F. (1986) Stereoselective amination of chiral enolates. A new approach to the asymmetric synthesis of α-hydrazino and α-amino acids derivatives. *J. Am. Chem. Soc.* **108,** 6395–6397.
37. Evans, D. A., Britton, T. C., Ellman, J. A., and Dorow, R. L. (1990) The asymmetric synthesis of α-amino acids. Electrophilic azidation of chiral imide enolates, a practical approach to the synthesis of (*R*)- and (*S*)-α-azido carboxylic acids. *J. Am. Chem. Soc.* **112,** 4011–4030.
38. Dharanipragada, R., Nicholas, E., Toth, G., and Hruby, V. J. (1989) Asymmetric synthesis of unusual amino acids: synthesis of optically pure isomers of β-methylphenylalanine. *Tetrahedron Lett.* **30(49),** 6841–6844.
39. Nicholas, E., Dharanipragada, R., Toth, G., and Hruby, V. J. (1989) Asymmetric synthesis of unusual amino acids: synthesis of optically pure isomers of β-methyltyrosine. *Tetrahedron Lett.* **30(49),** 6845–6848.
40. Nicholas, E., Dharanipragada, R., Russell, K. C., Van Hulle, K., Alarcon, A., and Hruby, V. J. (1991) Asymmetric synthesis of unusual amino acids: synthesis of Tyr and Phe analogues, in *Peptide 1990, Proceedings of the 21st European Peptide Symposium* (Giralt, E. and Andreu, D., eds.), Escom, Leiden, The Netherlands, pp. 368–369.
41. Dharanipragada, R., Bruck, M., and Hruby, V. J. (1992) The absolute configuration in the asymmetric synthesis of unusual amino acids. *Acta Crystallographica* **C48,** 1239–1241.
42. Dharanipragada, R., Van Hulle, K., Bannister, A., Bear, S., Kennedy, L., and Hruby, V. J. (1992) Asymmetric synthesis of unusual amino acids: an efficient synthesis of optically pure isomers of β-methylphenylalanine. *Tetrahedron* **48,** 4733–4748.
43. Evans, D. A. and Weber, A. E. (1986) Asymmetric glycine enolates aldol reactions: synthesis of cyclosporine's unusual amino acids. *J. Am. Chem. Soc.* **108,** 6757–6761.
44. Ariza, J., Font, J., and Ortuno, R. M. (1990) Enantioselective synthesis of hydroxy α-amino acids. (–)-erythro- and (–)-*threo*-γ-hydroxynorvalines. *Tetrahedron* **46(6),** 1931–1942.
45. Papageorgiou, C. and Benezra, C. (1984) A facile synthesis of optically pure valeroactone and β-hydroxy valeroactone from a common sugar-derived precursor. *Tetrahedron Lett.* **25(52),** 6041–6044.
46. Ariza, J., Font, J., and Ortuno, R. M. (1991) An efficient and concise entry to (–)-4,5-dihydroxy-D-*threo*-L-norvaline. Formal synthesis of clavalanine. *Tetrahedron Lett.* **32(17),** 1979–1982.
47. Inghardt, T., Frejd, T., and Svensson, G. (1991) Organoaluminium induced ring-opening of epoxypyranosides. IV. Synthesis and structure of γ-hydroxy-isoleucine stereoisomers and their corresponding lactones. *Tetrahedron* **47(32),** 6469–6482.

48. Fleming, P. R. and Sharpless, K. B. (1991) Selective transformations of the *threo*-2,3-dihydroxy esters. *J. Org. Chem.* **56,** 2869–2875.
49. Corey, E. J. and Chai, S. (1991) Highly enantioselective routes to Darzens and acetate aldol products from achiral aldehydes and *t*-butyl bromoacetate. *Tetrahedron Lett.* **32(25),** 2857–2860.
50. Strecker, A. (1850) Ueber die kunstliche bidung der milchsaure und einen neun, dem glycoll homologen korper. *Leibigs Ann. Chem.* **75,** 27–45.
51. Harada, K. and Okawara, T. (1973) Sterically controlled synthesis of optically active organic compounds XVIII. Asymmetric synthesis of amino acids by addition of hydrogen cyanide to Schiff base. *J. Org. Chem.* **38,** 707–710.
52. Kunz, H. and Sager, W. (1987) Diastereoselective Strecker synthesis of α-aminonitriles on carbohydrate templates. *Angew. Chem. Int. Ed. Engl.* **26,** 557–559.
53. Kunz, H. and Pfrengle, W. (1988) Carbohydrates as chiral templates: asymmetric Ugi-synthesis of α-amino acids using galactosylamines as the chiral matrices. *Tetrahedron* **44,** 5487–5494.
54. Kunz, H., Pfrengle, W., and Sager, W. (1989) Carbohydrates as chiral template: diastereoselective Ugi synthesis of (*S*)-amino acids using *O*-acylated D-arabinopyranosylamine as the auxilliary. *Tetrahedron Lett.* **30(31),** 4109–4110.
55. Kunz, H., Pfrengle, W., Ruck, K., and Sager, W. (1991) Stereoselective synthesis of L-amino acids via Strecker and Ugi reactions on carbohydrate templates. *Synthesis* **11,** 1039–1042.
56. Paulsen, H., Gyorgydeak, Z., and Friedmann, M. (1974) *exo*-Anomer effekt und circulardichroismus von glycopyranosylaziden. *Chem. Ber.* **107,** 1568–1578.
57. Chakraborty, T. K., Reddy, G. V., and Azhar Hussain, K. (1991) Diastereoselective Strecker synthesis using α-phenylglycinol as chiral auxiliary. *Tetrahedron Lett.* **32(51),** 7597–7600.
58. Cainelli, G. and Panunzio, M. (1991) Studies on *N*-metallo imines: synthesis of *N*-unsubstituted aziridines from *N*-trimethylsilyl imines and lithium enolates of α-halo esters. *Tetrahedron Lett.* **32(1),** 121.
59. Maurer, P. J., Takahata, H., and Rapoport, K. (1984) α-Amino acids as chiral enolates for asymmetric products. A general synthesis of D-α-amino acids from L-serine. *J. Am. Chem. Soc.* **106,** 1095–1098.
60. Dunn, P. J., Haner, R., and Rapoport, H. (1990) Stereoselective synthesis of 2,3-diamino acids. 2,3-diamino-4-phenylbutanoic acid. *J. Org. Chem.* **55,** 5017–5025.
61. Sasaki, N. A., Hashimoto, C., and Pauly, R. (1989) Stereoselective synthesis of optically pure β,γ-unsaturated α-amino acids in both L and D configurations. *Tetrahedron Lett.* **30(15),** 1943–1946.
62. Baldwin, J. E., Moloney, M. G., and North, M. (1989) Non-proteinogenic amino acid synthesis: synthesis of β,γ-unsaturated α-amino acids from aspartic acid. *Tetrahedron* **45(19),** 6319–6330.
63. Baldwin, J. E., Moloney, M. G., and North, M. (1989) Asymmetric α-amino acid synthesis: preparation of the β-anion derived from aspartic acid. *J. Chem. Soc.* **4,** *Perkin Trans.* **I,** 833–834.

64. Parry, R. J. and Lii, F.-L. (1991) Investigation of the biosynthesis of *trans*-(+)-*S*-1-propenyl-L-cysteine sulfoxide. Elucidation of the stereochemistry of the oxidative decarboxylation process. *J. Am. Chem. Soc.* **113,** 4704–4706.
65. Baldwin, J. E., North, M., and Flinn, A. (1989) Synthesis of nonproteinogenic amino acids part 2: preparation of a synthetic equivalent of the anion synthon for asymmetric amino acid synthesis. *Tetrahedron* **45(5),** 1453–1464.
66. Baldwin, J. E., North, M., Flinn, A., and Moloney, M. G. (1989) Synthesis of nonproteinogenic amino acids part 3: conversion of glutamic acid into γ,δ-unsaturated α-amino acids. *Tetrahedron* **45(5),** 1465–1474.
67. Hanson, G. J. and Russell, M. A. (1989) Stereoselective Δ^4-pipecolic acid synthesis via alkylation of a vinyl *N*-Boc-iminium ion derived from baikiain. *Tetrahedron Lett.* **30(42),** 5751–5754.
68. Hudlicky, M. and Merola, J. S. (1990) New stereospecific syntheses and X-ray diffraction structures of (–)-D-erythro- and (+)-L-*threo*-4-fluoroglutamic acid. *Tetrahedron Lett.* **31(51),** 7403–7406.
69. El Hadrami, M., Lavergne, J.-P., and Viallefont, P. (1991) Synthesis of (2S,4S,6S)-2-amino-6-hydroxy-4-methyl-8-oxodecanoic acid and (4S,E)-4-methylhex-2-enoic acid constituents of leucinostatines. *Tetrahedron Lett.* **32(32),** 3985–3988.
70. Seebach, D. and Steinmuller, D. (1968) Preparation of (S)-2-methylbutyl and (S)-*sec*-butyl ketones from optically active 2-methyl-1-butanol by the dithiane method. *Angew. Chem. Int. Edit.* **7,** 619–620.
71. Urbach, H. and Henning, R. (1989) Enantioselective synthesis of 1S,3S,5S- and 1R, 3S,5S-2-azabicyclo[3.3.0]octane-3-carboxylic acid starting from L-serine. *Heterocycles* **28(2),** 957–965.
72. Mulzer, J., Becker, R., and Brunner, E. (1989) Synthesis of hydroxylated 1-azabicyclo[3.1.0] hexane and prolinol derivatives by stereo- and regiocontrolled staudinger aminocyclization. Application to the nonproteinogenic amino acid (2S,3S,4S)-3-hydroxy-4-methylproline (HMP) and its enantiomer. *J. Am. Chem. Soc.* **111,** 7500–7504.
73. Hamada, Y., Iwai, K., and Shioiri, T. (1990) A new stereoselective synthesis of a γ-azetidinyl-β-hydroxy-α-amino acid moiety of mugineic acid—A formal synthesis of mugineic acid. *Tetrahedron Lett.* **31(35),** 5041–5042.
74. Hamada, Y. and Shioiri, T. (1986) Synthesis of mugineic acid through direct c-acylation using diphenyl phosphorazidate. *J. Org. Chem.* **51,** 5489–5490.
75. Schmidt, U., Beuttler, T., Lieberknecht, A., and Griesser, H. (1983) Aminosauren und peptide–XXXXII. Synthese von chlamydocin + epi-chlamydocin. *Tetrahedron Lett.* **24(34),** 3573–3576.
76. Schmidt, U., Meyer, R., Leitenberger, V., Stabler, F., and Lieberknecht, A. (1991) Total synthesis of the biphenomycins; II. Synthesis of protected (2S,4R)-4-hydroxyornithines. *Synthesis* **5,** 409–413.
77. Gasparski, C. and Miller, M. J. (1991) Synthesis of β-hydroxy-α-amino acids by aldol condensation using a chiral phase transfer catalyst. *Tetrahedron* **47(29),** 5367–5378.
78. O'Donnell, M. J., Bennett, W. D., and Wu, S. (1989) The stereoselective synthesis of α-amino acids by phase-transfer catalysis. *J. Am. Chem. Soc.* **111,** 2353–2355.

79. Yu, L. and Wang, F. (1987) Asymmetric synthesis of α-amino acids by phase-transfer catalysis. *Chem. J. Chin. Univ.* **8(4),** 336–340.
80. Barton, D. H., Herve, Y., Potier, P., and Thierry, J. (1987) Synthesis of novel α-amino acids and derivatives using radical chemistry: synthesis of L- and D-α-amino acids, L-α-aminopimelic acid and appropriate unsaturated derivatives. *Tetrahedron* **43(19),** 4297–4308.
81. Saito, K. and Harada, K. (1989) Asymmetric synthesis of amino acids by addition of cyanide to the Schiff base in the presence of cyanide modified hemin-copolymer. *Tetrahedron Lett.* **30(34),** 4535–4538.
82. Nagai, U. and Pavone, V. (1989) Preparation of all the four diastereomers of β-phenylcysteine methyl ester through chromatographic optical resolution of the 2,2-dimethylthiazolidine derivatives. *Heterocycles* **28(2),** 589–592.
83. Crich, D. and Davies, W. (1989) Asymmetric synthesis of α-alkylated tryptophan derivatives. *J. Chem. Soc., Chem. Comm.* **19,** 1418,1419.
84. Tietze, L. F. and Bratz, M. (1989) Diastereoselective formation of substituted cyclic nonproteinogenic α-amino acids by cyclization of activated imines. *Synthesis* **5,** 439–442.
85. Pearson, A. J., Bruhn, P. R., Gouzoules, F., and Lee, S.-H. (1989) Stereoselective synthesis of arylglycine derivatives using arene-manganese tricarbonyl complexes. *J. Chem. Soc., Chem. Comm.* **10,** 659–661.
86. Calmes, M., Daunis, J., Ismaili, H., and Jacquier, R. (1990) Supramolecular asymmetric induction: a new concept applied to the supported enantioselective synthesis of α-amino acids. *Tetrahedron* **46(17),** 6021–6032.
87. Reetz, M. T., Wunsch, T., and Harms, K. (1990) Stereoselective synthesis of α,γ-diamino-β-hydroxy amino acid esters: a new class of amino acids. *Tetrahedron: Asymmetry* **1(6),** 371–374.
88. Beaulieu, P. L. (1991) Diastereospecific synthesis of D- and L-allo-threonine and trichlorinated derivatives suitable for the preparation of tritium labeled material. *Tetrahedron Lett.* **32(8),** 1031–1034.
89. Jako, I., Uiber, P., Mann, A., Wermuth, C.-G., Boulanger, T., Norberg, B., Evrard, G., and Durant, F. (1991) Stereoselective synthesis of 3-alkylated glutamic acids: application to the synthesis of secokainic acid. *J. Org. Chem.* **56,** 5729–5733.
90. Amoroso, R., Cardillo, G., and Tomasini, C. (1991) 1,3-Asymmetric induction: highly enantioselective synthesis of α-amino acids via 2,5-*trans* disubstituted imidazolidin-4-ones. *Tetrahedron Lett.* **32(17),** 1971.
91. Kazmierski, W. M. and Hruby, V. J. (1991) Asymmetric synthesis of topographically constrained amino acids: synthesis of the optically pure isomers of α,β-dimethylalanine and α,β-dimethyl-1,2,3,4-tetrahydroisoquinoline-3-carboxylic acid. *Tetrahedron Lett.* **32(41),** 5769–5772.
92. Josien, H., Martin, A., and Chassaing, G. (1991) Asymmetric synthesis of L-diphenylalanine and L-9-fluorenylglycine via room temperature alkylations of a sultam-derived glycine imine. *Tetrahedron Lett.* **32(45),** 6547–6550.
93. Bailey, P. D., Brown, G. R., and Korber, F. (1991) Asymmetric synthesis of pipecolic acid derivatives using the Aza-Diels-Alder reaction. *Tetrahedron: Asymmetry* **2(12),** 1263–1282.

94. Easton, C. J., Hay, M. P., and Love, S. G. (1989) Regioselective formation of amidocarboxy- substituted free radicals. *J. Chem. Soc.* **2,** *Perkin Trans.* **I,** 265–268.
95. Bishop, J. E., Nagy, J. O., O'Connell, J. F., and Rapoport, H. (1991) Diastereo-selective synthesis of phycocycanobilin-cysteine adducts. *J. Am. Chem. Soc.* **113,** 8024–8035.
96. Jefford, C. W., Tang, Q., and Zaslona, A. (1991) Short, enantiogenic syntheses of (–)-Indolizidine 167B and (+)-monomorine. *J. Am. Chem. Soc.* **113,** 3513–3518.
97. Ueda, Y., Crast, L. B., Mikkilineni, A. B., and Partyka, R. A. (1991) Synthesis of phenoxyacetyl-*N*-(hydroxydioxocyclobutenyl)cycloserines. *Tetrahedron Lett.* **32(31),** 3767–3770.
98. Baldwin, J. E., Bradley, M., Adlington, R. M., Norris, W. J., and Turner, N. J. (1991) Further evidence for the involvement of a monocyclic β-lactam in the enzymatic conversion of δ-L-α-aminoadipoyl-L-cysteinyl-D-valine into isopenicilin N. *Tetrahedron* **47(3),** 457–480.
99. Ouazzani, F., Roumestant, M.-L., Viallefont, P., and El Hallaoui, A. (1991) Synthesis of enantiomerically pure phosphonic analogs of homoserine derivatives. *Tetrahedron: Asymm.* **2(9),** 913–917.
100. Hruby, V. J., Kazmierski, W., Matsunaga, T. O., Nikiforovich, G. V., and Prakash, O. (1991) Topographic consideration in the design of potent, receptor selective peptide hormones and neurotransmitters, in *Protein: Structure, Dynamics and Design* (Renugopalakrishnan, V., Carey, P. R., Smith, I. C., Huang, S. G., and Storer, A. C., eds.), Escom, Leiden, The Netherlands, pp. 271–276.
101. Kazmierski, W. M., Yamamura, H. I., and Hruby, V. J. (1991) Topographic design of peptide neurotransmitters and hormones on stable backbone templates: relation of conformation and dynamics to bioactivity. *J. Am. Chem. Soc.* **113,** 2275–2283.
102. Gehrig, C. A., Toth, G., Prakash, O., and Hruby, V. J. (1991) β-Me-Phe containing enkephalin analogs with high selectivity for the delta opioid receptor, in *Peptide Chemistry 1990* (Shimonishi, Y., ed.), Protein Research Foundation, Osaka, pp. 261–264.
103. Russell, K. C., Kazmierski, W. M., Nicolas, E., Ferguson, R., Knollenberg, J., Wegner, K., and Hruby, V. J. (1992) Asymmetric synthesis of unusual amino acids designed for topographic control of peptide conformation, in *Proceedings of the 12th American Peptide Symposium* (Smith, J. and Rivier, J., eds.), Escom, Leiden, The Netherlands, pp. 768–770.

CHAPTER 14

Synthesis of Fully Protected Peptide Fragments

Monika Mergler

1. Introduction

Development of appropriate resin linker combinations for solid-phase peptide synthesis (SPPS) has allowed rapid access to (fully) protected peptide fragments with a free C-teminal carboxyl moiety. These fragments may be assembled either in solution or on resin—an approach that has some intrinsic advantages compared to the stepwise methodology (*see* Chapter 15). Approaches to synthesize such fragments usually employ "orthogonal" (*see* Note 1) N^{α}-protection/side-chain protection. The orthogonal combination Fmoc/*t*Bu can, however, only be used if the peptide resin bond is cleavable either by acids weak enough to leave *t*Bu groups intact or by a method employing neither acids nor bases, such as photolysis *(1)* or catalysis by a Pd(0) complex *(2)*.

A few very acid-labile resins have been developed. Only a part of them is commercially available. Some are labile even toward acetic acid in TFE/DCM *(3)*, which means that the acidity of the protected amino acids and HOBt has to be taken into account (or else premature cleavage may decrease the yield). If, on the other hand, too strong an acid (or too high a concentration) is needed for cleavage, N^{ε}-Boc groups of lysine or *t*Bu ether groups of tyrosine may be partially cleaved, which is not acceptable. A good compromise is reached when cleavage can be carried out with 0.5–1% TFA in DCM *(4,5)* (*see* Chapter 5). Fully Boc/*t*-butyl-protected fragments can be obtained under these conditions, e.g., from the commercially available Sasrin™ resin *(6)*. This chapter will deal explicitly with the synthesis on and the conditions for cleavage of fragments

From: *Methods in Molecular Biology, Vol. 35: Peptide Synthesis Protocols*
Edited by: M. W. Pennington and B. M. Dunn

from Sasrin resin. Sample cleavage from Sasrin may also serve as a tool for SPPS monitoring (*see* Note 2).

Solid-phase synthesis using Fmoc strategy (e.g., on Wang resin) has already been dealt with in detail in preceding chapters of this book. The methods described there can also be applied to Sasrin resin, but, obviously, acids stronger than N-protected amino acids have to be avoided (just as prolonged treatment with piperidine) to avoid premature cleavage of the peptide from the resin. N^{α}- and side-chain protecting groups have to withstand treatment with 1% TFA in DCM. The most useful protecting groups for synthesis of protected fragments on Sasrin resin and their limitations are listed below:

for N^{α}:	Fmoc, Boc, Z
Asn, Gln:	Mtt, Trt, none (protection of the amide group is strongly recommended)
Asp, Glu:	O*t*Bu
Arg:	Pmc, Mtr (*see* Chapter 5)
Cys:	Trt, Acm (polar), S*t*Bu (*see* Note 3)
His:	none, Trt (*see* Note 4)
Lys:	Boc
Ser, Thr, Tyr:	*t*Bu
Met:	none, sulfoxide (polar, *see* Note 5)
Trp:	none, Boc (*see* Note 6)

All these protecting groups influence the solubility of the fragment; sufficient solubility is a crucial requirement for efficient fragment coupling (*see* Chapter 15). If necessary, various fragments with different combinations of protecting groups have to be prepared and checked for solubility.

If possible, Gly or Pro are chosen as C-terminal amino acid of the fragment with the free carboxyl moiety to avoid racemization during coupling later on, but when synthesizing fragments with C-terminal Pro, the formation of diketopiperazine (*see* Note 7) has to be circumvented by coupling, e.g., Fmoc dipeptides instead of the penultimate Fmoc amino acid at the risk of concomitant racemization. Recently, other deblocking procedures to suppress diketopiperazine formation have been published *(7)*. Unfortunately, nature has conceived most of its peptides not in the manner making the peptide chemist's work easy by putting glycine into appropriate positions. Therefore, coupling with other C termini cannot be avoided, and appropriate coupling conditions with minimal racemization have to be worked out. Coupling fully protected peptide azides is an old but, still valuable method known for minimal concomitant racemization (*see* Note 8). These acid azides are usually generated *in situ* from the corresponding hydrazides. Fully protected peptide hydrazides can also be obtained rapidly by SPPS on Sasrin resin followed by cleavage with hydrazine hydrate *(8)*. This chapter will deal exclusively with acidolytic cleavage leading to protected peptide fragments with a free C-terminal carboxyl group.

2. Materials

1. All reagents and solvents are commercially available. TFA and pyridine should be colorless or else they must be distilled. The 1% TFA/DCM solution can be stored in a tightly closed dark bottle for a few weeks. Piperidine solutions in DMF are not stable and may only be kept for a few days in a closed container.
2. DCM should be dry and acid-free, but the commercially available solvent usually fulfills these requirements, if it is properly stored. DMF has to be freed from basic impurities, such as dimethylamine, e.g., by treating it with acidic aluminum oxide or by distilling it *in vacuo* from ninhydrin. DMF should be kept in a dark bottle and the pH checked from time to time. Ethers must be free of peroxides especially when Cys- or Met-containing fragments are to be treated (*see* Note 9). Check, e.g., with Merckoquant Peroxid-Test strips 10.011 from Merck (Darmstadt, Germany). Methyl *t*-butyl ether is less prone to peroxide formation than diethyl or diisopropyl ether, and less volatile. The other solvents need no additional purification.
3. Fmoc-amino acid Sasrin resins are commercially available from BACHEM AG, Bubendorf, Switzerland.
4. The solutions for the Kaiser test are relatively stable; b and c have to be colorless:
   a. 5 g ninhydrin in 100 mL abs. ETOH;
   b. 80 g phenol in 20 mL abs. ETOH; and
   c. 2 mL 0.001*M* aq. KCN—add pyridine to 100 mL.
5. Equipment for TLC and HPLC analysis (*see* Chapters 1–5, *PAP*).
6. FAB-MS and other methods to confirm the structure of the fragment (*see* Chapters 6 and 7, *PAP*).

3. Methods

3.1. General Considerations

3.1.1. Instrumental

Synthesis and cleavage can be performed manually on a fritted-glass funnel or in any commercially available synthesizer suitable for Wang resin (some modifications may be necessary for the cleavage procedure). A shaker with a vessel equipped with a sintered-glass bottom is a simple, but versatile apparatus for synthesis and cleavage. It may be operated manually or automatically. Equipment specifically designed for continuous flow should not be used for SPPS on Sasrin resin. During the whole procedure, the peptide resin never needs to be transferred to another vessel or funnel. Filtration can be performed by suction or, preferentially, by inert gas pressure. This way the peptide resin is protected from oxygen and moisture, and the solvents are removed rapidly, but gently. Many fully automated synthesizers apply this method. On the other hand, the simple "fritted-glass funnel method" is suitable for the rapid manual synthesis and cleavage using small amounts of resin or, especially, for cleaving small samples to monitor a synthesis.

3.1.2. Solid-Phase Synthesis of Protected Fragments

Solid-phase synthesis according to the Fmoc strategy is described elsewhere in this book. For the convenience of the reader and to avoid any misunderstanding, a brief description of a standard protocol for SPPS on Sasrin resin is given in Section 3.2.1.

3.1.3. Pretreatment of the Peptide Resin Before Cleavage

The fully protected peptide fragment may be cleaved from the resin directly after the N-terminal amino acid has been coupled, but first the coupling reagents have to be washed out carefully with the solvent used for the preceding coupling step. To remove this solvent, the resin is washed with isopropanol (*see* Note 10), which will shrink it (this step is optional). Then the resin has to be washed **thoroughly** with DCM. These washes are extremely important to remove remainders of polar solvents (DMF and alike) completely. Polar contaminants "consume" TFA. They are protonated as well as the amide groups of the peptide under the anhydrous conditions of cleavage (compare *[5]*). Only after this "neutralization" will the cleavage proceed smoothly.

The peptide resin may now be cleaved or dried first to determine the weight gain of the peptide resin, or if one wishes to cleave only a part of the resin and store the remainder. Such a dried peptide resin has to be washed several times with DCM for proper swelling prior to cleavage.

3.1.4. The Cleavage

The prewashed peptide resin is now treated several times with 1% TFA/DCM (*see* Note 11). The amide moieties of the peptide fragment will "bind" TFA, thus reducing the amount of acid. The protonation of the amide bonds may be responsible for the good solubility of the fragment during cleavage (*see* Note 12). Repetitive short treatments with 1% TFA/DCM with immediate subsequent neutralization of the peptide-containing filtrates as described by Flörsheimer and Riniker *(5)* minimize the actual time of exposure to acid during cleavage. The first (with concomitant decrease of TFA concentration; *see* Section 3.1.3.) to fourth treatment will cleave most of the fragment, if no polar contaminants have been present, but treatments should be continued until no peptide is cleaved anymore (checked by TLC; *see* Note 13). Normally the resin will turn deeply violet as the cleavage proceeds (*see* Note 14). This color change may serve as an indicator for cleavage, but not for determining its "end point." Concomitantly, the volume of the resin decreases significantly. Inertization of the cleavage vessel is optional (*see* Note 15).

3.2. General Procedures

3.2.1. General Synthetic Procedure

1. Swell the dry Sasrin resin (with the appropriate Fmoc amino acid attached) by treating (*see* Note 16) it several times with DMF. Swelling will take some time. Other solvents, especially NMP, have been applied successfully in fully automated SPPS (*see* Chapter 3).
2. Treat the swollen resin with 20% piperidine/DMF for 5 min, suck off, and repeat the treatment for 10 min. If piperidine/DMF leads to incomplete Fmoc cleavage, it may be replaced by 20% piperidine/DMF containing 2–5% diazabicycloundecene.
3. Wash thoroughly with DMF, until the washes are neutral because the base has to be removed completely (traces of base may cause premature Fmoc cleavage during the coupling!). Washing can be sped up by shrinking, after a few DMF washes, with isopropanol (two to three washes) and swelling again with DMF (three to four washes; check pH). In most cases, the base will be removed after 10 washes.
4. Take a small sample. The Kaiser test *(9)* has to be positive, i.e., deeply blue or red (with proline) (*see* Chapter 8). With 2,4,6-trinitrobenzenesulfonic acid (TNBS) *(10)*, the beads turn red if free amino groups are present (*see* Notes 17 and 18). These two tests only show the presence of free amino groups in a qualitative manner. Quantitative information may be gained either via UV monitoring of the Fmoc cleavage or by completely cleaving a sample of peptide resin followed by HPLC analysis (*see* Chapter 3, *PAP*).
5. The coupling: First the Fmoc amino acid has to be activated, e.g., with DCC/HOBt (*see* Notes 19 and 20) or TBTU/DIPEA (*see* Note 21). Fmoc

amino acids and coupling reagents have to be used in excess (e.g., three-fold) to drive the coupling to completion. As HOBt is liberated during the coupling, the suspension will turn slightly acidic. Adjust to pH 7–7.5 by adding small amounts of DIPEA, but avoid an excess of base.

6. After 15–30 min take a small sample, wash it with DMF, and perform the Kaiser test (and, optionally, the TNBS test; *see* Chapter 8). In both cases, the beads should be nearly colorless and the supernatant yellowish. The sensitivity of the TNBS test can be increased by using a microscope to inspect the beads. Many other methods for monitoring SPPS have been developed, but none of them reached the popularity of the Kaiser test.
7. If the tests show (e.g., Kaiser test bluish or greenish) that the coupling is not yet completed even after 1 h, suck off and wash five times with DMF. Repeat the Kaiser test since impurities may have been the cause of a slightly positive test before. If not, repeat the coupling.
8. Eventually the tests will be virtually negative (*see* Note 17). Wash thoroughly with DMF to remove the reagents. Again, shrinking and reswelling are optional. To block free amino groups that have not been detected and thus to avoid deletions, the resin may be treated with a "capping" reagent, e.g., a large excess of acetic anhydride/pyridine in DMF for 10 min. Again wash carefully with DMF to remove the reagents.
9. Cleave the Fmoc group as described in step 2.
10. Coupling of the next Fmoc amino acid: follow 3–8 (*see* Note 22) and so on.
11. For washings after the last coupling step, *see* Section 3.1.3.

3.2.2. General Cleavage Procedure

Before cleavage, the peptide resin has been carefully washed and (optionally) dried (compare Section 3.1.3.).

1. The dry resin is weighed into a sufficiently large (mind the swelling!) fritted-glass funnel (preferably G4), where it remains throughout the whole procedure, e.g., a 100-mL funnel can be used for 5–10 g peptide resin.
2. The resin has to be washed at least five times with DCM (10–20 mL/g resin, contact time not <5 min). If the resin is not used in dried form, but directly after the last coupling step, it should be washed the same way, but contact time can be reduced to about 1 min. Prolonged suction should be avoided, and the resin must never be sucked to dryness. Slow stirring of the resin suspension is optional.
3. After placing the funnel on a clean vessel, the resin is treated with 1% TFA/DCM (10–15 mL/g) for 2–5 min. All the solvent is sucked into a vessel containing pyridine (>1.2 mL/mL TFA), and thus, the filtrate is immediately neutralized. Pyridine (200 mL/g resin) in methanol (2 mL/g resin) has also been recommended *(5)*. Again, prolonged suction and suck-

ing-through of air should be avoided (*see* Note 23). The resin should have changed its color slightly to distinctly (for exceptions, *see* Note 14).

4. This treatment is repeated with further portions of 1% TFA/DCM until it can be assumed that all the peptide has been cleaved from the resin, which by then should have turned deeply violet (*see* Note 14). Normally, three to six treatments will be sufficient.
5. The neutralized filtrates should be kept in separate vessels and analyzed, e.g., by TLC. Only the fractions containing a significant amount of peptide are pooled and subjected to work-up (*see* Note 24). Often the second and/or third fraction will contain most of the peptide.
6. To determine the cleavage yield (from weight loss of the resin), the resin has to be washed thoroughly with alcohol and ether, and dried to constant weight. The resin will be discolored rapidly by these washings.
7. When cleaving small amounts of peptide resin, e.g., for monitoring a synthesis (*see* Note 2), the filtrates need not be collected separately to avoid losses.

3.3. Work-Up Procedures

At first, the peptide fragment has to be freed from contaminating pyridinium trifluoroacetate (cf Section 3.2.2., step 3). Utmost care has to be taken to remove remainders of trifluoroacetic acid completely before using the fragment for subsequent coupling. During work-up, Met has to be protected from oxidation. Depending on the solubility of the fragment, work-up varies. A protected fragment may precipitate during cleavage or during neutralization, or it may be precipitated thereafter or extracted. The following paragraphs deal with the different work-up procedures.

1. Neutralization of the cleavage fractions already leads to precipitation of the peptide fragment (even though TLC checks should not be omitted since only a part of the peptide may have precipitated). Diethylether or methyl *t*-butyl ether is added to the pooled fractions to complete precipitation (at least the same volume, up to 5 vol of solvent). Stirring is recommended, and then the precipitate should be left to settle. Often the pyridinium trifluoroacetate crystallizes in long needles. The precipitate is filtered off and washed with ether. Since fully protected peptide fragments usually are hardly soluble in water, the precipitate may be triturated with water (until the washes are neutral) to remove the salt. Then it is washed with ether, dried, and weighed. The mother liquor and washes should be checked for peptide before discarding them. Polar impurities may be removed by dissolving the fragment in DMF (or another water-miscible solvent—the amount should be kept as low as possible) and precipitating it by adding water or

0.5*N* aqueous $KHSO_4$ (*see* Note 25). The precipitate is filtered off, washed with water until the washings are neutral, dried carefully, and weighed.

2. The neutralized fractions become extremely viscous, or a gel precipitates. Then ether (5–10 vol) has to be added in small portions. The gel should be stirred. Such precipitates may cause problems during filtration, and they are prone to form inclusions. Thus, they should be redissolved and precipitated if possible (*see* step 1).
3. Neutralization does not cause any visible effect.
   a. After pooling the fractions, the peptide may be precipitated with ether. Precipitation can be enhanced by removing a part of the DCM on a rotavap beforehand. A small sample is taken and treated with 10 vol of ether. If the peptide precipitates, it can be isolated by this method; the mother liquor has to be checked for peptide. For further treatment, *see* step 1.
   b. The peptide cannot be precipitated with an ether, so the DCM has to be removed *in vacuo*. It can be (discontinuously) replaced directly with ethyl acetate (*see* Note 26). High concentrations of pyridinium trifluoroacetate have to be avoided if the fragment contains very acid-sensitive moieties, such as Tyr(*t*Bu) or His(Trt). The resulting solution in ETOAc (which may contain small amounts of DCM) has to be extracted several times with water (the phase separation may take some time) and brine. The last aqueous washes should be neutral. All phases should be checked by TLC and reextracted with ETOAc, when necessary. The organic phase is dried with sodium sulfate and evaporated. The residue may be triturated with an appropriate solvent, e.g., an ether (*see* Note 27).
   c. The DCM has been removed, but the residue turns out to be insoluble in ethyl acetate (a rather unusual case). At first, a sample should be treated with DMF (*see* step 1) and, in case of dissolution, precipitated with water. Normally, dissolving and precipitation are more effective than just triturating the insoluble residue with water.
4. The fragment is scarcely soluble in 1% TFA/DCM, yet it is cleaved from the resin (as can be deduced from the color change). Therefore, samples of the resin should be extracted with DMF, DMA, NMP, and similar solvents, or mixtures thereof, until a suitable solvent system is found (*see* Note 28). (For further treatment, *see* step 1.) Such a behavior can be expected when cleaving fragments containing ions, e.g., short fragments containing the Ba-salt of sulfated tyrosine.

3.4. Purification Procedures

Normally, the fully protected peptide fragments cleaved from Sasrin resin and worked up as described turned out to be sufficiently pure according to TLC and HPLC. Their structures have to be confirmed by FAB-MS (*see* Chapter 7, *PAP*), amino acid analysis, and other methods. The formation of side products during cleavage seems to be rather unlikely because of the mild conditions, but it cannot be ruled out during synthesis. The well-established methods for the purification of unprotected peptides, such as RP-HPLC (*see* Chapter 3, *PAP*) and ion-exchange chromatography (*see* Chapters 2 and 5, *PAP*), usually cannot be applied because of the lack of fragment solubility in aqueous systems. The fragments are rather unpolar. Hence, they often are soluble in common organic solvents such as chloroform or methanol. Thus extraction, precipitation, and (flash) chromatography on silica may be suitable methods. Recently, purification methods for fully protected fragments using preparative chromatography have been developed by Lloyd-Williams et al. *(1)* and by Riniker et al. *(11)*.

3.5. Examples

1. Fmoc-Gly-Val-Val-Lys(Boc)-Asn(Trt)-Asn(Trt)-Phe-Val-Pro-Thr(*t*Bu)-Asn(Trt)-Val-Gly-OH—This fragment represents the sequence 21–33 of α-h CGRP. An example for a large-scale synthesis and cleavage is: SPPS starting with 125 g Fmoc-Gly-Sasrin (≈81.2 mEq). Peptide resin (after the last coupling step) is carefully washed with isopropanol and then with DCM. It is treated with 1% TFA/DCM (12 × 1000 mL, 10 min). Each fraction is neutralized with 15 mL pyridine (no precipitation). Fractions are pooled and DCM removed *in vacuo*. Oily residue is triturated with water (2 L) leaving a white precipitate which is filtered, washed with water (five times) and ether (eight times), and dried. Yield is 176 g (89%). Purity is (TLC) >90%
2. Boc-Ala-Cys(Acm)Asp(O*t*Bu)-Thr(*t*Bu)-Ala-Thr(*t*Bu)-Cys(Trt)-Val-Thr(*t*Bu)-His-Arg(Pmc)-Leu-Ala-Gly-OH—This fragment represents the sequence 1–14 of α-h CGRP. After cleavage, it is oxidized with I_2.

   An example for work-up by precipitation (cf Section 3.3., step 3a) is: SPPS starting with 70 g Fmoc-Gly-Sasrin (49 mEq). Peptide resin (after the last coupling step) is washed with isopropanol (seven times) and DCM (seven times). This is treated with 1% TFA/DCM (8 × 1000 mL, 10 min). Fractions are neutralized with pyridine and checked by TLC. The fragment does not precipitate. Fractions 1–7 are pooled and DCM removed *in vacuo* until ca 2000 mL are left. Five liters of diisopropyl ether are added to precipitate the peptide under stirring. The precipitate is filtered off, washed with ether (four times) and dried. Yield is 89.4 g (78%). Purity is (TLC) >80%.

3. Boc-His(Boc)-Lys(Boc)-Thr(*t*Bu)-Asp(O*t*Bu)-Ser(*t*Bu)-Phe-Val-Gly-OH—This fragment represents the sequence 1–8 of substance K. An example for work-up by trituration (cf Section 3.3., step 1), is: SPPS starting with 15 g Fmoc-Gly-Sasrin (10.5 mEq). Peptide resin (after the last coupling step and washes) is washed with DCM six times. This is treated with 1% TFA/DCM (5 × 300 mL, 5 min). Fractions are neutralized with pyridine, and fractions 2–5 are pooled. DCM is removed *in vacuo* leaving a gel, which was triturated with 200 mL diisopropyl ether. Precipitate is triturated with diisopropyl ether (3 × 100 mL), dried, and stirred in 300 mL water for 0.5 h, filtered off, washed with water, dried, triturated with ether, and dried. Yield is 11.68 g (82%). Purity is >96% (HPLC).
4. Fmoc-Ser(*t*Bu)-Pro-Lys(Boc)-Met-Val-Gln(Mtt)-Gly-OH—This fragment represents the sequence 1–7 of hBNP. An example for work-up by precipitation (cf Section 3.3., step 3c and Section 3.3., step 1) is: After SPPS, resin is washed and dried; 26.6 g peptide resin (9.4 mEq peptide) are prewashed with DCM. This is treated with 1% TFA/DCM (6 × 280 mL, 5 min). Each fraction is neutralized with 4.8 mL pyridine. All fractions are pooled and DCM removed in vacuo. Residue is dissolved in 140 mL DMA at ca. 40°C. The resulting solution is slowly poured into water (1.2 L). Precipitate is filtered off, washed with water (3 × 500 mL), and dried. Yield is 10.8 g (83%). Purity is >87% (deprotected, HPLC).
5. Fmoc-Asn(Mtt)-Lys(Boc)-Phe-His-Thr(*t*Bu)-Phe-Pro-Gln(Mtt)-Thr(*t*Bu)-Ala-Ile-Gly-OH—This fragment represents the sequence 17–28 of human calcitonin, a "small-scale" cleavage. An example for work-up by extraction (cf Section 3.3., step 3b) is: 1.67 g peptide resin (0.49 mEq) is prewashed with DCM (5 × 20 mL) and cleaved with 1% TFA/DCM (5 × 15 mL, 5 min). Peptide does not precipitate when neutralizing the fractions with pyridine. All fractions are pooled and DCM is removed *in vacuo*. Residue is triturated with diisopropyl ether and dried. It is dissolved again in EtOAc and extracted with water (5 × 10 vol%) and brine, EtOAc is removed, and residue is triturated with ether. Yield is 706 mg (67%). Purity is 65% (HPLC, after cleavage of the protecting groups).
6. Boc-Arg(Mtr)-Ser(*t*Bu)-Ser(*t*Bu)-Cys(Acm)-Phe-Gly-Gly-Arg(Mtr)-Met-Asp(O*t*Bu)-Arg(Mtr)-Ile-Gly-OH—This fragment represents the sequence 4–16 of α-h ANF. An example for a fragment forming a gel (cf Section 3.3., step 2) is: After SPPS, resin is washed and dried; 50 g peptide resin (14.2 mEq peptide) are prewashed with DCM. This is treated with 1% TFA/DCM (6 × 500 mL, 10 min). When neutralizing with pyridine, a part of the fractions turns into thick gels. Thus, they cannot be pooled directly: 500 mL ether are added under vigorous stirring to fraction 2–5 to precipitate the peptide. Fractions are pooled, left to settle, and filtered. The pre-

cipitate is washed with ether, water, and again ether, and dried. Yield is 30.7 g (89%). Purity is >90% (HPLC).

7. Pyr-Gln-Asp(O*t*Bu)-Tyr(*t*Bu)-Thr(*t*Bu)-Gly-OH—This fragment represents the sequence 1–6 of cerulein. An example for precipitation owing to neutralization (cf Section 3.3., step 1) is: SPPS starting with 20 g Fmoc-Gly-Sasrin (13 mEq). Peptide resin (after the last coupling step) is carefully washed with DCM. This is treated with 1% TFA/DCM (4 × 300 mL, 5 min). Each fraction is neutralized with 5 mL pyridine, causing the peptide to precipitate (the second fraction contains most of the peptide). Fractions 2 and 3 are pooled and diluted with ca 600 mL ether. The precipitate is left to settle, filtered off, triturated with ether and water, and dried. Yield is 7.82 g (95%). Purity is >85% (TLC).

4. Notes

1. Orthogonal: "an orthogonal system has been defined as a set of completely independent classes of protecting groups, such that each class of groups can be removed in any order and in the presence of all other classes" *(12)*.
2. The solid-phase synthesis on Sasrin resin can be monitored by cleaving samples at any stage. These samples, which only have to be washed carefully before cleavage, actually show what the resin-bound product looks like. This is an advantage of very acid-labile resins compared to, e.g., Wang resin, where protecting groups are concomitantly removed when cleaving from the resin.
3. When using Cys(S*t*Bu), piperidine treatment has to be kept as short as possible *(13)*.
4. His(Trt) is very acid-sensitive. Trt may be partially cleaved even with 1% TFA/DCM.
5. Met sulfoxide may be formed unintentionally during handling, especially during coupling of Met-containing fragments.
6. N^{in}-protection of Trp is strongly recommended when Trp is "exposed," e.g., C-terminal *(11,14)*.
7. Diketopiperazine formation is favored when Pro is the C-terminal amino acid (*see* Compound 1).
8. For further details, *see*, e.g., M. Bodanszky, *Principles of Peptide Synthesis*, Springer (1984).
9. Normally protected peptides are somewhat less sensitive toward oxidation than the corresponding free peptides, especially when many hydrophobic protecting groups are present.
10. Washes with isopropanol also remove dicyclohexyl urea (which is formed when coupling with DCC, DCC/HOBt, and so forth). This scarcely soluble compound also has to be removed carefully before cleavage since it "con-

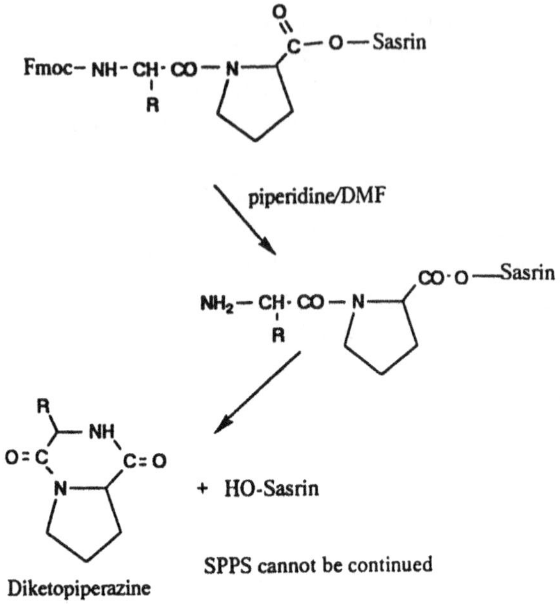

Compound 1.

sumes" TFA. Actually, it is dissolved by 1% TFA/DCM owing to salt formation, whereas its solubility in pure DCM is negligibly low. Dicyclohexyl urea can also be removed by treating the peptide resin with DCM/alcohol (1:1).

11. Higher yields of Trp and/or Met-containing fragments may be obtained by adding a neutral scavenger, e.g., 5% EDT. Nevertheless, indole protection, e.g., Trp(Boc), may be the best way to prevent the yield-decreasing alkylation.
12. Precipitation or low solubility of the fragment in TFA/DCM has been observed very rarely and was not found to obstruct the cleavage. On the other hand, neutralization of the TFA can cause precipitation of the protected fragment (*see* Section 3.3.).
13. TLC systems have to be less polar than systems for the analysis of free peptides. A few useful systems are:

$CHCl_3$/MeOH/AcOH	77.5:15:7.5
$CHCl_3$/MeOH/32% AcOH	15:4:1
ETOAc/pyridine/AcOH/H_2O	6:5:1:3
$CHCl_3$/MeOH/32% AcOH	5:3:1
$CHCl_3$/TFE/80% AcOH	6:2:1

Detection is by, e.g., Greig-Leaback.

14. The color change is probably caused by the stable carbocation formed by cleaving the peptide-resin bond. Minor color changes occur when cleaving Trp- or Met-containing peptides or when using scavenger.
15. Inertization is recommended, but not absolutely necessary when cleaving Met- or Cys-containing peptides.
16. "Treat" here means shake or stir slowly. The swollen beads may be damaged by mechanical forces.
17. The Kaiser ninhydrin test *(9)* is: Put a small sample of swollen beads in a test tube. Add three drops of each solution (*see* Section 2.), and keep the suspension at 100°C for 5–6 min. Free amino groups are detected by a deep-blue color (red: for proline). Unfortunately, the intensity of the color depends on the N-terminal amino acid. It may be rather weak, e.g., for Asp and Ser. On the other hand, all the coded amino acids (except Pro) seem to yield the same intensity of color with TNBS, making this test a valuable additional control (*see* Note 18). Special care has to be taken when coupling an Fmoc amino acid to Pro (which tends to be a sluggish reaction) since the Kaiser test gives somewhat ambiguous results.
18. The TNBS test *(10)*: Put a small sample of swollen beads into a test tube. Add a tiny amount of solid TNBA and three drops of 10% DIPEA in DMF. Vortex and wait 5 min. The supernatant should be yellow. Remove it with a Pasteur pipet or by centrifugation. The beads turn red when free amino groups are present.
19. Avoid skin contact with DCC, which is a strong allergen. When spilled, decompose with AcOH.
20. Activation with DCC yielding the Fmoc-amino acid OBt ester: Dissolve or suspend equimolar amounts of Fmoc amino acid and HOBt in DMF. Try to keep the amount of solvent low. Add a low excess (1.1X) of DCC, and stir for 0.5–1 h; the reaction is slightly exothermic, and dicyclohexyl urea will start to precipitate after a few minutes. Filter off and wash the precipiate (matted white needles) twice with small volumes of DMF. Add the combined filtrates to the resin. This method should not be applied to Fmoc-Arg derivatives. When activating Fmoc-Gly-OH, do not filter off the precipitate.
21. Activation with TBTU/DIPEA *(15)* is simpler. It can be performed directly before coupling (this is especially important when coupling Fmoc-Arg derivatives). Equimolar amounts of Fmoc amino acid and TBTU are suspended in DMF. On adding an equimolar amount of DIPEA, the TBTU will dissolve (the reaction is slightly exothermic), and the color may turn yellow or red. Stir and add to the resin. Try to keep the concentration of the reactands as high as possible.
22. The number of washings may have to be increased in the course of the synthesis, especially when synthesizing larger fragments. The volume of the

swollen resin should increase slowly. Sudden decreases indicate peptide aggregation. Then difficulties in Fmoc cleavage and coupling have to be expected.
23. Even water may condense on the resin surface because of strong sucking and cooling caused by the evaporation of DCM. From this point of view, pressure filtration is superior to suction.
24. When cleaving Trt-protected peptides, deeply yellow solutions may be obtained. The color disappears when neutralizing.
25. The peptide fragments may have been obtained as pyridinium salt that will be converted into the free acid by this procedure.
26. *n*-Butanol can also be used, when the residue cannot be dissolved in ethyl acetate.
27. An alternative "washing-protocol" removing most basic and acidic impurities follows:
    a. 5% Aqueous $NaHCO_3$;
    b. Water;
    c. 0.5–1*N* aqueous $KHSO_4$; and
    d. Water and brine until the washings are neutral. The peptide fragment is converted into the free acid.
28. The solvents TFE, HFIP, and their mixtures with DCM or chloroform are also excellent solvents for fully protected peptide fragments. Since they are weakly acidic, Sasrin will turn pink when treated with them.

References

1. Lloyd-Williams, P., Gairi, M., Albericio, F., and Giralt, E. (1991) Convergent SPPS X. Synthesis and purification of protected peptide fragments using the photolabile Nbb-resin. *Tetrahedron* **47,** 9867–9880.
2. Kunz, H. and Dombo, B. (1988) Solild phase synthesis of peptides and glycopeptides on polymeric supports with allylic anchor groups. *Angew. Chem. (Int. Engl. Ed.)* **27,** 711–713 (*Angew. Chem.* **100,** 732–734).

3a. Rink, H. (1987) Solid phase synthesis of protected peptide fragments using a trialkoxydiphenyl-methylester resin. *Tetrahedron Lett.* **28,** 3787–3790.

3b. Barlos, K., Gatos, D., Kallitsis, J., Papaphotiu, G., Sotiriu, P., Wenqing, Y., and Schäfer, W. (1989) Darstellung geschützter Peptidfragmente unter Einsatz substituierter Triphenylmethylharze. *Tetrahedron Lett.* **30,** 3943–3946.

3c. Albericio, F. and Barany, G. (1991) Hypersensitive acid-labile (HAL) tris(alkoxybenzyl) ester anchoring for solid-phase synthesis of protected peptide segments. *Tetrahedron Lett.* **32,** 1015–1018.

4. Sheppard, R. C. and Williams, B. J. (1982) A new protecting group combination for solid phase synthesis of protected peptides. *J. Chem. Soc., Chem. Commun.,* 587–589.
5. Flörsheimer, A. and Riniker, B. (1991) Solid phase synthesis of peptides with the highly sensitive HMPB-linker (4-(4-hydroxymethyl-3-methoxy phenoxy) butyric acid). *Peptides 1990, Proc 21st EPS,* Platja d'Aro, Escom, Leiden, pp. 131–133.

6. Mergler, M., Nyfeler, R., Tanner, R., Gosteli, J., and Grogg, P. (1988) Peptide synthesis by a combination of solid-phase and solution methods II. Synthesis of fully protected peptide fragments on 2-methoxy-4-alkoxy-4-alkoxybenzyl alcohol resin. *Tetrahedron Lett.* **29,** 4009–4012.
7. Kapurniotu, A., Ungermann, C., and Voelter, W. (1992) Optimized SPPS of the new stem cell prolifering inhibiting factor Ac-SDKP and derivation. *Proc. 2nd Int. Symp. on Innovation and Perspectives in Solid Phase Synthesis,* Canterbury 1991, Intercept, Andover, pp. 319–323.
8. Mergler, M. and Nyfeler, R. (1992) Easy synthesis of protected peptide hydrazides on solid support. *Peptides, Chemistry and Biology Proc. 12th APS,* Boston 1991, Escom, Leiden, pp. 551–552.
9. Kaiser, E., Colescott, R. L., Bossinger, C. D., and Cook, P. I. (1970) Color test for detection of free terminal amino groups in the solid-phase synthesis of peptides. *Anal. Biochem.* **34,** 595–598.
10. Hancock, W. S. and Battersby, J. E. (1976) A new method for the detection of incomplete coupling reaction in solid phase peptide synthesis using 2,4,6-trinitrobenzenesulfonic acid. *Analyt. Biochem.* **71,** 260–263.
11. Riniker, B., Fretz, H., and Kamber, B. (1993) *Peptides 1992 Proc. 22nd EPS,* Interlaken, Escom, Leiden, pp. 34,35.
12. Barany, G. and Merrifield, R. B. (1979) *Solid-Phase Peptide Synthesis. The Peptides,* vol. 2, Academic, New York, pp. 1–284.
13. Athertone, E., Pinori, M., and Sheppard, R. C. (1985) Peptide synthesis 6. Protection of the sulfhydryl group of cysteine in solid-phase synthesis using N_{α}-Fluorenylmethoxycarbonyl amino acids. Linear oxytocin derivatives. *J. Chem. Soc., Perkin Trans.* I, 2057–2064.
14. White, P. (1992) Fmoc-Trp (Boc)-OH: a new derivative for the synthesis of peptides containing tryptophan. *Peptides, Chemistry and Biology, Proc. 12th APS,* Boston 1991, Escom, Leiden, pp. 537–538.
15. Knorr, R., Trzeciak, A., Bannwarth, W., and Gillesen, D. (1989) New coupling reagents in peptide chemistry. *Tetrahedron Lett.* **30,** 1927–1930.

CHAPTER 15

Peptide Synthesis via Fragment Condensation

Rolf Nyfeler

1. Introduction

In the classical solution synthesis, fragment condensation has always been the way to build up peptide chains with more than approximately five amino acids' length. Since the early achievements in this field, like the synthesis of glucagon and secretin *(1)*, little has changed from a strategic point of view. For a recent example, see the synthesis of human epidermal growth factor *(2)*. However, taking the classical approach to synthesize larger peptides is cumbersome, and needs experience and time.

With recent developments of appropriate linker resin combinations in solid-phase peptide synthesis, protected peptide fragments have become readily available (*see* Chapter 14), usually in good yield and of high purity. These developments in solid-phase peptide synthesis opened up the way to new strategies: the combination of solid-phase and solution synthesis, and the fragment coupling onto resin. In the former approach, protected peptide fragments are synthesized on solid support, cleaved from the resin with full preservation of protecting groups, purified, and characterized. The fragments are then assembled in solution. In such a way, the advantages of solid phase are combined with the advantages of solution synthesis, allowing for the key steps, i.e., the fragment couplings, full control and monitoring, as well as isolation, purification, and characterization of the intermediates *(3,4)*.

The other strategy uses the protected fragments for coupling onto resin; the growing peptide chain is assembled on the polymer support. This

From: *Methods in Molecular Biology, Vol. 35: Peptide Synthesis Protocols*
Edited by: M. W. Pennington and B. M. Dunn

approach, if successful, allows very rapid preparation of larger peptides even compared to stepwise solid phase. The strategy is usually named convergent (*see* refs. *5,6* and literature cited therein). Assembling a peptide this way, using purified peptide fragments, is advantageous to the stepwise approach: side products from incomplete couplings are more easily removed, since they differ not by just one amino acid, as may be the case in stepwise solid-phase synthesis. However, truncated and deletion sequences cannot be fully excluded in the final product unless their absence in the fragments (obtained from stepwise solid-phase synthesis) can be shown. Fragment purity therefore is essential.

Whereas fragment coupling in solution is a very well-established procedure, fragment coupling to a resin-bound peptide chain constitutes a rather recent development, as can be deduced from the growing number of papers on the subject. However, the approach was discussed and compared to the combination approach quite some time ago, in 1981 *(7)*.

Both fragment coupling strategies require some experience because—unlike amino acid derivative coupling—no standard protocol can be given, and therefore, such couplings are usually not performed in an automated way, for the following reasons: The solubility of any designed protected fragment is not predictable with accuracy, but if there is insufficient solubility, coupling may become difficult. This holds especially for the convergent approach. Although in solution synthesis numerous fragment couplings in suspension or even in gels have been shown to proceed successfully *(8)*, such couplings will proceed sluggishly if at all in the convergent approach where diffusion of the reactants into the matrix is essential. The structure of the resin as well as the peptide loading on the resin clearly have their impact on the reaction rate of the coupling *(9)*. Low peptide loadings (0.1 mEq/g) usually improve difficult couplings *(10)*. Even so, taking into account that fragment coupling is slower compared to amino acid coupling and also that one rather tends to avoid large excesses of precious fragments, good solubility of a fragment is an important prerequisite for a successful fragment coupling on resin. Even then coupling may be hampered by polymer/peptide chain matrix interactions.

The synthesis of protected peptide fragments is described elsewhere in this book in full detail. There are numerous coupling methods described in the literature *(11–14)*; many of them may be applied for fragment couplings. It is beyond the scope of this chapter to mention or discuss them all. The aim of the following section is rather to give the reader an overview of all points to consider (*see* Section 3.1.) and to provide the reader with a practical guideline (*see* Section 3.2.).

2. Materials

1. All solvents and reagents mentioned are commercially available, and may be used as such. However, in slow couplings/diluted solutions, a high solvent quality is important. Ethers must be free of peroxides.
2. Fragment couplings on resin may be performed in any vessel recommended for solid-phase synthesis or in standard glass flasks equipped with a stirring device. These are also used for couplings in solution. Precipitated products may either be isolated by centrifugation or filtration.
3. Equipment for in-process analysis, such as TLC and HPLC (*see* Chapter 3, *PAP*) is available on the market in a great variety. HPLC equipment must include a UV detector.

3. Methods

3.1. General Considerations

3.1.1. When or Why to Choose a Fragment Approach

When your stepwise solid-phase synthesis failed to give a reasonable quantity and/or quality of the desired peptide, a fragment approach may be tried, especially in the case where the stepwise approach is hampered by β-sheet formation, which can occur at a peptide chain length of about 6–15 amino acids. These difficult couplings can be overcome by coupling a fragment that bridgespans over the troublemaker sequence. The same holds for any difficult (for whatever reason) coupling regions in individual couplings.

When purification of your peptide is tedious and you assume or know this to be the result of deletion sequences closely related in structure to the desired peptide, a fragment approach may help. The fragments used, however, have to be reasonably pure and should themselves not contain deletion sequences. Peptides from convergent synthesis usually show much less broadening of the main peak in RP-HPLC and are easier to purify. Side products from incomplete couplings differ from the peptide by the length of a fragment and are thus usually easy to remove.

When your peptides to be synthesized either contain repetitive sequences or share common sequences, such sequences have to be synthesized only once, purified, and either coupled several times (for repetitive sequences) or to different peptides. The approach seems especially useful for synthesis of multiple antigenic peptide matrices (MAPs *[15]*) or templates *(16)*.

Whenever you have chosen a classical solution synthesis of your peptide (of let us say more than four to six amino acids), you would use fragment condensation. Depending on structure, even a pentapeptide may be preferentially assembled by a 3 + 2 coupling.

3.1.2. How to Choose the Synthetic Strategy: In Solution vs on Resin

Whether you choose a fragment approach in solution or on resin depends on your experience and on the time available; for large-scale syntheses, economy has also to be taken into consideration. The solution approach needs more experience and time, but is more generally applicable. Numerous examples are described in the literature. Furthermore, C-terminal fragments usually have to be synthesized in solution. This again needs more practical knowledge than making them on resin. Unfortunately, rapid synthesis of fragments protected at their carboxy terminus by *tert.butyl*, or *benzyl*esters via solid phase is not (yet) possible. Protected fragments obtained from solid-phase synthesis may serve as intermediates for this purpose and may be esterified *(17)* in a subsequent step. Special attention has to be paid to the risk of epimerization. Protected amides may be obtained via aminolysis directly from the peptide resin.

Fragment coupling onto resin is more rapid since all components are made via solid phase. However, restrictions, such as the solubility of the fragment and the interaction of the fragment with the peptide–polymer matrix, may slow down the coupling or even prevent it from proceeding. Coupling fragments onto resin may often not be as straightforward as coupling amino acid derivatives.

3.1.3. How to Choose the Coupling Sites

For both approaches, you have to define your coupling sites, whether you use Boc or Fmoc strategy. To define reasonable fragments, consider the following points:

> Length of the fragment: Typical fragment length would be from 5–10 and maybe up to 15 amino acids. Even some larger peptides of up to 22 amino acids length (e.g., fragment 1–22 of CRF) have been prepared in the author's laboratories and successfully coupled. However, fragments of this length usually are difficult to purify, and may contain deletion and truncated sequences. These also will couple and thus, unfortunately, the advantage of fragment coupling is partially lost. With fragments of the

recommended length, there is a good chance for sufficient solubility. Hence, purification can be easily achieved and, finally, coupling proceeds rapidly and efficiently.

Epimerization: Whenever possible, Gly is chosen as the C-terminal amino acid for the obvious reason that it cannot racemize. Second choice would be Pro since at least the azlactone mechanism of racemization is impossible. The disadvantage of C-terminal Pro fragments is that their synthesis is somewhat more complicated because of diketopiperazine formation (*see* Chapter 14).

Reaction rate and side reactions: Gly again would be preferred for fast reactions. In case there is no Gly or Pro available at appropriate positions, you should avoid sterically hindered amino acids, such as Ile or Val, as coupling sites (both N- and C-terminal). As N-terminus you should select neither a secondary amino acid (*N*-methylamino acid or Pro) nor Gln. The latter may lead to pyroglutamate formation.

3.1.4. How to Choose the Strategy: Protecting Groups

Both synthetic strategies well known from stepwise solid-phase synthesis can be applied to the fragment approach. Boc strategy would use N-Boc as temporary, Bzl type as (semi) permanent side-chain protection; Fmoc strategy would use Fmoc as temporary protection and *tert.butyl* type protecting for side chains. Details concerning the choice of protecting groups are available elsewhere in this book (*see* Chapter 14). A third possibility, restricted to the coupling in solution approach, is the use of carbobenzoxy (Z) as temporary protecting group and *tert.butyl*-type protection for side chains. However, once the peptide chain contains Met and/or Cys, hydrogenation of Z may not be possible and alternative orthogonal protecting groups, such as Fmoc, Adpoc, Bpoc or the like, have to be used. For a definition of orthogonality, *see* Chapter 14, Note 1.

The three protecting-group strategies mentioned (Boc/Bzl, Fmoc/*tert.butyl*, Z/*tert.butyl*) are the most widely accepted and used ones. Numerous other strategies/protecting-group combinations are possible. The only requirement is orthogonality between temporary and permanent protecting groups.

Those familiar with Fmoc strategy would choose Fmoc-protected fragments, and those familiar with Boc strategy, Boc-protected fragments. For cleavage of the temporary protecting group and also for washing procedures, the standard protocol could be used. Those not familiar with solid-phase synthesis should consult Chapter 14 first and then perform coupling in solution with the Fmoc-protected fragments obtained according to the general protocol given there. The C-terminal fragment is used as *tert.butyl*ester, unless the peptide to be synthesized is an amide (no protection required).

3.1.5. How to Choose the Appropriate Fragment Coupling Method

Undoubtedly, many of the peptide coupling methods described in the literature up to now may be used for fragment couplings. Usually one prefers either preactivating the carboxylic moiety or a one-pot reaction with carboxylic acids (rather than synthesizing active esters of fragments) with one notable exception: Acyl azides, well known for the low epimerization risk, are prepared from hydrazides, which themselves are usually prepared from methyl esters, but also via hydrazinolysis of appropriate peptide resins, e.g., Sasrin *(18)*. An alternative, however, to this procedure may be the use of diphenylphosphoryl azide (DPPA) and free carboxylic acids. It is beyond the scope of this chapter to enumerate all the suitable methods. Protocols for some of the most useful standard methods, like mixed anhydride or the abovementioned azide coupling, are to be found in the literature *(19)*.

DCC coupling methodology with all its variations in structure of the reagent (water-soluble carbodiimides and the like) or in terms of additives (HOSu, HOBt, 3-hydroxy-1,2,3-benzotriazin-4[3H]-one and the like) still can be considered as generally applicable. It can be used in most of the common solvents for peptide synthesis.

TBTU-type reagents may also be used for fragment couplings, with or without addition of HOBt. For fragment couplings with epimerization risk, use of TPTU is recommended rather than TBTU *(20)*. BOP is known to give very fast coupling reactions, also in fragment couplings. However, these couplings may proceed with concomitant partial epimerization.

3.1.6. How to Handle the Epimerization Risk

If there is no Gly at the C-terminus of the fragment, there always is a possibility (risk) of racemization of the C-terminal amino acid, and no method can be considered absolutely safe. There are a few general rules and a lot of data available from literature; nevertheless, it is absolutely necessary to take appropriate care for each individual case of fragment coupling with respect toward epimerization. Products obtained from such fragment couplings should always be checked for epimerization either by hydrolysis and subsequent analysis for D/L amino acid, or by preparing the unwanted diastereomer and developing a suitable analytical method capable of separating the pair of diastereomers (such as RP-HPLC). Contamination with a small amount of diastereomer may be acceptable if appropriate purification methods are at hand, such as countercurrent distribution or preparative HPLC (Chapter 4, *PAP*).

In other cases, there is no other way than running a series of experiments not only with various coupling reagents, but also with appropriate variations in reaction conditions, such as solvent, excess of reagent, concentration of components, temperature, the nature of additives, and so on, followed by analysis of the products obtained with regard to epimerization. These experiments can be carried out on a small scale (10–100 mg of peptide fragment or peptide resin). According to our experience, DCC/HOSu and DCC/HOBt are still the methods that—generally speaking—are the most promising ones. Recently, addition of copper(II)chloride was claimed to prevent racemization *(21)*. For couplings in solution in a classical synthesis, azide couplings may still be highly recommended, although some cases of epimerization, especially for C-terminal His, Tyr, and Phe, are known from the literature.

3.2. General Procedures: The Experimental Approach

3.2.1. The Choice of the Solvent

Determine the solubility of the fragment(s) (*see* Note 1). Start with DMF, DMA, or NMP or mixtures of these. HMPA is a good solvent in mixtures. However, it should not be used because of its cancerogenicity. DMSO also is a very good solvent. However, its oxidative potential restricts its general use, especially when Met is present in the peptide sequence. Methylene chloride and/or trifluoroethanol may be other choices. The latter may require laborious optimization of reaction conditions (Note 2). Addition of Li salts, such as LiCl or $LiClO_4$, may enhance solubility of fragments *(21)*. Addition of the coupling reagent and/or the additives may also facilitate dissolution.

Attempt to obtain a concentration of the fragment in the range from 20 to 5 g/100 mL of solvent (Note 3). Below a 5 g/100 mL concentration, coupling still may take place, but rather slowly. If there is no other choice, one may try couplings in the 1–5 g/100 mL range. When performing a coupling onto a peptide resin, the coupling solvent must be capable of swelling that resin. The peptide resin has to be washed with the coupling solvent before coupling.

When performing a coupling in solution, solubility of both fragments has to be checked. If the amine component consists of a salt, an appropriate base, such as DIPEA or NMM, has to be added (1 Eq) to liberate the amine. This also may have an influence on solubility.

3.2.2. The Coupling Conditions

Usually the carboxylic component is used in excess when performing a coupling onto resin. The excess may be chosen from 1.5 Eq to whatever is needed. A good starting point is 2–3 Eq.

In solution couplings, you generally use less excess. A good choice for starting is in the range of 1.0–1.5 Eq of the carboxy component.

Reagents and additives, such as DCC and HOBt, are usually used in equimolar amounts or in slight excess with regard to the carboxy component. A good starting point here is to use 1 Eq of reagent and additive (*see* Note 4).

The simplest way of performing a fragment coupling is the "one-pot" procedure. All components/additives/base are dissolved in the solvent, and the reagent, e.g., DCC or TBTU (*see* Notes 5 and 6), is added last. For couplings onto resin, the resin has to be prewashed with the coupling solvent. For TBTU couplings, at least equimolar amounts of DIPEA have to be added, preferentially 1.5 Eq with regard to TBTU. For DCC/HOBt couplings, the apparent pH should be between 6.5 and 7. Addition of base, such as DIPEA (up to a pH of 8), may speed up reaction. Be careful, however, with elevated pH values when there is a risk of epimerization. Faster reaction rates may be obtained via preactivation of the carboxylic component. The latter is dissolved. The additive (HOBt) and the reagent (DCC or TBTU) are added, and the mixture left for 0.5–2 h. Dicyclohexyl urea may be filtered off in the case of DCC preactivation before addition. The preactivation mixture is then added to the neutralized solution of the amino component. In solid phase, the resin has to be washed with the coupling solvent, which subsequently has to be filtered off. The preactivated carboxylic acid solution is then added.

3.2.3. The Coupling Time: The Monitoring of the Coupling

Coupling time may vary from a few hours to a few days. Usual coupling time for fragments is overnight.

For monitoring coupling reactions in solution, TLC is a very convenient method. Conditions leading to a separation of both starting materials and the product have to be worked out; specific development of TLC plates with ninhydrin (for NH_2[NH]), Greig-Leaback (—CONH—), and UV helps in interpreting the chromatograms.

Monitoring coupling reactions on resin is a more tedious case. Standard tests, such as the Kaiser-test and the TNBS-test (both described in Chapter 14) may only give a first indication. Their results, especially when negative, quite often turn out not to be reliable. The best way to monitor a coupling is to cleave a sample from the resin and analyze it by RP-HPLC. Cleaved samples also have to be prepared from the fragments for reference purpose. RP-HPLC illustrates the coupling both in a qualitative and quantitative manner. Other methods for monitoring are amino acid analyses, peptide sequencing, or mass spectroscopy.

Should the test(s) reveal incomplete coupling, one may add more reagent or, preferentially, more reagent and carboxy component. The amounts to add depend on the turnover of the coupling. One may use up to the same amounts already used in the first instance.

In solution couplings, the components may be added in solid form or in concentrated solutions. In solid-phase couplings, either more reagent is added or, preferentially, the resin is filtered and the coupling repeated all over again with fresh solutions of all components as described above (*see* Note 7). If coupling does not go to completion, but the turnover is acceptable, the amino function may be blocked, e.g., by acetylation, and the synthesis is continued (*see* Note 8).

3.2.4. The Work-Up

Work-up of couplings performed on resin is carried out as described for stepwise solid-phase synthesis elsewhere in this book (*see* Chapter 4). All unused components, reagents, additives, and so on are simply washed out, and the peptide resin is ready either for final cleavage or for cleavage of the temporary N^{α}-protecting group according to the protocols described in the corresponding chapters in this book. Thereafter, synthesis is continued by either stepwise coupling or a further fragment coupling. Work-up of couplings performed in solution follow the general protocols used in solution chemistry. The work-up procedure strongly depends on the properties of the specific peptide synthesized. No standard work-up protocol can therefore be given.

Work-up may proceed via precipitation (*see* Note 9) either by adding water (for lipophilic products) or a hydrophobic solvent, such as ethyl acetate or ether. If the product does not precipitate, the solution is either diluted with, e.g., ethyl acetate (followed by an extraction procedure), or evaporated. The residue obtained may then be treated in an appropriate manner: dissolution and precipitation, dissolution and extraction, trituration, dissolution followed by purification with countercurrent distribution, or any suitable type of chromatography. *See* the following section for some examples.

3.3. Examples

3.3.1. Fragment Coupling on Resin Using DCC/HOBt

- I Fragment: Fmoc-Ala-Gln(Mtt)-Ser(*t*Bu)-Gly-Leu-Gly-OH
- II Peptide resin: H-Cys(Acm)-Asn(Mtt)-Ser(*t*Bu)-Phe-Arg(Pmc)-Tyr(*t*Bu)-O-Wang

I represents position 17–22, and II position 23–28 of human Atrial Natriuretic Peptide. The loading of the starting Fmoc-Tyr(*t*Bu)-O-Wang resin was 0.6 mEq/g. Compare with Section 3.3.5. for a corresponding coupling in solution.

3.3.1.1. Coupling Conditions

Peptide resin (2.7 g) and 2.2-g fragment (2 Eq) were coupled in 22 mL DMF using 4 Eq of HOBt and 4 Eq of DCC. After 5 min, DIPEA (2 Eq) was added. The reaction mixture was gently agitated. After an overnight coupling, a sample was cleaved with TFA/H_2O/DTT (95:5:5) and analyzed by HPLC. Content of unreacted peptide II was found to be <0.5%. Addition of base accelerates the reaction, but it is not absolutely necessary.

The fragment can also be preactivated using amounts of solvent and reagents as given above for approx 30 min. The resulting solution is then added to the resin previously swollen with DMF.

3.3.2. Fragment Coupling on Resin Using TBTU

- I Fragment: Fmoc-Ser(*t*Bu)-Asn(Trt)-Lys(Boc)-Gly-Ala-Ile-Ile-Gly-OH
- II Peptide resin: H-Leu-Met-Val-Gly-Gly-Val-Val-O-Wang

I represents position 26–33, and II position 34–40 of β-Amyloid protein (1-40). The loading of the starting Fmoc-Val-O-Wang resin was 0.6 mEq/g.

3.3.2.1. Coupling Conditions

Preactivation of fragment I: 3.6 g of the fragment (2 Eq) were dissolved in a 1:1 mixture of DMA and NMP (60 mL), 2 Eq of TBTU and 3 Eq of DIPEA were added, and the mixture was gently agitated for 30 min and then added to the peptide resin II previously washed with the coupling solvent mixture. Coupling time was overnight, and turnover as determined by HPLC after cleaving was over 96%.

3.3.3. Fragment Coupling in Solution Using DCC/HOSu

- Fragment I: Z-Gly-Val-Val-Lys(Boc)-Asn(Trt)-Asn(Trt)-Phe-Val-Pro-Thr(*t*Bu)-Asn(Trt)-Val-Gly-OH
- Fragment II: H-Ser(*t*Bu)-Lys(Boc)-Ala-Phe-NH_2 HCl

Fragment I represents the sequence 21–33, and fragment II the sequence 34–37 of α-human CGRP; 4 g of fragment I (1.7 mmol) and 1.1 g of fragment II (1 Eq) were dissolved in DMF (40 mL). After addition of NMM (1 Eq), HOSu (3 Eq), and DCC (3 Eq), the reaction mixture was left at room temperature for 2 d. Monitoring by TLC showed disappearance of both starting materials. The product was precipitated with ether, and treated with ethyl acetate and isopropanol. Yield was 75%. Purification was achieved by countercurrent distribution, using the solvent system: MeOH/1N AcOH/1.2-Dichlorethane/$CHCl_3$ (10:3:8:4). After 1300 cycles, fractions containing pure product were pooled, evaporated, and the product precipitated with water. Yield (purification) was 88%.

3.3.4. Fragment Coupling in Solution via Azide

- Fragment I: Z-Ile-Phe-Thr(*t*Bu)-Asn-Ser(*t*Bu)-Tyr(*t*Bu)-$NHNH_2$
- Fragment II: H-Arg(HCl)-Lys(Boc)-Val-Leu-Gly-OH

Fragment I represents the sequence 5–10, and fragment II sequence 11–15 of human GRF; 10 g fragment I (10 mmol) were converted into the azide in DMF/DMA (3:2, 60 mL) using 4 Eq of HCl and 2.5 Eq of *tert.butyl*nitrite at –12°C. After 30 min reaction time, the mixture was neutralized with DIPEA. A solution of 8 g fragment II (1 Eq) in DMSO/DMF/water (50/40/4.5 mL) was added to the azide solution at –10°C, and the pH was adjusted to 8. The mixture was kept at room temperature for 3 d, concentrated *in vacuo*, and the product precipitated with water containing 10 mmol HCl. Further purification was done by dissolution in $CHCl_3$/MeOH (1:1) at 45°C, concentrating *in vacuo* and addition of MeOH. Yield was 70%.

3.3.5. Fragment Coupling in Solution Using DCC/HOBt

- Fragment I: Fmoc-Ala-Gln(Mtt)-Ser(*t*Bu)-Gly-Leu-Gly-OH
- Fragment II: H-Cys(Acm)-Asn-Ser(*t*Bu)-Phe-Arg(HCl)-Tyr(*t*Bu)-O*t*Bu HCl

Fragment I represents the sequence 17–22, and fragment II the sequence 23–28 of human Atrial Natriuretic Peptide. Compare this to Section 3.3.1.

for a corresponding coupling on resin; 2.4 g fragment I (2.2 mmol) and 2.4 g fragment II (1 Eq) were dissolved in DMF/DMA (3:2, 50 mL). HOBt (1 Eq), DCC (1 Eq), and NMM (1 Eq) were added at 0°C and the reaction mixture left at room temperature overnight. Monitoring by TLC showed absence of starting materials. DCU was filtered off and the product precipitated with ethyl acetate/ether (1:1). Yield was 70%.

4. Notes

1. Fragments containing carboxylic acids, such as TFA or acetic acid, either as contaminants or as counterions may not be used for fragment couplings with DCC, since they themselves may couple to the amine. Even if present in low percentage, they may interfere to a great deal because they have a low molecular weight compared to the fragment, and hence, they may be present in quite high a concentration based on molarity. For couplings with TBTU, the presence of TFA may be tolerated as long as the desired reaction proceeds smoothly.
2. Trifluoroethanol is rather acidic and requires some "neutralization" with base. Furthermore trifluorethyl esters may be formed as (by)products *(8)*.
3. Keep in mind when performing these solubility experiments that you may well wish to use an excess of the carboxy component in the coupling. Therefore, requirements for solubility should not be set too low.
4. HOBt and HOSu may be used in excess (1.5–2 Eq with respect to the coupling reagent); sometimes, the presence of basic moieties like unprotected Arg side chain may "consume" additive.
5. For fast reaction, BOP may also be used, usually in the presence of a base, such as NMM, or DIPEA, and of HOBt (optional).
6. For literature references, *see* ref. *23* for DCC/HOBt, ref. *20* for TBTU, and ref. *24* for BOP.
7. It may be worth trying somewhat different conditions for a second coupling in solid phase, e.g., changing the solvent and/or the reagent; furthermore, addition of some structure-breaking or chaotropic agents or cosolvents may be tried ($LiClO_4$, ethylene carbonate, DMSO, trifluoroethanol, HFIP, and the like [*see* Chapter 1]).
8. Acetylation may facilitate final purification of the peptide. Acetylation blocks the amino group forever and prevents any further reaction.
9. If couplings were performed using DCC, the dicyclohexyl urea formed (and precipitated) during the coupling reaction would be filtered off first.

References

1. Wuensch, E. and Wendlberger, G. (1972) Zur Synthese des Sekretins V. *Chem. Ber.* **105,** 2508–2514.

2. Shin, S. Y., Kaburaki, Y., Watanabe, M., and Munekata, E. (1992) Total solution synthesis of human epidermal growth factor by the assembly of nine building blocks. *Biosci. Biotech. Biochem.* **56,** 404–408.
3. Nyfeler, R., Wixmerten, U., Seidel, C., and Mergler, M. (1992) Peptide synthesis by a combination of solid phase and solution methods, in *Peptides, Proc. 12th APS 1991* (Smith, J. A. and Rivier, J. E., eds.), Escom, Leiden, pp. 661–663.
4. Riniker, B., Fretz, H., and Kamber, B. (1993) *Peptides 1992. Proc. 22nd EPS,* (Schneider, C. H. and Eberle, A. N., eds.) Escom, Leiden, pp. 34,35.
5. Nokihara, K. and Hellstern, H. (1990) Synthesis of cardiodilatin related peptides by fragment assembly on a polymer support, in *Peptide Chemistry 1989* (Yanaihara, N., ed.), Protein Research Foundation, Osaka, pp. 315–320.
6. Albericio, F., Lloyd-Williams, P., Gairi, M., Jou, G., Celma, C., Kneib-Cordonnier, N., Grandas, A., Eritja, R., Pedroso, E., Van Rietschoten, J., Barany, G., and Giralt, E. (1992) Convergent solid phase peptide synthesis, in *Proc. 2nd Int. Symp. on Innovation and Perspectives in Solid Phase Synthesis,* Canterbury 1991, Intercept, Andover, pp. 39–47.
7. Atherton, E., Brown, E., Priestley, G., Sheppard, R. C., and Williams, B. J. (1981) Exploratory studies on solid phase segment condensation synthesis, in *Peptides, Proc 7th APS* (Rich, D. H. and Gross, E., eds.), Pierce, Rockford, pp. 163–175.
8. Felix, A. M., Wang, C.-T., and Lambros, J. (1985) Coupling of large protected peptide fragments in trifluoroethanol: synthesis of Thymosin α 1, in *Peptides, Proc. 9th APS Toronto* (Deber, C. M., Hruby, V. J., and Kopple, K. D., eds.), Pierce, Rockford, pp. 389–396.
9. Albericio, F., Pons, M., Pedroso, E., and Giralt, E. (1989) Comparative study of supports for solid-phase couplings of protected peptide segments. *J. Org. Chem.* **54,** 360–366.
10. Barlos, K., Gatos, D., and Schaefer, W. (1991) Synthese von Prothymosin α. *Angew. Chem.* **103,** 572–575.
11. Wuensch, E. (ed.) (1974) Houben-Weyl, *Methoden der organischen Chemie,* vol. 15, part II, Thieme, Stuttgart.
12. Gross, E. and Meienhofer, J. (1979) Major methods of peptide bond formation. *The Peptides,* vol. 1, Academic, Orlando, FL.
13. Bodanszky, M. (1984) *Principles of Peptide Synthesis.* Springer, Berlin.
14. Hudson, D. (1988) Methodological implications of simultaneous solid-phase peptide synthesis. 1. Comparison of different coupling procedures. *J. Org. Chem.* **53,** 617–624.
15. Posnett, D. N., Mc Grath, H., and Tam, J. P. (1988) A novel method for producing anti-peptide antibodies. *J. Biol. Chem.* **263,** 1719–1725.
16. Doerner, B., Carey, R. I., Mutter, M., Labhardt, A. M., Steiner, V., and Rink, H. (1992) New routes to artificial proteins applying the TASP concept, in *Proc. 2nd Int. Symp. on Innovation and Perspectives in Solid Phase Synthesis,* Canterbury 1991, Intercept, Andover, pp. 163–170.
17. Kamber, B. and Riniker, B. (1992) The solid phase synthesis of protected peptides combined with fragment coupling in solution, in *Peptides, Proc. 12th APS 1991* (Smith, J. A. and Rivier, J. E., eds.), Escom, Leiden, pp. 525–526.

18. Mergler, M. and Nyfeler, R. (1992) Easy synthesis of protected peptide hydrazides on solid support, in *Peptides, Proc. 12th APS 1991* (Smith, J. A. and Rivier, J. E., eds.), Escom, Leiden, pp. 551–552.
19. Bodanszky, M. and Bodanszky, A. (1984) *The Practice of Peptide Synthesis, Reactivity and Structure, Concepts in Organic Chemistry,* vol. 21, Springer, Berlin.
20. Knorr, R., Trzeciak, A., Bannwarth, W., and Gillessen, D. (1989) New coupling reagents in peptide chemistry. *Tetrahedron Lett.* **30,** 1927–1930.
21. Miyazawa, T., Otomatsu, T., Fukui, Y., Yamada, T., and Kuwata, S. (1992) Effect of copper(II)chloride on suppression of racemization in peptide synthesis by the carbodiimide method. *Int. J. Peptide Prot. Res.* **39,** 237–244.
22. Thaler, A., Seebach, D., and Cardinaux, F. (1991) Lithium salt effects in peptide synthesis (a) Part I *Helv.* **74,** 617–627; (b) Part II *Helv.* **74,** 628–643.
23. Koenig, W. and Geiger, R. (1970) Eine neue Methode zur Synthese von Peptiden. *Chem. Ber.* **103,** 788–789.
24. Castro, B., Dormoy, J. R., Evin, G., and Selve, C. (1975) Reactifs de couplage peptidique IV. *Tetrahedron Lett.* **14,** 1219–1222.

Index

GPSR Compliance

The European Union's (EU) General Product Safety Regulation (GPSR) is a set of rules that requires consumer products to be safe and our obligations to ensure this.

If you have any concerns about our products, you can contact us on ProductSafety@springernature.com

In case Publisher is established outside the EU, the EU authorized representative is:

Springer Nature Customer Service Center GmbH
Europaplatz 3
69115 Heidelberg, Germany

Batch number: 10761599

Printed by Printforce, the Netherlands